Limnology and Marine Biology

NIPA® GENX ELECTRONIC RESOURCES & SOLUTIONS P. LTD.
New Delhi-110 034

About the Editors

Dr. V. B. Sakhare is Professor at Post Graduate and Research Department of Zoology, Yogeshwari Mahavidyalaya, Ambajogai (Maharashtra). He has 26 years' experience as an outstanding teacher and researcher and has done pioneering work in the field of Reservoir Fisheries, Limnology and Fish Biology. Prof. Sakhare has successfully organized more than 7 conferences and workshops.

Prof. Sakhare is member of Board of Studies in Zoology of Dr. Babasaheb Ambedkar Marathwada University, Chhatrapati Sambhajinagar and Walchand College of Arts and Science, Solapur (Autonomous college). Prof. Sakhare has authored/edited more than 45 books. He is a recognized research guide of Dr. Babasaheb Ambedkar Marathwada University, and Solapur University, Solapur.

His exemplary qualifications and experience in the field of 'Inland Fisheries' has benefited to several students as well as fish farmers and produced 4 Doctoral candidates, as also led to publications to spread his knowledge across diaspora.

Dr. Santoshkumar Abujam obtained Master's degree in Life Sciences (Zoology) from Dibrugarh University (DU), Assam. He was awarded the degree of Doctor of Philosophy by the same University. Dr. Abujam has gathered long experience in diverse areas of fisheries research through several pre-doctoral and post-doctoral assignments respectively as a Project Fellow (ICAR-NBFGR Lucknow sponsored); Junior Research Fellow (MoEF, New Delhi sponsored); Post Doctoral Research Associate (DBT & DST New Delhi sponsored) and a Guest Faculty in the Department of Geology (Zoology-a subsidiary subject), RGU from August, 2017 to May, 2019. Presently, he has been working as an Assistant Professor in the Department of Zoology, Pravabati College, Mayang Imphal West, Manipur (India). He has been able to publish more than 80 research articles; 05 books and 15 articles in book chapter/proceedings by virtue of his deep involvement in fishery research.

Dr. V. Ravi is an Associate Professor of Marine Biology at Annamalai University with a focus on Fish Biology and Ecology. After earning a Ph.D. from the Centre of Advanced Study in Marine Biology, Annamalai University, Tamil Nadu in 2000, Dr. Ravi was recognized with the Prof. R.V. Seshaiya's Gold Medal for the University's Best Thesis. He has over twenty years of teaching experience and over twenty-five years of research experience. To his credit, he has had more than 50 research papers published in peer-reviewed national and international journals and he has also authored a book titled "Practical Manual on Keys to Marine Fish Identification". In addition, he has guided five students for their M.Phil. and six for their Ph.D. He has successfully completed the major research project funded by UGC and MOEFCC. In 2016, he traveled to Australia, where he visited the Great Barrier Reef, the largest coral reef system in the world, and noted its biodiversity. Besides, he has held a deputation position in the Zoology Department at the Govt. College for Women in Kumbakonam since 2017. Currently, he is serving in the Zoology Department at Periyar Arts College (affiliated to Annamalai University) in Cuddalore, Tamil Nadu.

Limnology and Marine Biology

Editors

V.B. Sakhare
Professor
Post Graduate Department of Zoology
Yogeshwari Mahavidyalaya
Ambajogai, Maharashtra, India

Santoshkumar Abujam
Assistant Professor
Department of Zoology
Pravabati College
Mayang Imphal, Manipur, India

V. Ravi
Associate Professor
Department of Zoology
Periyar Arts College
Devanampattinam, Cuddalore
Tamil Nadu, India

NIPA® GENX ELECTRONIC RESOURCES & SOLUTIONS P. LTD.
New Delhi-110 034

NIPA® GENX ELECTRONIC RESOURCES & SOLUTIONS P. LTD.

101,103, Vikas Surya Plaza, CU Block
L.S.C. Market, Pitam Pura, New Delhi-110 034
Ph : +91-11-43860225, Mob.: +91 9717133558, 9540816132
E-mail: newindiapublishingagency@gmail.com
Website: www.nipaersources.com

Print ISBN: 978-93-58873-09-2
ebook ISBN: 978-93-58872-05-7

Composed and Designed by NIPA®.

Preface

This volume is the compilation of twenty-six research and review articles contributed by senior faculty members of different universities and scientific institutions. With its application oriented and interdisciplinary approach, we hope that the students, teachers, researchers, scientists, policy makers and environmental lawyers will find this volume much more useful.

Editors are grateful to the following under mentioned distinguished scientists and other fellow colleagues for their constant encouragements, suggestions, valuable guidelines and necessary help. These distinguished, beloved scientists and well-wishers are Dr. Anurag Kr. Singh of IMS-BHU, Varanasi (Uttar Pradesh), Dr. Anurag Protim Das of Gargaon College, Simaluguri (Assam), Dr.S. R. Sitre and Dr.A. K. Pimpalshende of Nilkanthrao Shinde Science and Arts College, Bhadrawati (Maharashtra), Dr. S. M. Hussain and Dr. B.N.Hazarika of College of Horticulture and Forestry, Pasighat (Arunchal Pradesh), Dr. T. H. Dave, Dr.Hardik Sikotariya, Dr. Ketan V. Tank, Dr.Ketan Makwana and Dr. Pinak Bamaniya, Dr. H. L. Parmar, Dr.Sayan Biswas, Dr.Neha Kharadi and Dr. Mayur Bhadarka, Dr.Mohit Bamaniya and Dr. D. T. Vaghela of Kamdhenu University, Veraval (Gujarat), Dr. Devashish Ka of Department of Life Science and Bioinformatics, Silchar (Assam), Dr. Gora Shiva Prasad of West Bengal University of Animal and Fishery Sciences, Kolkata (West Bengal), Dr.H. Puinyabati of Pravabati College, Mayang Imphal (Manipur), Dr.Tribeni Hazarika, Dr. Januka Sharma, Dr. Monjeet Sonowal of Arunachal University of Studies, Namsai (Arunachal Pradesh), Dr.Kardivel Govindasamy, ICAR-ATARI, Guwahati (Assam), Dr. H.S. Praveenjoshi, Dr. Hariyala Sirisha and Dr. Kranthi Kumar Dhande, Andhra Pradesh Fisheries University, Narasapuram (Andhra Pradesh), Dr. M. Shomorendra of Thambal Marik College, Oinam (Manipur), Dr. M. Sivagangai of Periyar Arts College, Devanampattinam, Cuddalore (Tamil Nadu), Dr. Mostafa A. M. Soliman and Dr. Safwat Gobran and Dr. Mohammed A. A. Abo-Taleb of Agricultural Research Centre, Giza (Egypt), Dr. Pravin U. Meshram of Sevadal Mahila Mahavidyalaya, Nagpur (Maharashtra), Dr. R. Anandhan and V. Kavitha of Government Arts College, Kumbakonam (Tamil Nadu), Dr. S. Borthakur of College of Fisheries, Raha, Nagaon, Dr. Sanyogta Meshram of Shree Ramdeobaba Engineering College, Nagpur (Maharashtra), Dr. Sanyogita R. Verma and Dr. Vaishali A. Meshram of Anand Niketan College, Anandwan, Warora (Maharashtra), Dr.Sharda Dhase and

Shahnoor Khan of CSIR-National Environmental Engineering Research Institute, Nagpur (Maharashtra), Dr. Shreya D. Indurkar, Seevadal Mahila Mahavidyalaya, Nagpur (Maharashtra), Dr. Vaishali P. Meshram, M.P. Deo Dharampeth College, Nagpur (Maharashtra), Dr. Aasidhara Darvekar, K.Z.S. Science College, Bramhani-Kalmeshwar (Maharashtra), Dr. Shyama Prasad Biswas of Dibrugarh University, Dibrugarh (Assam), Dr. Budhin Gogoi, Moran College, Moranhat, Charaideo (Assam), Dr. Vivekanand Safi and Dr. Tadang Meena of KVK, Karsingsa (Arunachal Pradesh), Dr. Tilling Tayo, KVK Longding, ICAR for NEH Region (Arunachal Pradesh), Dr. Shivendra Kumar, College of Fisheries, Dholi (Bihar), Dr. Ram Kumar, Koshi College, Khagaria (Bihar), Dr. Parmanand Prabhakar, College of Fisheries, Kishanganj (Bihar), Dr. Mukesh Kumar Singh, College of Fisheries, Dholi (Bihar), Dr. Nabam Gama, KVK, Karsingsa, (Arunachal Pradesh), and Dr. T. Neeraja, Andhra Pradesh Fisheries University, Narasapuram (Andhra Pradesh).

We thank to Publisher M/s NIPA GENX Electronic Resources and Solutions Pvt. Ltd., New Delhi for taking pains in bringing out the book.

We are grateful to our family members for helping us by several ways and encouraging us for publication of this volume.

Editors

List of Contributors

A.K. Pimpalshende
Nilkanthrao Shinde Science & Arts College
Bhadrawati, Maharashtra

Aasidhara Darvekar
Department of Zoology
K.Z.S. Science College, Bramhani
Kalmeshwar, Nagpur, Maharashtra

A.I. Shaikh
Department of Zoology
Yogeshwari Mahavidyalaya, Ambajogai
Maharashtra

Anurag Kr. Singh
IMS-BHU, Varanasi, Uttar Pradesh

Anurag Protim Das
Department of Zoology
Gargaon College, Simaluguri, Assam

B.N. Hazarika
KVK East Siang
College of Horticulture and Forestry
CAU (I), Pasighat, Arunachal Pradesh

Budhin Gogoi
Department of Zoology, Moran College
Moranhat, Charaideo, Assam

D.T. Vaghela
College of Fisheries Science
Kamdhenu University, Veraval Gujarat

Devashish Ka
Division of Wetlands, Fishery Science and Aquaculture, Department of Life Science and Bioinformatics,
Assam (Central) University, Silchar, Assam

Gora Shiva Prasad
Faculty of Fishery Sciences
West Bengal University of Animal & Fishery Sciences, Kolkata, West Bengal

H. Puinyabati
Department of Zoology
Pravabati College, Mayang
Imphal, Manipur

H. L. Parmar
Department of Fisheries Resource Management, College of Fisheries Science
Kamdhenu University
Veraval, Gujarat

Hardik Sikotariya
Department of Fisheries Resource Management, College of Fisheries Science
Kamdhenu University
Veraval, Gujarat

Hariyala Sirisha
College of Fishery Science
Andhra Pradesh Fisheries University
Narasapuram, Andhra Pradesh

Januka Sharma
Department of Zoology, Faculty of Science & Technology, Arunachal University of Studies, Namsai Arunachal Pradesh

Kardivel Govindasamy
ICAR-ATARI, Guwahati, Zone-VI
Assam

Ketan Makwana
Department of Fisheries Resource Management, College of Fisheries Science,
Kamdhenu University
Veraval, Gujarat

Ketan V. Tank
College of Fisheries Science
Kamdhenu University, Veraval, Gujarat

Kranthi Kumar Dhande
College of Fishery Science
Andhra Pradesh Fisheries University
Muttukuru, Andhra Pradesh

M. Shomorendra
Department of Zoology
Thambal Marik College, Oinam Manipur

M. Sivagangai
Department of Zoology
Periyar Arts College, Devanampattinam
Cuddalore, Tamil Nadu

Mayur Bhadarka
College of Fisheries Science
Kamdhenu University, Veraval, Gujarat

Mohammed A.A. Abo-Taleb
Regional Center for Food and Feed
Agricultural Research Centre
Giza, Egypt

Mohit Bamaniya
College of Fisheries Science
Kamdhenu University, Veraval, Gujarat

Monjeet Sonowal
Department of Zoology
Arunachal University of Studies Namsai
Arunachal Pradesh

Mostafa A.M. Soliman
Utilization of by-Products Department
Animal Production Research Institute
(APRI), Agricultural Research Center
ElDokki, Giza, Egypt

Mukesh Kumar Singh
Department of Aquaculture
College of Fisheries, Dholi, Bihar

Nabam Gama
KVK, Karsingsa, Papum Pare District
Arunachal Pradesh

Neha Kharadi
College of Fisheries Science
Kamdhenu University, Veraval, Gujarat

Parmanand Prabhakar
Department of Fish Processing Technology
College of Fisheries
Kishanganj, Bihar

Pinak Bamaniya
College of Fisheries Science
Kamdhenu University, Veraval, Gujarat

Praveenjoshi H.S.
College of Fishery Science
Andhra Pradesh Fisheries University
Narasapuram, Andhra Pradesh

Pravin U. Meshram
Dept. of Environmental Science
Sevadal Mahila Mahavidyalaya
Nagpur, Maharashtra

R. Anandhan
Department of Zoology, Government Arts
College (A), Kumbakonam
Tamil Nadu

Ram Kumar
Koshi College (Munger University)
Khagaria, Bihar

S. Borthakur
College of Fisheries, AAU, Raha
Nagaon, Assam

S.M. Hussain
KVK East Siang
College of Horticulture and Forestry CAU
(I), Pasighat, Arunachal Pradesh

S.R. Sitre
Nilkanthrao Shinde Science & Arts College
Bhadrawati, Maharashtra

Safwat Gobran
Central Laboratory for Aquaculture Research
(CLAR), Agricultural Research Center,
Egypt

Sanyogta Meshram
Department of Chemistry
Shree Ramdeobaba Engineering College,
Nagpur, Maharashtra

Sanyogita R. Verma
Department of Zoology, Anand Niketan
College, Anandwan Warora, Maharashtra

Sayan Biswas
Department of Aquaculture
College of Fisheries Science
Kamdhenu University, Veraval, Gujarat

Shahnoor Khan
CSIR-National Environmental Engineering Research Institute
Nagpur, Maharashtra

Sharda Dhadse
CSIR-National Environmental Engineering Research Institute
Nagpur, Maharashtra

Shivendra Kumar
Department of Aquaculture
College of Fisheries, Dholi, Bihar

Shreya D. Indurkar
Department of Environmental Science
Seevadal Mahila Mahavidyalaya
Nagpur, Maharashtra

Shyama Prasad Biswas
Department of Life Sciences
Dibrugarh University, Dibrugarh Assam

Sikha Sonowal
Department of Zoology
Faculty of Science and Technology
Arunachal University of Studies Namsai, Arunachal Pradesh

T. Neeraja
College of Fishery Science
Andhra Pradesh Fisheries University
Narasapuram-534275, Andhra Pradesh

T. H. Dave
College of Fisheries Science
Kamdhenu University, Veraval, Gujarat

Tadang Meena
KVK, Karsingsa, Papum Pare District
Arunachal Pradesh

Tilling Tayo
KVK Longding, AP Center Basar
ICAR for NEH Region, A.P.

Tribeni Hazarika
Department of Zoology
Faculty of Science & Technology Arunachal University of Studies Namsai, Arunachal Pradesh

V. Kavitha
Department of Zoology
Government College for Women (A)
Kumbakonam, Tamil Nadu

V. Ravi
Department of Zoology, Periyar Arts College, Devanampattinam
Cuddalore, Tamil Nadu

V. B. Sakhare
Yogeshwari Mahavidyalaya
Ambajogai, Maharashtra

Vaishali A. Meshram
Anand Niketan College
Anandwan Warora District Chandrapur
Maharashtra

Vaishali P. Meshram
Department of Chemistry, M P Deo
Dharampeth Col., Nagpur, Maharashtra

Vivekanand Safi
KVK, Karsingsa, Papum Pare District
Arunachal Pradesh

Contents

List of Colour Plates

1

Fish Diversity in Indian Mangrove Ecosystems and Adjacent Waters A Checklist

V. Ravi

INTRODUCTION

Ecosystem services and biodiversity conservation are well recognized as essential to human well-being and have become a policy priority. India is gifted with a large coastline and vast stretches of coastal wetlands. The estuaries and back waters which may include mangroves, mud flats, bays etc. extend over a large part of the Indian coasts. The mangrove ecosystem functions as a habitat for nurseries, thus benefiting fishers in the food chain (Mumby *et al.*, 2004; Venkatachalam *et al.*, 2018). Mangrove habitats increase the survival ratio of young fish that migrate to other offshore habitats when they mature, resulting in the enrichment of fish biomass in adjacent mangrove areas (Mumby *et al.*, 2004; Aburto-Oropeza *et al.*, 2008). Mangroves enhance fish production by providing food and shelter to fishery resources. Mangrove forests are highly productive, with average primary productivity level similar to that of tropical terrestrial forests (Alongi, 2009). Therefore, mangrove loss could have a devastating effect on fishery outcomes. The estimated fish catch is up to 70% higher in mangrove areas than in non-mangrove areas (Das, 2017).

NATURE OF MANGROVE FORESTS

Mangroves are amongst the most productive marine ecosystems on the Earth, providing a unique habitat opportunity for many species and key goods and significant services for human beings. They are salt tolerant plants, and are found distributed growing in estuaries, lagoons, backwaters and bays at the land-sea interface in tropical and subtropical areas of the world. Mangroves thrive on low-oxygen through the breathing roots or pneumatophores that are used for respiration. They need freshwater for their growth and so they live at the sea edge, they must remove the salts from the water that is taken in by the roots. Majority of the salts are taken in by the roots and the salt glands on the leaves of some species excrete salts. Mangrove species have a unique way of reproduction through viviparity as the seeds germinate and develop into seedlings while the seeds are still attached to the parent tree.

MANGROVES AS KEYSTONE SPECIES

A keystone species is a species that has a disproportionately large effect on its natural environment relative to its abundance that helps define an entire ecosystem. Mangroves are a keystone species providing essential services that act as the base for the entire estuarine community.

ECOSYSTEM SERVICES OF MANGROVE FORESTS

- Mangroves provide natural infrastructure to help protect nearby populated areas by reducing erosion and absorbing storm surge impacts during extreme weather events such as cyclones.
- The complex root systems of mangroves filter nitrates, phosphates and other pollutants from the water, improving the water quality flowing from rivers and streams into the estuarine and ocean environment.
- Mangrove forests capture massive amounts of carbon dioxide emissions and other greenhouse gases from the atmosphere, and then trap and store them in their carbon-rich flooded soils for millennia.
- Mangrove forests also provide habitat and refuge to a wide array of wildlife such as birds, fish, invertebrates, mammals and plants.
- Mangroves act as an important breeding and nursery ground for juvenile marine species including prawn, crabs, and many commercial fish species thereby enhancing fishery productivity and contributing to livelihood and local food security.

MANGROVE ECOSYSTEMS IN INDIA

The mangrove cover in India is 4,975 sq km, which is 0.15% of the country's total geographical area according to the India State of Forest Report, 2019. The deltas of the Ganges, Mahanadi, Krishna, Godavari, and the Cauvery rivers contain mangrove forests. West Bengal has the highest percentage of area under total Mangrove cover followed by Gujarat and Andaman Nicobar Islands. Sundarbans in West Bengal are the largest mangrove forest regions in the world. It is listed as a UNESCO World Heritage Site. and this forest is home to the Royal Bengal tiger, Gangetic dolphins and Estuarine crocodiles. The Bhitarkanika Mangroves of Odisha is the second largest mangrove forest in India which is protected by the two river deltas of River Brahmani and Baitarani. It is one of the most significant Ramsar wetlands in India. The Coringa mangroves form an important mangrove in Godavari-Krishna riverine system in Andhra Pradesh. The backwaters in Kerala have a high density of mangrove forest. Pichavaram in Tamil Nadu has a vast expanse of water covered with mangrove forests and home to many aquatic bird species. It is designated as Ramsar site. The details of Mangrove Forest cover (Area in sq.km) in different States/UT in India are given in Table 1.

Table 1: Mangrove Forest cover (Area in sq.km) in different States/UT in India

State/UT	Name of the mangrove forest	Area in sq. km (2017)
East coast		
West Bengal	Sundarban	2114
A&N islands	North Andaman Nicobar	617
Andhra Pradesh	Coringa	404
Odisha	Bhitarkanika	243
Tamil Nadu	Pichavaram and Muthupet	49
Puducherry	Puducherry	2
West coast		
Gujarat	Gulf of Kachchh and Gulf of Khambhat	1140
Maharashtra	Malwan	304
Goa	Goa	26
Daman & Diu	Daman & Diu	3
Kerala	Vembanad	9

MANGROVE SPECIES OF INDIA

The list of various mangrove species distributed in India are presented in Table 2.

Table 2: Mangrove species of India- checklist

Mangrove species	
1.	*Acanthus ebracteatus*
2.	*Acanthus ilicifolius*
3.	*Acanthus volubilis*
4.	*Acrostichum aureum*
5.	*Acrostichum speciosum*
6.	*Aegialitis rotundifolia*
7.	*Aegiceras corniculatum*
8.	*Aglaia cucullata*
9.	*Ardisia elliptica*
10.	*Arthrocnemum indicum*
11.	*Avicennia alba*
12.	*Avicennia marina*

13.	*Avicennia officinalis*
14.	*Brownlowia tersa*
15.	*Bruguiera cylindrica*
16.	*Bruguiera gymnorrhiza*
17.	*Bruguiera parviflora*
18.	*Bruguiera sexangula*
19.	*Calotropis gigantea*
20.	*Carallia brachiata*
21.	*Cerbera manghas*
22.	*Ceriops decandra*
23.	*Ceriops tagal*
24.	*Citrullus colocynthis*
25.	*Clerodendrum inerme*
26.	*Cynometra iripa*
27.	*Cynometra ramiflora*
28.	*Derris trifoliata*
29.	*Dolichandrone spathaceae*
30.	*Excoecaria agallocha*
31.	*Flagellaria indica*
32.	*Heliotropium curassavicum*
33.	*Heritiera fomes*
34.	*Heritiera littoralis*
35.	*Ipomoea pes-caprae*
36.	*Kandelia candel*
37.	*Lumnitzera littorea*
38.	*Lumnitzera racemosa*
39.	*Nypa fruticans*
40.	*Pemphis acidula*
41.	*Phoenix paludosa*
42.	*Pongamia pinnata*
43.	*Rhizophora annamalayana*
44.	*Rhizophora apiculata*
45.	*Rhizophora lamarckii*
46.	*Rhizophora mucronata*
47.	*Rhizophora stylosa*
48.	*Salicornia brachiata*
49.	*Sarcolobus globosus*
50.	*Scyphiphora hydrophyllaceae*
51.	*Sesuvium portulacastrum*

52.	*Sonneratia alba*
53.	*Sonneratia apetala*
54.	*Sonneratia caseolaris*
55.	*Sonneratia griffithi*
56.	*Sonneratia ovata*
57.	*Spinifex littoreus*
58.	*Suaeda maritima*
59.	*Suaeda monoica*
60.	*Xylocarpus gangeticus*
61.	*Xylocarpus granatum*
62.	*Xylocarpus meckongensis*
63.	*Xylocarpus moluccensis*

FISH DIVERSITY IN THE MANGROVE ECOSYSTEM AND ADJACENT WATERS IN INDIA

The present checklist of diversity of fishes of Indian mangrove ecosystems and adjacent waters are presented in Table 3. Totally 331 species of fishes are listed presently under the two Classes (Chondrichthyes and Osteichthyes), 35 Orders, 85 Families and 201 Genera. Among the Order-wise contribution, Gobiiformes dominated with 60 species, followed by Perciformes (46 species), Clupeiformes (42 species), Pleuronectiformes (22 species), Carangiformes (19 species), Acanthuriformes (19 species), Siluriformes (17 species) and Anguilliformes (13 species). The species number of other Orders are presented in Table 4.

Table 3: Checklist of fish diversity in the mangrove ecosystem and adjacent waters in India

Sl.No	Order	Family	Species Name	IUCN status
1.	Carcharhiniformes	Carcharhinidae	*Carcharhinus limbatus*	VU
2.	Carcharhiniformes	Carcharhinidae	*Carcharhinus melanopterus*	VU
3.	Carcharhiniformes	Carcharhinidae	*Glyphis gangeticus*	CR
4.	Torpediniformes	Narcinidae	*Narcine brunnea*	LC
5.	Pristiformes	Rhinobatidae	*Rhinobatos annandalei*	CR
6.	Myliobatifomes	Myliobatidae	*Aetobatus narinari*	EN
7.	Myliobatifomes	Dasyatidae	*Himantura imbricata*	DD
8.	Myliobatifomes	Dasyatidae	*Himantura uarnak*	EN
9.	Myliobatifomes	Dasyatidae	*Pastinachus sephen*	NT
10.	Myliobatifomes	Dasyatidae	*Pateobatis bleekeri*	EN
11.	Myliobatifomes	Dasyatidae	*Telatrygon zugei*	VU

12.	Elopiformes	Elopidae	*Elops machnata*	LC
13.	Elopiformes	Megalopidae	*Megalops cyprinoides*	DD
14.	Anguilliformes	Anguillidae	*Anguilla bengalensis*	NT
15.	Anguilliformes	Anguillidae	*Anguilla bicolor*	NT
16.	Anguilliformes	Congridae	*Uroconger lepturus*	LC
17.	Anguilliformes	Moringuidae	*Moringua raitaborua*	DD
18.	Anguilliformes	Muraenidae	*Strophidon sathete*	LC
19.	Anguilliformes	Muraenesocidae	*Congresox talabonoides*	LC
20.	Anguilliformes	Muraenesocidae	*Congresox talabon*	LC
21.	Anguilliformes	Muraenesocidae	*Muraenesox bagio*	LC
22.	Anguilliformes	Muraenesocidae	*Muraenesox cinereus*	LC
23.	Anguilliformes	Ophicthidae	*Bascanichthys deraniyagalai*	DD
24.	Anguilliformes	Ophicthidae	*Cirrhimuraena playfairii*	LC
25.	Anguilliformes	Ophicthidae	*Pisodonophis boro*	LC
26.	Anguilliformes	Ophicthidae	*Pisodonophis cancrivorus*	LC
27.	Osteoglossiformes	Notopteridae	*Notopterus notopterus*	LC
28.	Clupeiformes	Chirocentridae	*Chirocentrus dorab*	LC
29.	Clupeiformes	Clupeidae	*Anodontostoma chacunda*	LC
30.	Clupeiformes	Clupeidae	*Anodontostoma thailandiae*	LC
31.	Clupeiformes	Clupeidae	*Dussumieria acuta*	LC
32.	Clupeiformes	Clupeidae	*Dussumieria elopsoides*	LC
33.	Clupeiformes	Clupeidae	*Escualosa thoracata*	LC
34.	Clupeiformes	Clupeidae	*Gonialosa manmina*	LC
35.	Clupeiformes	Clupeidae	*Hilsa kelee*	LC
36.	Clupeiformes	Clupeidae	*Nematalosa nasus*	LC
37.	Clupeiformes	Clupeidae	*Sardinella albella*	LC
38.	Clupeiformes	Clupeidae	*Sardinella fimbriata*	LC
39.	Clupeiformes	Clupeidae	*Sardinella longiceps*	LC
40.	Clupeiformes	Clupeidae	*Sardinella sindensis*	LC
41.	Clupeiformes	Clupeidae	*Tenualosa ilisha*	LC
42.	Clupeiformes	Clupeidae	*Tenualosa toli*	VU
43.	Clupeiformes	Engraulidae	*Coilia boreensis*	DD
44.	Clupeiformes	Engraulidae	*Coilia dussumieri*	LC
45.	Clupeiformes	Engraulidae	*Coilia neglecta*	LC
46.	Clupeiformes	Engraulidae	*Coilia ramacarati*	DD
47.	Clupeiformes	Engraulidae	*Coilia reynaldi*	LC
48.	Clupeiformes	Engraulidae	*Encrasicholina punctifer*	LC
49.	Clupeiformes	Engraulidae	*Setipinna phasa*	LC

50.	Clupeiformes	Engraulidae	*Setipinna taty*	LC
51.	Clupeiformes	Engraulidae	*Setipinna tenuifilis*	DD
52.	Clupeiformes	Engraulidae	*Stolephorus commersonii*	LC
53.	Clupeiformes	Engraulidae	*Stolephorus indicus*	LC
54.	Clupeiformes	Engraulidae	*Thryssa baelama*	LC
55.	Clupeiformes	Engraulidae	*Thryssa dussumieri*	LC
56.	Clupeiformes	Engraulidae	*Thryssa hamiltoni*	LC
57.	Clupeiformes	Engraulidae	*Thryssa malabarica*	DD
58.	Clupeiformes	Engraulidae	*Thryssa mystax*	LC
59.	Clupeiformes	Engraulidae	*Thryssa purava*	DD
60.	Clupeiformes	Engraulidae	*Thryssa vitrirostris*	LC
61.	Clupeiformes	Pristigasteridae	*Ilisha elongata*	LC
62.	Clupeiformes	Pristigasteridae	*Ilisha kampeni*	LC
63.	Clupeiformes	Pristigasteridae	*Ilisha megaloptera*	LC
64.	Clupeiformes	Pristigasteridae	*Ilisha melastoma*	NE
65.	Clupeiformes	Pristigasteridae	*Ilisha novacula*	DD
66.	Clupeiformes	Pristigasteridae	*Opisthopterus tardoore*	LC
67.	Clupeiformes	Pristigasteridae	*Pellona ditchela*	LC
68.	Clupeiformes	Pristigasteridae	*Raconda russeliana*	LC
69.	Gonorynchiformes	Chanidae	*Chanos chanos*	LC
70.	Cypriniformes	Cyprinidae	*Puntius sarana subnusutus*	LC
71.	Cypriniformes	Cyprinidae	*Puntius sophore*	LC
72.	Cypriniformes	Cyprinidae	*Cirrhinus cirrhosus*	VU
73.	Cypriniformes	Cyprinidae	*Labeo catla*	LC
74.	Cypriniformes	Cyprinidae	*Labeo calbasu*	LC
75.	Cypriniformes	Cyprinidae	*Labeo fimbriatus*	LC
76.	Cypriniformes	Cyprinidae	*Labeo rohita*	LC
77.	Cypriniformes	Cyprinidae	*Danio aequipinnatus*	DD
78.	Siluriformes	Ariidae	*Arius africanus*	DD
79.	Siluriformes	Ariidae	*Arius arius*	LC
80.	Siluriformes	Ariidae	*Arius gagora*	NT
81.	Siluriformes	Ariidae	*Arius maculatus*	DD
82.	Siluriformes	Ariidae	*Arius subrostratus*	NE
83.	Siluriformes	Ariidae	*Nemapteryx caelata*	NE
84.	Siluriformes	Ariidae	*Plicofollis platystomus*	LC
85.	Siluriformes	Bagridae	*Mystus cavasius*	LC
86.	Siluriformes	Bagridae	*Mystus gulio*	LC
87.	Siluriformes	Bagridae	*Mystus vittatus*	LC
88.	Siluriformes	Bagridae	*Sperata aor*	LC

89.	Siluriformes	Pangasiidae	*Pangasius pangasius*	LC
90.	Siluriformes	Plotosidae	*Plotosus canius*	NE
91.	Siluriformes	Siluridae	*Ompok pabda*	NT
92.	Siluriformes	Siluridae	*Ompok bimaculatus*	NT
93.	Siluriformes	Siluridae	*Wallago attu*	NT
94.	Siluriformes	Heteropneustidae	*Heteropneustes fossilis*	LC
95.	Aulopifomes	Synodontidae	*Harpadon nehereus*	NT
96.	Aulopiformes	Synodontidae	*Saurida tumbil*	LC
97.	Holocentriformes	Holocentridae	*Sargocentron cornutum*	LC
98.	Batrachoidiformes	Batrachoididae	*Allenbatrachus grunniens*	NE
99.	Kurtiformes	Apogonidae	*Apogonichthyoides niger*	LC
100.	Kurtiformes	Kurtidae	*Kurtus indicus*	NE
101.	Gobiiformes	Ambassidae	*Ambassis gymnocephalus*	LC
102.	Gobiiformes	Ambassidae	*Ambassis ambassis*	LC
103.	Gobiiformes	Ambassidae	*Chanda nama*	LC
104.	Gobiiformes	Ambassidae	*Parambassis baculis*	LC
105.	Gobiiformes	Ambassidae	*Parambassis ranga*	LC
106.	Gobiiformes	Butidae	*Butis humeralis*	LC
107.	Gobiiformes	Butidae	*Butis koilomatodon*	LC
108.	Gobiiformes	Butidae	*Butis melanostigma*	NE
109.	Gobiiformes	Butidae	*Odonteleotris canina*	NE
110.	Gobiiformes	Eleotrididae	*Butis butis*	LC
111.	Gobiiformes	Eleotrididae	*Bunaka gyrinoides*	LC
112.	Gobiiformes	Eleotrididae	*Eleotris fusca*	LC
113.	Gobiiformes	Gobiidae	*Acentrogobius caninus*	LC
114.	Gobiiformes	Gobiidae	*Acentrogobius audax*	LC
115.	Gobiiformes	Gobiidae	*Acentrogobius cyanomus*	LC
116.	Gobiiformes	Gobiidae	*Acentrogobius madraspatensis*	NE
117.	Gobiiformes	Gobiidae	*Aulopareia cyanomos*	LC
118.	Gobiiformes	Gobiidae	*Awaous gutum*	DD
119.	Gobiiformes	Gobiidae	*Bathygobius fuscus*	LC
120.	Gobiiformes	Gobiidae	*Bathygobius ostreicola*	DD
121.	Gobiiformes	Gobiidae	*Favonigobius reichei*	LC
122.	Gobiiformes	Gobiidae	*Glassogobius callidus*	LC
123.	Gobiiformes	Gobiidae	*Glassogobius giuris*	LC
124.	Gobiiformes	Gobiidae	*Mahidolia mystacina*	LC
125.	Gobiiformes	Gobiidae	*Parachaeturichthys polynema*	LC

126.	Gobiiformes	Gobiidae	*Psammogobius biocellatus*	LC
127.	Gobiiformes	Oxudercidae	*Amblyotrypauchen arctocephalus*	LC
128.	Gobiiformes	Oxudercidae	*Apocryptes bato*	LC
129.	Gobiiformes	Oxudercidae	*Apocryptodon madurensis*	NE
130.	Gobiiformes	Oxudercidae	*Boleophthalmus boddarti*	LC
131.	Gobiiformes	Oxudercidae	*Boleophthalmus dussumieri*	LC
132.	Gobiiformes	Oxudercidae	*Brachyamblyopus brachysoma*	NE
133.	Gobiiformes	Oxudercidae	*Brachygobius nasus*	LC
134.	Gobiiformes	Oxudercidae	*Odontamblyopus rubicundus*	LC
135.	Gobiiformes	Oxudercidae	*Oligolepis acutipennis*	LC
136.	Gobiiformes	Oxudercidae	*Oxuderces dentatus*	DD
137.	Gobiiformes	Oxudercidae	*Oxyurichthys microlepis*	LC
138.	Gobiiformes	Oxudercidae	*Oxyurichthys tentacularis*	DD
139.	Gobiiformes	Oxudercidae	*Parapocryptes rictuosus*	DD
140.	Gobiiformes	Oxudercidae	*Parapocryptes serperaster*	LC
141.	Gobiiformes	Oxudercidae	*Paratrypauchen microcephalus*	LC
142.	Gobiiformes	Oxudercidae	*Periophthalmodon schlosseri*	LC
143.	Gobiiformes	Oxudercidae	*Periophthalmus argentilineatus*	LC
144.	Gobiiformes	Oxudercidae	*Periophthalmus chrysospilos*	LC
145.	Gobiiformes	Oxudercidae	*Periophthalmus novemradiatus*	DD
146.	Gobiiformes	Oxudercidae	*Periophthalmus variabilis*	NE
147.	Gobiiformes	Oxudercidae	*Periophthalmus walailakae*	NE
148.	Gobiiformes	Oxudercidae	*Periophthalmus waltoni*	LC
149.	Gobiiformes	Oxudercidae	*Pseudapocryptes elongatus*	LC
150.	Gobiiformes	Oxudercidae	*Pseudapocryptes lanceolatus*	LC
151.	Gobiiformes	Oxudercidae	*Pseudapocryptes macrolepis*	LC
152.	Gobiiformes	Oxudercidae	*Scartelaos gigas*	NE
153.	Gobiiformes	Oxudercidae	*Scartelaos histophorus*	LC
154.	Gobiiformes	Oxudercidae	*Stigmatogobius sadanundio*	NE
155.	Gobiiformes	Oxudercidae	*Taenioides anguillaris*	LC
156.	Gobiiformes	Oxudercidae	*Taenioides buchanani*	DD

157.	Gobiiformes	Oxudercidae	*Taenioides cirratus*	DD
158.	Gobiiformes	Oxudercidae	*Yongeichthys nebulosus*	LC
159.	Gobiiformes	Oxudercidae	*Yongeichthys criniger*	LC
160.	Gobiiformes	Trypauchenidae	*Trypauchen vagina*	LC
161.	Mugiliformes	Mugilidae	*Crenimugil seheli*	LC
162.	Mugiliformes	Mugilidae	*Ellochelon vaigaiensis*	LC
163.	Mugiliformes	Mugilidae	*Mugil cephalus*	LC
164.	Mugiliformes	Mugilidae	*Osteomugil cunnesius*	NE
165.	Mugiliformes	Mugilidae	*Planiliza macrolepis*	LC
166.	Mugiliformes	Mugilidae	*Planiliza melanoptera*	LC
167.	Mugiliformes	Mugilidae	*Planiliza parsia*	LC
168.	Mugiliformes	Mugilidae	*Rhinomugil corsula*	LC
169.	Mugiliformes	Mugilidae	*Sicamugil hamiltonii*	NE
170.	Mugiliformes	Mugilidae	*Valamugil speigleri*	NE
171.	Cichliformes	Cichlidae	*Etroplus suratensis*	LC
172.	Cichliformes	Cichlidae	*Oreochromis mossambicus*	VU
173.	Cichliformes	Cichlidae	*Pseudetroplus maculatus*	LC
174.	Blenniformes	Blennidae	*Omobranchus ferox*	LC
175.	Beloniformes	Hemiramphidae	*Hemiramphus far*	NE
176.	Beloniformes	Hemiramphidae	*Hyporhamphus quoyi*	NE
177.	Beloniformes	Hemiramphidae	*Hyporhamphus dussumieri*	LC
178.	Beloniformes	Hemiramphidae	*Hyporhamphus limbatus*	LC
179.	Beloniformes	Hemiramphidae	*Rhynchorhamphus georgii*	LC
180.	Beloniformes	Hemiramphidae	*Zenarchopterus striga*	LC
181.	Beloniformes	Horaichthyidae	*Oryzias setnai*	LC
182.	Beloniformes	Belonidae	*Strongylura leiura*	LC
183.	Beloniformes	Belonidae	*Strongylura strongylura*	LC
184.	Beloniformes	Belonidae	*Xenentodon cancila*	LC
185.	Beloniformes	Adrianichthyidae	*Oryzias melastigma*	LC
186.	Beloniformes	Adrianichthyidae	*Oryzias dancena*	LC
187.	Cyprinodontiformes	Aplocheilidae	*Aplocheilus panchax*	LC
188.	Cyprinodontiformes	Aplocheilidae	*Aplocheilus blockii*	LC
189.	Synbranchiformes	Synbranchidae	*Ophichthys cuchia*	LC
190.	Carangiformes	Carangidae	*Alectis ciliaris*	LC
191.	Carangiformes	Carangidae	*Alectis indica*	LC
192.	Carangiformes	Carangidae	*Alepes kleinii*	LC
193.	Carangiformes	Carangidae	*Atropus atropus*	LC
194.	Carangiformes	Carangidae	*Atule mate*	LC
195.	Carangiformes	Carangidae	*Carangoides ciliarius*	LC

196.	Carangiformes	Carangidae	*Carangoides malabaricus*	LC
197.	Carangiformes	Carangidae	*Carangoides preustus*	LC
198.	Carangiformes	Carangidae	*Caranx heberi*	LC
199.	Carangiformes	Carangidae	*Caranx ignobilis*	LC
200.	Carangiformes	Carangidae	*Caranx sexfasciatus*	LC
201.	Carangiformes	Carangidae	*Gnathanodon speciosus*	LC
202.	Carangiformes	Carangidae	*Megalaspis cordyla*	LC
203.	Carangiformes	Carangidae	*Parastromateus niger*	LC
204.	Carangiformes	Carangidae	*Scomberoides lysan*	LC
205.	Carangiformes	Carangidae	*Scomberoides tol*	LC
206.	Carangiformes	Carangidae	*Trachinotus blochii*	LC
207.	Carangiformes	Carangidae	*Trachinotus mookalee*	LC
208.	Carangiformes	Menidae	*Mene maculata*	LC
209.	Istiophoriformes	Sphyraenidae	*Sphyraena barracuda*	LC
210.	Istiophoriformes	Sphyraenidae	*Sphyraena jello*	LC
211.	Istiophoriformes	Sphyraenidae	*Sphyraena obtusata*	LC
212.	Anabantiformes	Anabantidae	*Anabas testudineus*	LC
213.	Anabantiformes	Nandidae	*Nandus nandus*	LC
214.	Pleuronectiformes	Soleidae	*Solea heinii*	DD
215.	Pleuronectiformes	Cynoglossidae	*Cynoglossus arel*	DD
216.	Pleuronectiformes	Cynoglossidae	*Cynoglossus cynoglossus*	LC
217.	Pleuronectiformes	Cynoglossidae	*Cynoglossus lida*	LC
218.	Pleuronectiformes	Cynoglossidae	*Cynoglossus lingua*	LC
219.	Pleuronectiformes	Cynoglossidae	*Cynoglossus macrostomus*	VU
220.	Pleuronectiformes	Cynoglossidae	*Cynoglossus puncticeps*	LC
221.	Pleuronectiformes	Paralichthyidae	*Pseudorhombus arsius*	LC
222.	Pleuronectiformes	Paralichthyidae	*Pseudorhombus elevatus*	LC
223.	Pleuronectiformes	Paralichthyidae	*Pseudorhombus triocellatus*	LC
224.	Pleuronectiformes	Bothidae	*Bothus myriaster*	LC
225.	Pleuronectiformes	Soleidae	*Brachirus orientalis*	LC
226.	Pleuronectiformes	Soleidae	*Brachirus pan*	LC
227.	Pleuronectiformes	Soleidae	*Dagetichthys albomaculatus*	LC
228.	Pleuronectiformes	Soleidae	*Solea ovata*	LC
229.	Pleuronectiformes	Soleidae	*Zebrias synopturoides*	LC
230.	Pleuronectiformes	Cynoglossidae	*Cynoglossus arel*	DD
231.	Pleuronectiformes	Cynoglossidae	*Cynoglossus bilineatus*	DD
232.	Pleuronectiformes	Cynoglossidae	*Cynoglossus puncticeps*	LC
233.	Pleuronectiformes	Cynoglossidae	*Cynoglossus lingua*	LC
234.	Pleuronectiformes	Cynoglossidae	*Cynoglossus cynoglossus*	LC

235.	Pleuronectiformes	Cynoglossidae	*Paraplagusia bilineata*	LC
236.	Sygnathiformes	Fistulariidae	*Fistularia commersonii*	LC
237.	Callionymiformes	Callionymidae	*Callionemus carebares*	LC
238.	Scombriformes	Scombridae	*Rastrelliger kanagurta*	LC
239.	Scombriformes	Scombridae	*Scomberomorus commerson*	NT
240.	Scombriformes	Stromateidae	*Pampus argentius*	NE
241.	Scombriformes	Stromateidae	*Pampus chinensis*	NE
242.	Scombriformes	Trichiuridae	*Eupleurogrammus muticus*	DD
243.	Scombriformes	Trichiuridae	*Lepturacanthus pantului*	DD
244.	Scombriformes	Trichiuridae	*Lepturacanthus savala*	NE
245.	Scombriformes	Trichiuridae	*Trichiurus gangeticus*	NE
246.	Scombriformes	Trichiuridae	*Trichiurus lepturus*	LC
247.	Perciformes	Centropomidae	*Lates calcarifer*	LC
248.	Perciformes	Cepolidae	*Acanthocepola abbreviata*	LC
249.	Perciformes	Gerreidae	*Gerres filamentosus*	LC
250.	Perciformes	Gerreidae	*Gerres erythrourus*	LC
251.	Perciformes	Gerreidae	*Gerres limbatus*	LC
252.	Perciformes	Gerreidae	*Gerres oyena*	LC
253.	Perciformes	Haemulidae	*Pomadasys argenteus*	LC
254.	Perciformes	Haemulidae	*Pomadasys kaakan*	LC
255.	Perciformes	Haemulidae	*Pomadasys maculatus*	LC
256.	Perciformes	Leiognathidae	*Eubleekeria bindus*	LC
257.	Perciformes	Leiognathidae	*Eubleekeria blochii*	LC
258.	Perciformes	Leiognathidae	*Eubleekeria splendens*	LC
259.	Perciformes	Leiognathidae	*Gazza minuta*	LC
260.	Perciformes	Leiognathidae	*Leiognathus berbis*	NE
261.	Perciformes	Leiognathidae	*Leiognathus daura*	NE
262.	Perciformes	Leiognathidae	*Leiognathus equulus*	LC
263.	Perciformes	Leiognathidae	*Leiognathus ruconius*	NE
264.	Perciformes	Leiognathidae	*Photopectoralis bindus*	DD
265.	Perciformes	Leiognathidae	*Secutor insidiator*	DD
266.	Perciformes	Lutjanidae	*Lutjanus argentimaculatus*	LC
267.	Perciformes	Lutjanidae	*Lutjanus fulviflamma*	LC
268.	Perciformes	Lutjanidae	*Lutjanus johnii*	LC
269.	Perciformes	Lutjanidae	*Lutjanus lemniscatus*	LC
270.	Perciformes	Lutjanidae	*Lutjanus russelli*	LC
271.	Perciformes	Lutjanidae	*Lutjanus sebae*	LC
272.	Perciformes	Mullidae	*Upeneus taeniopterus*	LC
273.	Perciformes	Mullidae	*Upeneus tragula*	LC

274.	Perciformes	Mullidae	*Upeneus sulphureus*	LC
275.	Perciformes	Toxotidae	*Toxotes jaculatrix*	LC
276.	Perciformes	Polynemidae	*Eleutheronema tetradactylum*	EN
277.	Perciformes	Polynemidae	*Leptomelanosoma indicum*	NE
278.	Perciformes	Polynemidae	*Polydactylus sextarius*	NE
279.	Perciformes	Polynemidae	*Polynemus paradiseus*	LC
280.	Perciformes	Polynemidae	*Polynemus plebeius*	NE
281.	Perciformes	Scatophagidae	*Scatophagus argus*	LC
282.	Perciformes	Serranidae	*Epinephelus bleekeri*	NT
283.	Perciformes	Serranidae	*Epinephelus coioides*	NT
284.	Perciformes	Serranidae	*Epinephelus lanceolatus*	VU
285.	Perciformes	Serranidae	*Epinephelus malabaricus*	NT
286.	Perciformes	Serranidae	*Epinephelus chlorostigma*	LC
287.	Perciformes	Serranidae	*Epinephelus tauvina*	DD
288.	Perciformes	Siganidae	*Siganus canaliculatus*	LC
289.	Perciformes	Siganidae	*Siganus javus*	LC
290.	Perciformes	Siganidae	*Siganus virgatus*	LC
291.	Perciformes	Teraponidae	*Terapon jarbua*	LC
292.	Perciformes	Terapontidae	*Terapon puta*	LC
293.	Scorpaeniformes	Platycephalidae	*Cociella crocodila*	LC
294.	Scorpaeniformes	Platycephalidae	*Grammoplites scaber*	NE
295.	Scorpaeniformes	Platycephalidae	*Platycephalus indicus*	DD
296.	Scorpaeniformes	Scorpaenidae	*Minous monodactylus*	LC
297.	Moroniformes	Drepaneidae	*Drepane longimana*	LC
298.	Moroniformes	Drepaneidae	*Drepane punctata*	LC
299.	Acanthurifomes	Acanthuridae	*Acanthurus mata*	LC
300.	Acanthurifomes	Acanthuridae	*Acanthurus xanthopterus*	LC
301.	Acanthurifomes	Sciaenidae	*Chrysochir aureus*	LC
302.	Acanthurifomes	Sciaenidae	*Daysciaena albida*	LC
303.	Acanthurifomes	Sciaenidae	*Dendrophysa russelli*	LC
304.	Acanthurifomes	Sciaenidae	*Johnius amblycephalus*	LC
305.	Acanthurifomes	Sciaenidae	*Johnius belangeri*	LC
306.	Acanthurifomes	Sciaenidae	*Johnius carutta*	LC
307.	Acanthurifomes	Sciaenidae	*Johnius coiter*	LC
308.	Acanthurifomes	Sciaenidae	*Johnius dussumieri*	LC
309.	Acanthurifomes	Sciaenidae	*Johnius sina*	DD
310.	Acanthurifomes	Sciaenidae	*Kathala axillaris*	LC
311.	Acanthurifomes	Sciaenidae	*Nibea soldado*	LC

312.	Acanthurifomes	Sciaenidae	*Otolithoides pama*	DD
313.	Acanthurifomes	Sciaenidae	*Otolithoides biauritus*	DD
314.	Acanthurifomes	Sciaenidae	*Otolithus ruber*	LC
315.	Acanthurifomes	Sciaenidae	*Panna microdon*	LC
316.	Acanthurifomes	Sciaenidae	*Pennahia macrophthalmus*	LC
317.	Acanthurifomes	Sciaenidae	*Protonibea diacanthus*	NT
318.	Spariformes	Sparidae	*Acanthopagrus berda*	LC
319.	Spariformes	Sparidae	*Argyrops spinifer*	LC
320.	Spariformes	Lethrinidae	*Lethrinus nebulosus*	LC
321.	Spariformes	Lobotidae	*Lobotes surinamensis*	LC
322.	Spariformes	Sparidae	*Rhabdosargus sarba*	LC
323.	Spariformes	Sillaginidae	*Sillaginopsis panijus*	LC
324.	Spariformes	Sillaginidae	*Sillago sihama*	LC
325.	Tetraodontiformes	Tetraodontidae	*Chelonodontops potoca*	LC
326.	Tetraodontiformes	Tetraodontidae	*Chelonodon pleurospilus*	EN
327.	Tetraodontiformes	Tetraodontidae	*Dichotomycetere fluviatilis*	LC
328.	Tetraodontiformes	Tetraodontidae	*Lagocephalus lunaris*	LC
329.	Tetraodontiformes	Tetraodontidae	*Leiodon cutcutia*	LC
330.	Tetraodontiformes	Tetraodontidae	*Takifugu oblongus*	LC
331.	Tetraodontiformes	Triacanthidae	*Triacanthus biaculeatus*	LC

Table 4: The Order-wise fishes and their numbers

Sl.No	Order	Number
1.	Carcharhiniformes	3
2.	Torpediniformes	1
3.	Pristiformes	1
4.	Myliobatifomes	6
5.	Elopiformes	2
6.	Anguilliformes	13
7.	Osteoglossiformes	1
8.	Clupeiformes	42
9.	Gonorynchiformes	1
10.	Cypriniformes	8
11.	Siluriformes	17
12.	Aulopifomes	2
13.	Holocentriformes	1
14.	Batrachoidiformes	1
15.	Kurtiformes	2
16.	Gobiiformes	60

17.	Mugiliformes	9
18.	Cichliformes	3
19.	Blenniformes	1
20.	Beloniformes	12
21.	Cyprinodontiformes	2
22	Synbranchiformes	1
23.	Carangiformes	19
24.	Istiophoriformes	3
25.	Anabantiformes	2
26.	Pleuronectiformes	22
27.	Sygnathiformes	1
28.	Callionymiformes	1
29.	Scombriformes	9
30.	Perciformes	46
31.	Scorpaeniformes	4
32.	Moroniformes	2
33.	Acanthurifomes	19
34.	Spariformes	7
35.	Tetraodontiformes	7
	Total species	331

IUCN status

The IUCN (International Union for Conservation of Nature) Red List Categories and Criteria are intended to be an easily and widely understood system for classifying species at high risk of global extinction. It divides species into nine categories: Not Evaluated, Data Deficient, Least Concern, Near Threatened, Vulnerable, Endangered, Critically Endangered, Extinct in the Wild and Extinct.

1. **Extinct (EX)** – beyond reasonable doubt that the species is no longer extant.
2. **Extinct in the wild (EW)** – survives only in captivity, cultivation and/or outside native range, as presumed after exhaustive surveys.
3. **Critically endangered (CR)** – in a particularly and extremely critical state.
4. **Endangered (EN)** – very high risk of extinction in the wild, meets any of criteria A to E for Endangered.
5. **Vulnerable (VU)** – meets one of the 5 Red List criteria and thus considered to be at high risk of unnatural (human-caused) extinction without further human intervention.

6. **Near threatened (NT)** – close to being endangered in the near future.
7. **Conservation Dependent (CD)** - a category whose documentation previously was absent - inferred to simply be an oversight
8. **Least concern (LC)** – unlikely to become endangered or extinct in the near future.
9. **Data deficient (DD)** – No sufficient data to assess their risk of extinction
10. **Not evaluated (NE)** – Not yet having been assessed

IUCN Red List Categories were verified with the IUCN's online database and the details of number of fishes fall under these categories along with percentage are given in Table 5-6 and Fig.1. Among the IUCN Red List Category, 2 Critically Endangered (CR) species, 4 Endangered (EN) species, 8 Vulnerable (VU) species and 13 Near Threatened (NT) species are found observed and they indicate the high-risk category based on the IUCN status. Further, 239 species of Least Concern (LC) species, 34 species of Data Deficient (DD) and 32 Not Evaluated (NE) are noticed.

Table 5: IUCN status of Red list Category of Fish diversity – their number and percentage

IUCN Red list Category	**Number**	**%**
Critically Endangered (CR)	2	1
Endangered (EN)	4	1
Vulnerable (VU)	8	2
Near Threatened (NT)	13	4
Least Concern (LC)	238	72
Data Deficient (DD)	34	10
Not Evaluated (NE)	32	10
Total species	**331**	**100**

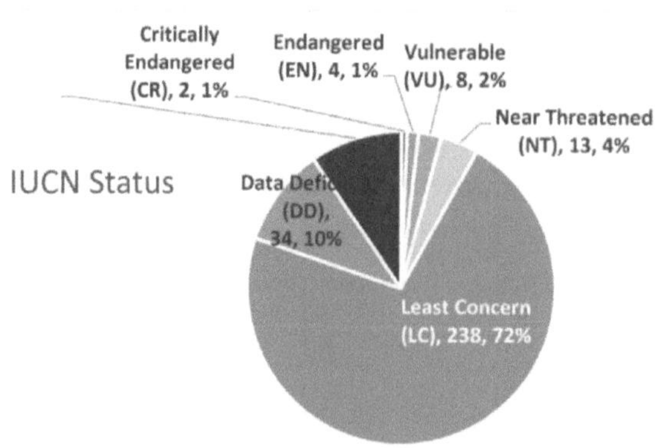

Fig. 1: IUCN status of Red list Category of Fish diversity – their number and percentage

Table 6: Species under high risk in the red list category

Sl.No	Order	Family	Species Name	IUCN Status
1.	Carcharhiniformes	Carcharhinidae	*Glyphis gangeticus*	CR
2.	Pristiformes	Rhinobatidae	*Rhinobatos annandalei*	CR
3.	Myliobatifomes	Dasyatidae	Pateobatis bleekeri	EN
4.	Myliobatifomes	Dasyatidae	*Himantura uarnak*	EN
5.	Myliobatifomes	Myliobatidae	*Aetobatus narinari*	EN
6.	Perciformes	Polynemidae	*Eleutheronema tetradactylum*	EN
7.	Carcharhiniformes	Carcharhinidae	Carcharhinus limbatus	VU
8.	Carcharhiniformes	Carcharhinidae	*Carcharhinus melanopterus*	VU
9.	Cichliformes	Cichlidae	*Oreochromis mossambicus*	VU
10.	Clupeiformes	Clupeidae	*Tenualosa toli*	VU
11.	Cypriniformes	Cyprinidae	Cirrhinus cirrhosus	VU
12.	Myliobatifomes	Dasyatidae	*Telatrygon zugei*	VU
13.	Perciformes	Serranidae	*Epinephelus lanceolatus*	VU
14.	Pleuronectiformes	Cynoglossidae	*Cynoglossus macrostomus*	VU
15.	Acanthurifomes	Sciaenidae	Protonibea diacanthus	NT
16.	Anguilliformes	Anguillidae	*Anguilla bengalensis*	NT
17.	Anguilliformes	Anguillidae	*Anguilla bicolor*	NT
18.	Aulopifomes	Synodontidae	*Harpadon nehereus*	NT
19.	Myliobatifomes	Dasyatidae	Pastinachus sephen	NT
20.	Perciformes	Serranidae	*Epinephelus bleekeri*	NT
21.	Perciformes	Serranidae	*Epinephelus coioides*	NT
22.	Perciformes	Serranidae	*Epinephelus malabaricus*	NT
23.	Scombriformes	Scombridae	Scomberomorus commerson	NT
24.	Siluriformes	Ariidae	*Arius gagora*	NT
25.	Siluriformes	Siluridae	*Ompok pabda*	NT
26.	Siluriformes	Siluridae	*Ompok bimaculatus*	NT
27.	Siluriformes	Siluridae	*Wallago attu*	NT

CONSERVATION OF MANGROVES

Mangrove ecosystems provide a wide range of benefits to mankind, including flood and cyclone damage mitigation (Menendez *et al.*, 2020), coastal protection (Hochard *et al.*, 2019), carbon sequestration (Kristensen *et al.*, 2008), and recreation (Spalding and Parrett, 2019). Similarly, mangroves play a multifaceted and indispensable role in supporting a multitude of ecosystem services. They contribute to soil formation, harness the power of photosynthesis for primary production, facilitate nutrient cycling, and regulate water movement within coastal regions. Beyond these ecological services, they serve as a vital source of natural products, providing proteins, tannins, and timbers to local communities (Yuki Yamamoto, 2023). Additionally, mangroves actively participate in regulating crucial ecosystem processes, including biological control, nutrient cycling, air quality maintenance, and the preservation of biodiversity (Yuki Yamamoto, 2023).

Large-scale destruction of mangroves is a relatively recent phenomenon. The pressures of increasing populations, food production and industrial and urban development threaten mangrove habitats (Field 2000). Conversion of mangroves for commercial aquaculture, agricultural, industrial and tourist facilities and industrial-scale forestry practices have led to dramatic losses of mangroves (Valiela *et al.* 2001). The immense socio-economic importance of mangroves for fisheries, offers a strong incentive to conserve mangroves for long term through proper management of both mangroves and fisheries associated with mangroves. In order to conserve mangrove forests with respect to fisheries, certain broad management needs to be considered (Bijoy Nandan *et al.*, 2021) avoiding mangrove loss, restoring natural mangroves and managing fisheries.

Most strikingly, mudskippers are strongly associated with mangrove ecosystems and tropical mudflats (Ansari *et al.*, 2014). Sundarbans mangroves support high density and abundance of mudskippers as it has a vast area of potential mudflats. However, the clayey dominated periphery/bunds of the riverine system of Sundarbans face damage through soil erosion during floods, which replace the soft muddy area where mudskippers inhabit. Mudflats of coastal regions are inhabited by mudskippers and thus mudflats act not only as feeding ground but also as breeding and nursery ground. Mudskipper's diversity helps to understand the health of coastal regions (Ravi *et al.*, 2021), hence needs adequate conservation of the resource.

REFERENCES

Aburto-Oropeza, O., Ezcurra, E., Danemann, G., Valdez, V., Murray, J., Sala, E., 2008. Mangroves in the Gulf of California increase fishery yields. Proc. Natl. Acad. Sci., 105 (30), 10456–10459.

Alongi, D.M. (2009). The Energetics of Mangrove Forests, Springer Verlag, New York. Ansari, Abid, A., Subrata Trivedi, Shalini Saggu and Hasibur Rehman. 2014. Mudskipper: A biological indicator for environmental monitoring and assessment of coastal waters. J. Entomol. Zool. Stud., 2(6): 22-33.

Bijoy Nandan, S, S. Sreelekshmi and M. Harikrishnan, 2021. Fishery Production Potential of the Mangroves in India - A Review. J. Indian Soc. Coastal Agric. Res., 39(2): 151-160.

Chandrasekharan, V.S. 2000. Relationship between plankton and finfish and shellfish juveniles in Pichavaram mangrove waterways, Southeast coast of India. Seaweed Res. Utiln 22 (1&2): 199-207.

Das, S., 2017. Ecological restoration and livelihood: contribution of planted mangroves as nursery and habitat for artisanal and commercial fishery. World Dev. 94, 492–502.

Field, C.D. (2000). Mangroves. In: Seas at the Millennium: An Environmental Evaluation. Volume III, Global Issues and Processes (ed. C. Sheppard), Ch. 108, pp. 17–30.

FSI, 2017. State of Forest Report, Forest Survey of India, Dehradun, India

Hochard, J.P., Hamilton, S., Barbier, E.B., 2019. Mangroves shelter coastal economic activity from cyclones. Proc. Natl. Acad. Sci., 116 (25): 12232–12237.

Krishnamurthy, K., et al. 1984. The Distribution of Fishes in The Global Mangroves. Current Science, 53(17): 901–06.

Kristensen, E., Bouillon, S., Dittmar, T., Marchand, C., 2008. Organic carbon dynamics in mangrove ecosystems: a review. Aquat. Bot.,89 (2), 201–219.

Menendez, P., Losada, I.J., Torres-Ortega, S., Narayan, S., Beck, M.W., 2020. The global flood protection benefits of mangroves. Sci. Rep., 10 (1), 1–11.

Mumby, P.J., Edwards, A.J., Arias-González, J.E., Lindeman, K.C., Blackwell, P.G., Gall, A., Gorczynska, M.I., Harborne, A.R., Pescod, C.L., Renken, H., et al., 2004. Mangroves enhance the biomass of coral reef fish communities in the Caribbean. Nature, 427 (6974), 533–536.

Paromita Ray , Giridhar Malla, J.A. Johnson and K. Sivakumar 2022. An overview of the fish diversity and their threats in the Gowthami-Godavari Estuary in Andhra Pradesh, India. Journal of Threatened Taxa,14(8): 21588–21604.

Ravi,V. G. Mahadevan and R. Bharathi, 2021. Diversity of Mudskippers along the selective mudflats of mangroves of east coast of India. J. Mar. Biol. Ass. India, 63 (1): 110-116.

Spalding, M., Parrett, C.L., 2019. Global patterns in mangrove recreation and tourism. Mar. Policy, 110, 103540.

Valiela, I., Bowen, J.L. and York, J.K. (2001). Mangrove forests: one of the world's threatened major tropical environments. Bioscience, 51(10): 807–815.

Venkatachalam, S., Kandasamy, K., Krishnamoorthy, I., Narayanasamy, R., 2018. Survival and growth of fish (Lates calcarifer) under integrated mangrove-aquaculture and open-aquaculture systems. Aquac. Rep., 9, 18–24.

Venkatraman, C., P. Padmanaban S. Shrinivaasu and G. Sivaleela, 2016. Faunal Diversity of Bhitarkanika Mangroves, Odisha. Rec. zool. Surv. India, 116(Part-4) : 407-430.

Yuki Yamamoto, 2023. Living under ecosystem degradation: Evidence from the mangrove–fishery linkage in Indonesia. Journal of Environmental Economics and Management,118 (2023)102788.

2

Diversity of Commercially Important Decapod Crustaceans from Mudasalodai Tamil Nadu, Southeast Coast of India

V. Ravi

INTRODUCTION

Biodiversity can be useful as an ecosystem indicator for conservation and monitoring, through continuous assessment of its main properties including stability, primary productivity, exploitation tolerance and even global environmental changes (Douglas *et al.*, 2012). Crustaceans are found one of the most morphologically diverse taxonomic groups on the planet (Martin and Davis, 2001). Among them, decapods are the most studied taxon, mainly on account of the commercial interest of some species and their great diversity (Isarch and Munoz, 2015). Decapods are chiefly composed of marine species that live in waters depths ranging from shallow to deeper than 5000 m. Isarch and Munoz (2015) described the importance of this group lies in several factors such as the great biomass they represent, their significant role in marine food webs and the commercial interest of many decapod species. Order Decapoda comprises the commercially important shrimps, crabs and lobsters. Out of 649 species of marine crabs so far recorded from Indian waters, only 12 species of edible crabs namely *Portunus sanguinolentus, P. plagicus, Charybdis feriatus, C. lucifera, C. annulata, C. natator, Scylla serrata, S. tranquebarica, Matuta lunaris, Sesarma tetragonum* and *Varuna litterata*, which inhabit the coastal waters and adjoining brackishwater environments, support commercial fisheries (Menon and Pillai, 1996). However substantial quantities of crabs are landed every year as by-catches of shrimp trawlers and indigenous fishing units throughout the county.

India represents one of the major contributors to the world production of marine crustaceans. Shrimps constitute as the main seafood export industry for the major foreign exchange earner as well as a source of livelihood for millions of fish workers. As shrimps are targeted for fishing, information on the present status of the fishery is critical for implementation of proper management. Some of the important penaeid shrimps that support commercial fisheries along the Indian seas are *Fenneropenaeus indicus*, *P.semisulcatus P.monodon*, *P.merguiensis*, *P.japonicus*, *P.penicillatus*, *Metapenaeus dobsoni*, *M.monoceros*, *M. affinis*, *M. kutchensis* , *M. brevicornis*, *Parapenaeopsis stylifera*, *P.hardwickii*, *P.sculptilis*, *P.maxillipedo*, *P.uncta*, *Trachypenaeus curvirostris*, *Metapenaeopsis stridulans*, *Parapenaeus longipes*, *Solenocera crassicornis* and *S.choprai* (Mohan Joseph and Jayaprakash, 2003). Rajkumar *et al.* (2015) carried out molecular identification of shrimp species, *Penaeus semisulcatus, Metapenaeus dobsoni, Metapenaeus brevicornis, Fenneropenaeus indicus, Parapenaeopsis stylifera* and *Solenocera crassicornis* from Tamil Nadu, India.

Being the more diverse group of brachyuran crabs, they are found at even 6000 m depths to seas shore and they are dominant in many estuarine habitats where salinity and temperatures can fluctuate dramatically daily (Ng *et al.,* 2008). Decapod crustaceans are one of the important faunal communities serving as very good food source to human beings in the marine and brackish water ecosystems. The brachyuran crabs, are found reported about 700 genera and 5000 to 10,000 species worldwide (Kaestner, 1970; Sternberg and Cumberlidge, 2001; Ng *et al.,* 2008; Yeo *et al.,* 2008). About 990 species of marine brachyuran crabs belonging to 281 genera and 36 families are recorded in Indian waters (Myla Chakravarty *et al.,* 2015). Of which, 404 species of crabs belonging to 26 families and 152 genera are reported in Tamil Nadu (Kathirvel, 2008). There were number of studies about the diversity of brachyuran crabs near our study area reported. On an average, about 25000 tons of crabs are exploited annually from marine sector, of which over 50% is landed in Gujarat and Tamil Nadu alone. Maximum percentages of crab catches are landed from Gulf of Mannar, Palk Bay, Nagapattinam and Puducherry landings of Tamil Nadu coast in India (Rao *et al.,* 1973). The crabs were reported from Chennai coast (Thangaraj Subramanian, 1998) Akkaraipettai (Nagapattinam) landing centres with 3 families (Portunidae, Sesarmidae and Ocypodidae) 6 genera and 13 species (Rajaraman *et al.,* 2016). In the present study, an attempt to assess the present status of biodiversity of decapod crustaceans (shrimps and crabs) were made on marine captured and landed at Mudasalodai, Tamil Nadu, India.

MATERIALS AND METHODS

Study area

Mudasalodai(11°29′N 79°46′E) is a well-known fish landing centre, located near Parangipettai Marine Biological station, Cuddalore district, Tamil Nadu, south east coast of India. About 300 mechanized boats (trawlers) and non-mechanized are engaged for fishing activities regularly. Trawlers are operated for more than half day and approximately 500 families are involving in fishing activities.

Decapod crustacean

In the present study, samplings of the decapod crustacean specimens (commercially important shrimps and crabs) were collected from the study area during July 2022- June 2023. Crustaceans were identified to species level and their numbers and individual wet weights recorded. The crustaceans were identified by following Chhapghar (1957), Sakai (1976), Galil (1994, 1997), (Ng *et al.,* 2008) for brachyuran crabs, Sethuramalingam and Ajmal Khan (1991) and Perez Farfante and Kinsley (1967) for shrimps. Month-wise and season-wise trawl landing data were subjected to univariate analysis for species diversity, species richness and evenness in the crustacean population using BioDiversityPro software (McAleece, *et al.* 1997). Margaleff M richness, Shannon evenness (J') and Shannon diversity (H') were used to evaluate relative abundance and evenness of species diversity. For cluster analysis, Bray-Curtis similarity was performed to assess natural grouping of crustaceans during different months.

Biodiversity indices

a. Shannon – Wiener index

In the present study, the data were analyzed for diversity index ($H' \log^2$) using Shannon – Wiener's (1949) formula:

$$H' = -\sum^{S} Pi \log 2\, Pi$$

$$i = 1$$

$$H' = \frac{3.3219(N \log N - \sum ni - \log ni}{N}$$

which can be rewritten as

H' =

where, H'= species diversity in bits of information per individual

ni = proportion of the samples belonging to the ith species (number of individuals of the ith species)

N = total number of individuals in the collection and

$\sum$ = sum

b. Species richness

Species richness (S) was calculated using two formulae given by Margalef's diversity index (Margalef, 1958).

Margalef's index = $\frac{(S-1)}{\ln N}$

S = total number of species

N = total number of individuals in the sample

In = natural logarithm

c. Species evenness index

The equitability (J') was computed using the following formula of Pielou (1966):

$$J' = \frac{H'}{\log_2 S} \text{ or } \frac{H'}{\ln S}$$

where, J' = evenness, H' = species diversity in bits of information per individual and S = total number of species.

In addition, statistical analysis was performed using Pearson's correlation coefficient (r), mean and percentage calculations for the landings of shrimps and crabs.

RESULTS

Checklist and Systematic position of the commercially important decapod crustaceans recorded in Station

Phylum Arthropoda

Class Crustacea

Order Decapoda

Family Penaeidae (Shrimps)

Genus *Fenneroppenaeus*

Fenneropenaeus indicus (H.Milne Edwards, 1837)

Fenneropenaeus merguiensis (De Man, 1888)

Fenneropenaeus penicillatus (Alcock, 1905)

Genus *Penaeus*

Penaeus monodon (Fabricius, 1798)

Penaeus semisulcatus (De Man, 1888)

Genus *Metapenaeus*

Metapenaeus affinis (H.Milne Edwards, 1837)

Metapenaeus brevicornis (H.Milne Edwards, 1837)

Metapenaeus dobsoni (Miers, 1878)

Metapenaeus monoceros (Fabricius, 1798)

Genus Parapenaeopsis

Parapenaeopsis stylifera (H.Milne Edwards, 1837)

Family Portunidae (Marine crabs)

Genus *Portunus*

Portunus. pelagicus (Linnaeus, 1758)

Portunus sanguinolentus (Herbst, 1783)

Genus *Scylla*

Scylla serrata (Forskal, 1775)

Scylla tranqubarica (Fabricius, 1798)

Genus *Charybdis*

Charybdis feriatus (Linnaeus, 1758)

Charybdis granulata (De Haan, 1833)

Charybdis lucifera (Fabricius, 1798)

Charybdis natator (Herbst, 1794)

Charybdis truncata (Fabricius, 1798)

Family Palinuridae (Lobster)

Genus *Panulirus*

Panulirus homarus (Linnaeus, 1758)

Panulirus ornatus (Fabricius, 1798)

Panulirus penicillatus (Oliver, 1791)

Panulirus polyphagus (Herbst, 1793)

Panulirus versicolor (Latrielle, 1804)

Genus *Puerulus*

Puerulus sewelli Ramadan, 1938

Family Scyllaridae

Genus *Thenus*

Thenus orientalis (Lund, 1793)

Species composition of Decapod crustaceans in the study areas

The list of species recorded Family-wise (decapod crustaceans) in the study areas are presented Fig 1. In the present study, four families belongs the order Decapoda (Class Crustacea) in the Phylum Arthropoda. The families include Penaeidae, Portunidae, Palinuridae and Scyllaridae. Altogether, 26 species belonging to 10 genera were identified as commercial importance in the study area. Family Penaeidae dominated with 10 species of shrimps including *Fenneropenaeus indicus, F.merguiensis, F.penicillatus, Penaeus monodon, P.semisulcatus, Metapenaeus affinis,M.brevicornis, M.dobsoni, M.monoceros* and *Parapenaeopsis stylifera.* Followed by Penaeidae, family Portunidae constituted 9 species of crabs including *Portunus pelagicus, P.sanguinolentus, Scylla serrata, S. tranqubarica, Charybdis feriatus, C. granulata, C. natator, C. lucifera* and *C. truncata.* Similarly lobsters were identified in two families such as Palinuridae (5 species- *P.homarus, P. ornatus, P. pencillatus, P. polyphagus, P.versicolor* and *P.sewelli*) and Scyllaridae (1 species- *Thenus orientalis*).

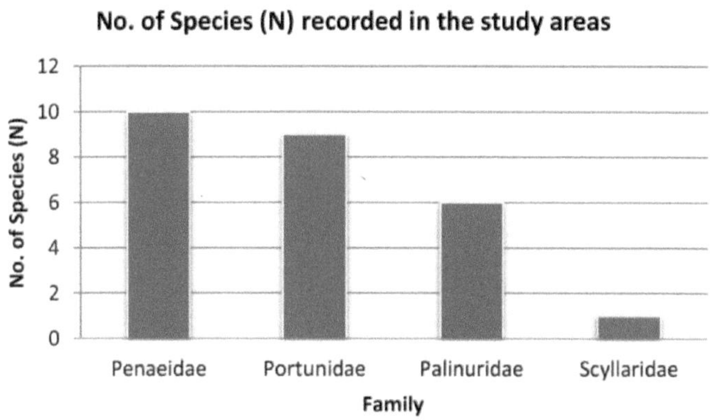

Fig. 1. No. of species recorded Family-wise in the study areas

Biodiversity indices of the Decapod crustaceans in the study area

The data collected in all the Stations were approached to various Univariate Diversity Indices and the results are presented in Table 1.

The species richness (Margaleff richness index M) ranged from *P. monodon* (0.844) and *Puerulus sewelli* (1.899) with mean value of 1.2765 in the study area. The Species evenness (Shannon evenness J') ranged from *S. tranquebarica* (0.725) and *T. orientalis* (0.99) with mean value of 0.94 in the study area. The Species diversity (Shannon H' Diversity) ranged from *S. tranquebarica* (0.437) and *P. versicolor* & *P. sewelli* (0.601 each) with mean value of 0.567 in the study area.

Table 1: Biodiversity indices for decapod crustaceans recorded in the study area

Species	Margaleff M Richness	Shannon J' Evenness	Shannon H' Diversity
Shrimps			
Fenneropenaeus indicus	0.856	0.997	0.6
F. merguiensis	0.887	0.969	0.584
F. penicillatus	0.885	0.984	0.592
Penaeus monodon	0.844	0.996	0.6
P. semisulcatus	0.851	0.979	0.589
Metapenaeus affinis	0.849	0.976	0.588
M. brevicornis	0.909	0.985	0.593
M. dobsoni	0.875	0.971	0.585
M. monoceros	0.867	0.964	0.581
Parapenaeopsis stylifera	0.879	0.991	0.596
Crabs			
Portunus pelagicus	0.916	0.876	0.527
P.sanguinolentus	0.891	0.913	0.55
Scylla serrata	0.88	0.781	0.47
S. tranqubarica	0.885	0.725	0.437
Charybdis feriatus	0.909	0.869	0.523
C. granulate	1.163	0.944	0.568
C. natator	0.938	0.842	0.507
C. lucifera	1.038	0.911	0.549
C. truncate	1.182	0.926	0.557
Lobsters			
Panulirus Homarus	1.576	0.977	0.588
P. ornatus	1.649	0.984	0.592
P. pencillatus	1.732	0.975	0.587
P. polyphagus	1.732	0.996	0.599
P. versicolor	1.784	0.997	0.601
Puerulus sewelli	1.899	0.999	0.601
Thenus orientalis	1.443	0.99	0.596

DISCUSSION

Gathering information on species composition is the basis for the understanding of processes affecting the equilibrium of communities or ecosystems (Bertini *et al.*, 2004). Biodiversity and community structures are now well-known to be important determinants of ecosystem functioning (Raghukumar and Anil, 2003). Species diversity monitoring is a useful technique for assessing damage to the system and maintenance of good species diversity, which is a positive management objective. Decapod crustacean occupies a significant portion of aquatic food resources of the world and hence conservation of these resources is important from commercial as well as ecological point of view (Dineshbabu *et al.,* 2011).

Marine crabs, one of the seafood items in great demand, formed about 11% of the edible crustacean landing in the year 2005. Although about 700 species of brachyuran crabs have been reported from the Indian waters, the Family Portunidae contributes maximum to the commercially important crab fishery of the country. The fishery for edible crabs in the coastal waters is predominated by the Cross crab, Charybdis feriatus, the Spotted crab *Portunus sanguinolentus* and the Reticulate crab *P. pelagicus. C. lucifera* is emerging as a fishery of importance in Andhra Pradesh and Kerala and is slowly becoming popular as an edible item. Crabs are caught as by-catch and more than 80% of the total landing is by trawlers. Indigenous gears such as gill nets and traps are also used in selected areas targeting individual species, especially *P.pelagicus*. Crabs are usually caught from a depth of about 10 m to 60 m. Trawlers occasionally go up to 80 m during the post-monsoon months, along the southwest coast. It is the recent advances in fishing technology that has enabled fishermen to venture into deeper waters engaging themselves in multi-day fishing. This has resulted in increased landing of edible crabs, especially *C. feriatus* Lobsters are one of the highly priced crustaceans in India and are in great demand as a delicacy in the internal market and as a foreign exchange earner in export market. In the present study, lobsters were identified in two families such as Palinuridae (5 species- *P.homarus, P. ornatus, P. pencillatus,P. polyphagus,P.versicolor* and *P.sewelli*) and Scyllaridae (1 species- *Thenus orientalis*). Landings of the lobsters *P. ornatus* and *P. homarus* are reported from Kayapattinam and Tharuvaikulam (Nair *et al.,* 1973; Rajamani and Manickaraja, 1995).

Decapod crustaceans are among the most diverse and abundant animal groups in marine environments. Crustaceans have a significant role in the marine ecosystem, especially in marine food webs, where they link high and low trophic levels (Cartes, 1998). Protecting benthic habitats as a whole would involve the conservation of their decapod communities (Valdes and Gonzalez, 2015). Human exploitation may disastrously impact this fauna, both through fisheries and through the aquarium trade (Calado *et al.* 2003; Costa *et al.* 2007; Balaji *et al.* 2009; Gregati *et al.* 2010). For evaluation of degree of conservation of these resources, continuous assessment programs are needed for sustainable utilization and these findings are helpful in the future.

REFERENCES

Balaji, K., G. Thirumaran, R. Arumugan, K.P. Kumaraguruvasagam and Anantharaman, P. 2009. A review on marine ornamental invertebrates. World Applied Sciences Journa, 17(8): 1054-1059.

Bertini, Giovana, Adilson Fransozo and Gustavo A.S. De Melo, 2004. Biodiversity of brachyuran crabs (Crustacea: Decapoda) from non-consolidated sublittoral bottom on the northern coast of Sa˜ o Paulo State,Brazil.Biodiversity and Conservation 13: 2185–2207.

Boschi E.E. 2000a. Biodiversity of marine decapod brachyurans of the Americas. Journal of Crustacean Biology, 2: 337–342.

Calado, R., J. Lin, A.L. Rhyne, R. Araujo and L. Narciso. 2003. Marine ornamental decapods – popular, pricey and poorly studied. Journal of Crustacean Biology 23(4): 963-973.

Cartes, J. E. 1998. Feeding strategies and partition of food resources in deep-water decapod crustaceans (between 400-2300 m). Journal of the Marine Biological Association of the United Kingdom,78:509-524.

Chapghar, B. F. 1957. On the marine crabs (Decapoda: Brachyura) of Bombay State. J. Bom. Nat. Hist. Soc., 54: 503-549.

Costa, R.C., A. Fransozo, F.A.M. Freire and A.L. Castilho. 2007. Abundance and ecological distribution of the 'sete-barbas' shrimp Xiphopenaeus kroyeri (Heller, 1862) (Decapoda, Penaeoidea) in three bays of the Ubatuba region, southeastern Brazil. Gulf and Caribbean Research, 19: 33-41.

Dineshbabu A.P., Durgekar R.N. and Zacharia P.U., 2011. Estuarine and marine decapods of Karnataka. Inventory. Fishing Chimes, 30(10&11): 20-24.

Douglas Fernandes Rodrigues Alves, Samara de Paiva Barros-Alves, Valter Jose Cobo, Daniel JoseMarcondes Lima and Adilson Fransozo, 2012. Checklist of the brachyuran crabs (Crustacea: Decapoda)in the rocky subtidal of Vitoria Archipelago, Southeast coast of Brazil. Check List, 8(5): 940–950.

Galil, B. S. 1997. Crustacea Decapoda: A revision of the Indo Pacific species of genus Calappa Weber, 1795 (Calappidae). In: Crosnier, A. (Ed.), Resultats desCampagnes MUSORSTOM, Volume 18. Memoires du Museum National d' Histoire naturelle, Paris, 176: 271-335.

Gregati, R.A., V. Fransozo, L.S. Lopez-Greco and M.L. Negreiros-Fransozo. 2010. Reproductive cycle and ovarian development of the marine ornamental shrimp Stenopus hispidus in captivity. Aquaculture, 306: 185-190.

Hooper, D.U.,F.S. Chapin, J.J .Ewel, A. Hector, P. Inchausti, S. Lavorel, J.H. Lawton, D.M Lodge, M.. Loreau and S. Naeem, 2005. Effects of biodiversity on ecosystem functioning: a consensus of current knowledge. Ecol.Monogr., 75: 3-36.

Isarch, Eva García and Isabe Munoz, 2015. Biodiversity and biogeography of decapod crustaceans in the canary current large marine ecosystem.ioc technical series, No. 115, pp. 257-271.

Kaestner, A. 1970. Invertebrate Zoology. III. Crustacea. (Translated and Adopted from the second German, edited by H.W. Levi). Intersciences Publishers, New York, USA.

Kathirvel, M. 2008. Biodiversity of Indian marine brachyuran crabs. Rajiv Gandhi Chair Special Publication, 7: 67 – 78.

Mace, G., Masundire, H., Baillie, J., Ricketts, T., Brooks, T., 2005. Biodiversity. In: Hassan, R., Scholes, R., Ash, N. (Eds.), Ecosystems and Human Well- Being: Current State and Trends (Findings of the Condition and Trends Working Group). Island, p. 77-122.

Margalef, R., 1958. Temporal succession and spatial heterogeneity in phytoplankton. In: Perspectives in Marine biology, Buzzati-Traverso (ed.), Univ. Calif. Press, Berkeley, pp. 323-347.

Martin, J. W and G. E. Davis, 2001. An Updated Classification of the Recent Crustacea. Natural History Museum of Los Angeles County. Science Series, 39: 1 – 124.

McAleece, N., Gage, J.D.G., Lambshead, P.J.D., Paterson, G.L.J. (1997). BioDiversity Professional statistics analysis software. Jointly developed by the Scottish Association for Marine Science and the Natural History Museum London.

Mohan Joseph, M. and A.A. Jayaprakash (Eds), 2003. Status of Exploited Marine Fishery Resources of India. Central Marine Fisheries Research Institute, Kochi.

Myla, Sarvabhowma Chakravarty , Vijaya Bharathi Tirumani and Rama Chandra Ganesh Paturi, 2017. Body Weight, Carapace Length and Width Relationship and Condition Factor of the Mud Crab Scylla serrata (Forskal) in Mangrove Ecosystem , Not. Sci. Biol., 9(3):338-343.

Perez Farfante, I. and Kensley, B. 1997. Penaeoid and sergestoid shrimps and prawns of the world. Keys and diagnosis for the families and genera. Memoires Museum National D'HistoireNaturelle, p. 1-223.

Pielou, E.C. 1966. The measurement of diversity in different types of biological collections. Journal of Theoretical Biology, 13, pp. 131–144.

Raghukumar S. and Anil A.C, 2003. Marine biodiversity and ecosystem functioning: A perspective. Curr.Sci., 84(7):884-892.

Rajamani, M. and M. Manickaraja, 1991. On the collection of spiny lobsters by skin divers in the Gulf of Mannar off Tuticorin. Marine Fisheries Information Service, Technical and Extension Series No. 113: 17-18.

Rajaraman,P. K. Durairaj, D. Prabhakaran and K. Priyadharshini, 2016. Diversity of Brachyuran crabs captured and landed by trawl at Akkaraipettai landing station, Nagapattinam, Tamil Nadu, India. Indo – Asian Journal of Multidisciplinary Research,2(2):576 – 579 2015.

Rajkumar,G. P. Saravana Bhavan, R. Udayasuriyan, C. Vadivalagan, Ramamoorthy,N., P. K. Karuppasamy and R. Sri Sakthi Priyadarshini, 2016. Proximate, amino acid and fatty acid composition the marine crabs from the Southeast coast of India. Journal of Marine Biosciences,2 (1), 91-98.

Rao, P. V., M. Vedavyasa and M. M. Thomas. 1973. Crab fishery resources of India. In: Proceeding Symposium on Living Resources of the Seas around India. 581-591.

Sakai, T. 1976. Crabs of Japan and the adjacent seas. Kodansha Ltd., Tokyo. Japan, 773 pp.

Sethuramalingam, S and Ajmal Khan S. 1991. Brachuran crabs of Parangipettai coast. CAS in Marine Biology, Annamalai University, Parangipettai, India.

Shannon, C. E. and W. Wiener, 1949. The mathematical theory of communication. University of Illinois Press, 177 p.

Sternberg R Von and N. Cumberlidge. 2001. On the hetrotreme - thoracotreme distinction in the Eubrachyurade Saint Laurent, 1980 (Decapods, Brachyura). Crustaceana, 74: 321 – 338.

Thangaraj Subramanian, V. 1998. An assessment off crab resources of Chennai (Madras) Marine Fisheries Information Service, Technical and Extension Series, 152: 2 - 6.

Valdes, L. and Deniz-Gonzalez, I. (eds). 2015. Oceanographic and biological features in the Canary Current Large Marine Ecosystem. IOC-UNESCO, Paris. IOC Technical Series, No. 115: 383.

Yeo, D.C.J., Ng. P.K.L and Cumberlidge, N. 2008. Global diversity of crabs (Crustacea: Decapoda: Brachyura) in fresh water. Hydrobiologia, 595: 275 - 286.

3

Biodiversity of The Clupeiform Fishes from Cuddalore, Tamil Nadu

V. Ravi and M. Sivagangai

INTRODUCTION

Species of the Clupeiformes (Teleostei), commonly known as herrings, shads, menhadens, sardines, anchovies, and their relatives, are a major component of forage fishes in coastal ecosystems and dominate worldwide forage fish landings (Tacon and Metian, 2009a; FAO, 2020). Clupeiformes are globally distributed fishes with having nearly 419 marine, freshwater, and diadromous species (Bloom and Egan, 2018) and they are filter-feeding fishes that form large schools having a diverse group of trophic guilds and habitats (Bloom and Egan, 2018). Additional to providing ecological and economic support, clupeiforms contribute to global food security given their abundance, easy access, and exceptionally high nutrient content (FAO, 2018). In some human communities, clupeiforms comprise the major or the sole protein source (Alder *et al*., 2008; Mohanty *et al*., 2019). Historically, clupeiform presence has been associated with persistent human settlement, growth, and survival (Thornton *et al*., 2010; Levin *et al*., 2016).

Clupeiform fishes are ecologically diverse and span all aquatic habitats, including coastal and open marine environments, oceanic islands, estuaries, and freshwater rivers and lakes (Whitehead, 1985; Lavoue *et al*., 2013). Species can be restricted to marine, estuarine, or fresh waters, or they can be euryhaline, where a subset exhibit diadromy (Whitehead, 1985). Strictly marine clupeiforms (33.7% of all species) are distributed in every ocean, except for the Southern Ocean (Whitehead, 1985), while strictly freshwater species (17.8% of all species) are found on every continent except for Antarctica (Bloom and Egan, 2018). Major threats to clupeiforms are similar to those found for other groups of fishes (e.g., Roberts and Hawkins, 1999; Reynolds *et al*., 2005), with overexploitation as the leading threat for all clupeiforms in all habitats. While overexploitation may be the most prolific threat by impacting the highest number of clupeiforms, pollution may be the most detrimental, as it affects greater numbers of CR species. When compared to other economically and ecologically important fish groups globally assessed using the IUCN Red List methodology, clupeiforms have the lowest estimated percentage of threatened and NT species overall. Using the midpoint of species evaluated as elevated concern, roughly 11% are currently at high risk compared to approximately 22% of tunas and billfishes (Collette *et al.*, 2011), 19% of sparids (Comeros-Raynal *et al.*, 2016), and 19% of groupers (Sadovy de Mitcheson *et al.*, 2020).

Order Clupeiformes comprises five families such as Denticipitidae (denticle herrings), Pristigasteridae (longfin herrings), Engraulidae (anchovies), Chirocentridae (wolf herrings) and Clupeidae (herrings:shads, sprats, sardines, pilchards, and menhadens) (Nelson *et al.*, 2016). and they inhabit about half the species are Indo-West Pacific, and almost one-quarter are in the Western Atlantic. The Indian oil sardine, *Sardinella longiceps* is an important pelagic fish species contributing to about 15% of the total marine fish production in India. Cuddalore district of Tamil Nadu, India has a coastal Length of 57.5 Km. comprising of 49 fishing villages with total marine fishermen population of 47000. Among them 23840 are active fishermen and nearly 15000 fisherwomen are engaged in fishing related activities. There are 235 Mechanized fishing boats and 2345 FRP boats, Non motorised boats 1906 are operating in this district. In the present study, the diversity of Clupeiform fishes is identified in Cuddalore coastal region.

MATERIALS AND METHODS

Description of the Study Area

Cuddalore (Lat.11^{0}43 N., Long. 79^0 49 E) (Fig. 1) is one of Tamil Nadu's most significant fishing districts. Its coastline stretches 57.5 km from Nallavadu in the north to Thandavaraya Sholaganpettai in the south, and it is home to 46 fishing villages with a combined population of over 47,000 fishermen. The Cuddalore fishing harbor is among Tamil Nadu's oldest minor harbors. High-value seafood exports, including as crabs, lobster, cuttlefish, octopus, mollusks, shrimp, and edible fish, are a significant source of economy in this region. As soon as fresh fish are landed in the wee hours of the morning, this landing center will be operational. Regular fish landing hours are from 3 to 5 a.m., and the fish market is open from 4 a.m. until full.

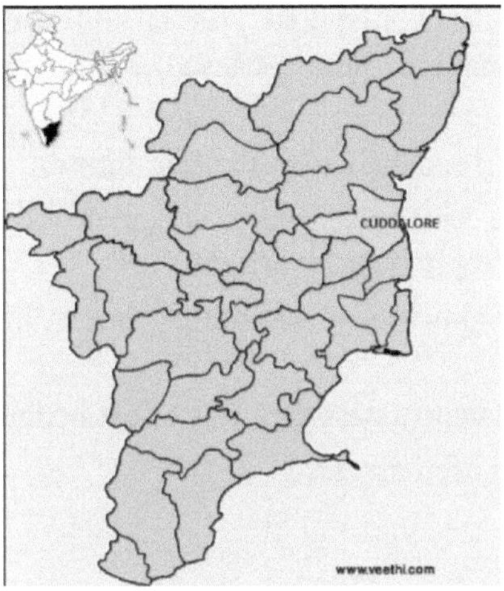

Fig. 1: Study area- Cuddalore, Tamil Nadu, Southeast coast of India

Identification of Clupeiform fishes

In the present study, the clupeiform fishes were collected from the Cuddalore fish landing centre, Tamil Nadu during July 2023 to February 2024. Fish were kept for additional examination in 5% formalin. Fish were identified later after Mohsin and Ambak (1996), Nelson *et al.* (2016), Jayaram (2002), and Compagno (1997).

IUCN status

The IUCN (International Union for Conservation of Nature) Red List Categories and Criteria are intended to be an easily and widely understood system for classifying species at high risk of global extinction. It divides species into nine categories: Not Evaluated, Data Deficient, Least Concern, Near Threatened, Vulnerable, Endangered, Critically Endangered, Extinct in the Wild and Extinct. In the present study, IUCN status of the fishes were verified from the official website of the IUCN organization.

1. Extinct (EX) – beyond reasonable doubt that the species is no longer extant.
2. Extinct in the wild (EW) – survives only in captivity, cultivation and/or outside native range, as presumed after exhaustive surveys.
3. Critically endangered (CR) – in a particularly and extremely critical state.
4. Endangered (EN) – very high risk of extinction in the wild, meets any of criteria A to E for Endangered.

5. Vulnerable (VU) – meets one of the 5 Red List criteria and thus considered to be at high risk of unnatural (human-caused) extinction without further human intervention.
6. Near threatened (NT) – close to being endangered in the near future.
7. Conservation Dependent (CD) - a category whose documentation previously was absent - inferred to simply be an oversight
8. Least concern (LC) – unlikely to become endangered or extinct in the near future.
9. Data deficient (DD) – No sufficient data to assess their risk of extinction
10. Not evaluated (NE) – Not yet having been assessed

RESULTS

Diversity of Clupeiform fishes recorded from Cuddalore coast

Table 1 and Figs. 2 and 3 exhibit the species of the Order Clupeiformes fish classification records from the Cuddalore coast. A total of 26 species (Fig 4-6) were identified in four distinct families, including the Clupeidae, Pristigasteridae, Engraulidae, and Chirocentridae. With 12 species spread across 5 genera, the Family Engraulidae was the most dominant, followed by the Family Clupeidae with 8 species spread across 5 genera. Table 1 and Figs. 2&3 provide information on the other families.

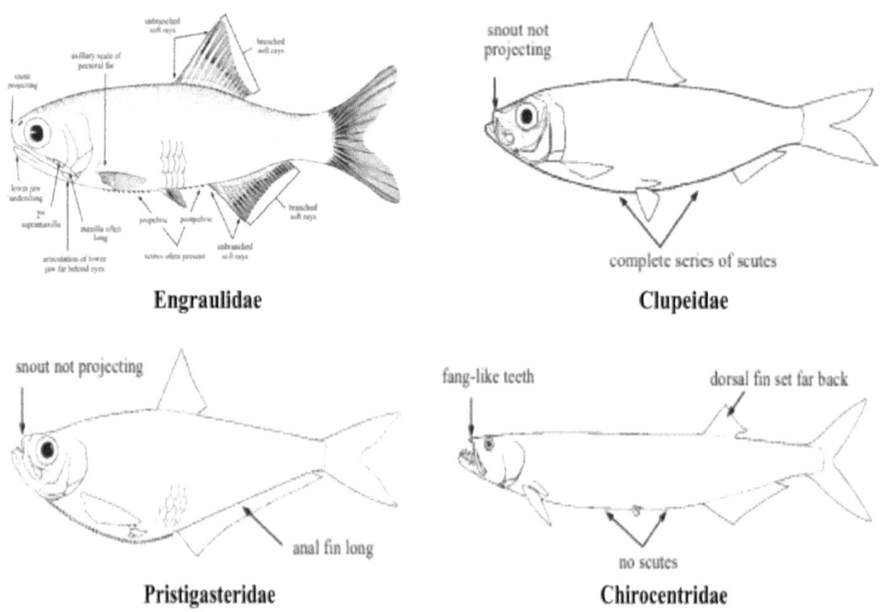

Fig. 2: Key characters of Clupeiform fish Families

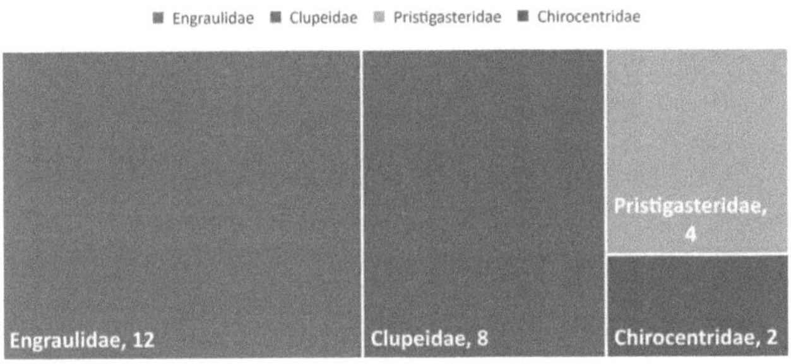

Fig. 3: Key characters of Clupeiform fish Families (See colour version on page 341)

Table 1: Classification of the Order Clupeiformes fishes recorded in Cuddalore coast

Sl.No	Order	Family	Species name	IUCN
1.	Clupeiformes	Pristigasteridae	*Ilisha megaloptera*	LC
2.	Clupeiformes	Pristigasteridae	*Ilisha melastoma*	NE
3.	Clupeiformes	Pristigasteridae	*Opisthopterus tardoore*	LC
4.	Clupeiformes	Pristigasteridae	*Pellona ditchela*	LC
5.	Clupeiformes	Engraulidae	*Coilia dussumieri*	LC
6.	Clupeiformes	Engraulidae	*Encrasicholina devesi*	LC
7.	Clupeiformes	Engraulidae	*Encrasicholina heteroloba*	LC
8.	Clupeiformes	Engraulidae	*Encrasicholina punctifer*	LC
9.	Clupeiformes	Engraulidae	*Setipinna phasa*	LC
10.	Clupeiformes	Engraulidae	*Setipinna taty*	LC
11.	Clupeiformes	Engraulidae	*Setipinna tenuifilis*	DD
12.	Clupeiformes	Engraulidae	*Stolephorus commersonii*	LC
13.	Clupeiformes	Engraulidae	*Stolephorus indicus*	LC
14.	Clupeiformes	Engraulidae	*Stolephorus insularis*	LC
15.	Clupeiformes	Engraulidae	*Thryssa malabarica*	DD
16.	Clupeiformes	Engraulidae	*Thryssa mystax*	LC
17.	Clupeiformes	Chirocentridae	*Chirocentrus dorab*	LC
18.	Clupeiformes	Chirocentridae	*Chirocentrus nudus*	LC
19.	Clupeiformes	Clupeidae	*Amblygaster sirm*	LC
20.	Clupeiformes	Clupeidae	*Anodontostoma chacunda*	LC
21.	Clupeiformes	Clupeidae	*Dussumieria acuta*	LC
22.	Clupeiformes	Clupeidae	*Escualosa thoracata*	LC
23.	Clupeiformes	Clupeidae	*Sardinella albella*	LC
24.	Clupeiformes	Clupeidae	Sardinella fimbriata	LC
25.	Clupeiformes	Clupeidae	Sardinella gibbosa	LC
26.	Clupeiformes	Clupeidae	Sardinella longiceps	LC

Ilisha megaloptera

Ilisha melastoma

Opisthopterus tardoore

Pellona ditchela

Coilia dussumieri

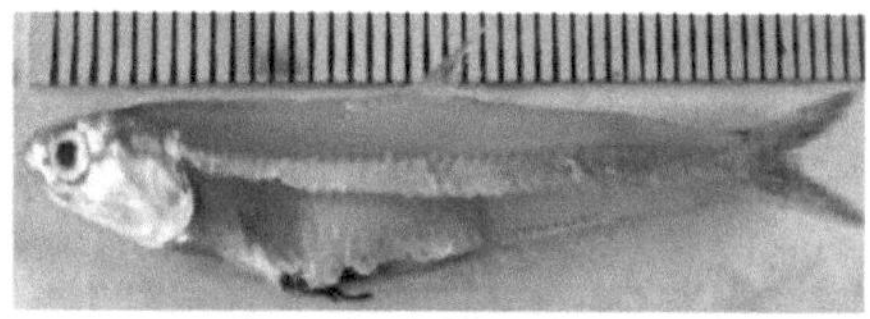

Encrasicholina devesi

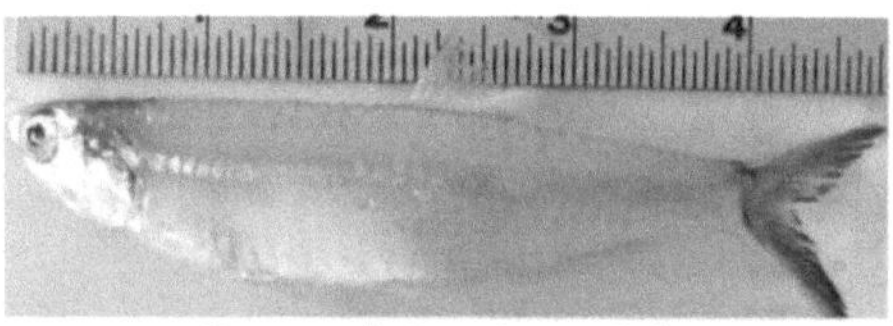

Encrasicholina heteroloba

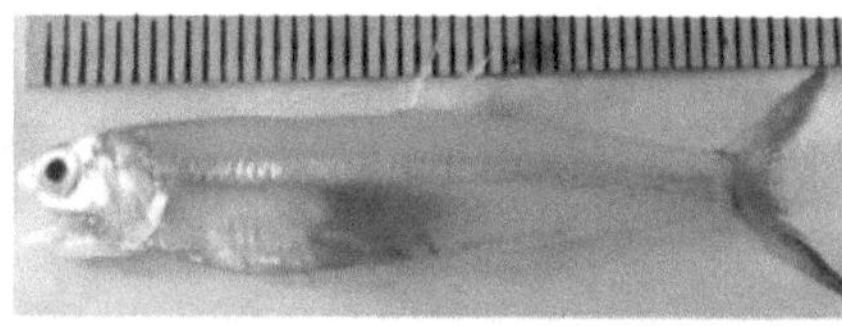

Encrasicholina punctifer

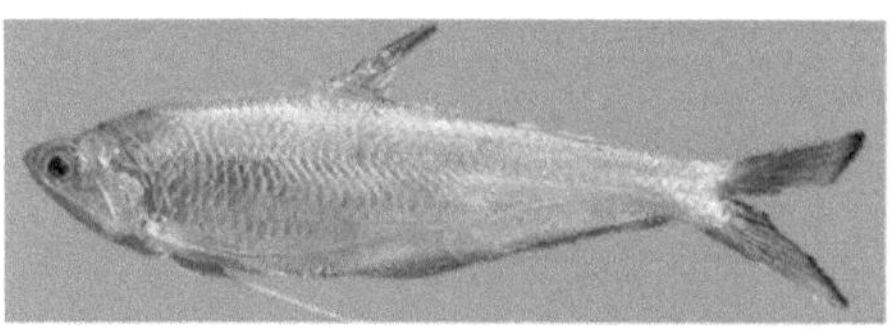

Setipinna phasa

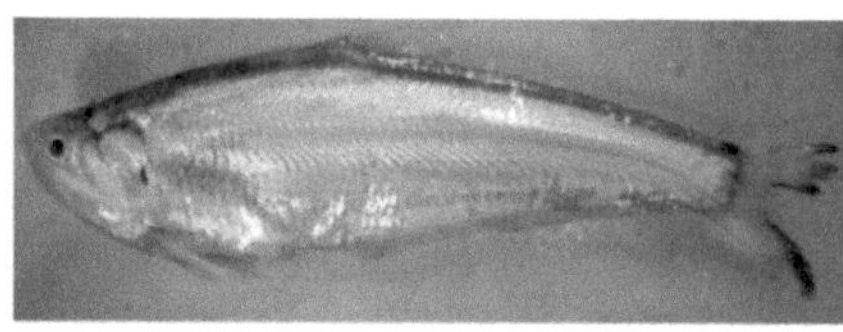

Setipinna taty

Fig. 4: Clupeiform fishes recorded from Cuddalore coast

Setipinna tenuifilis

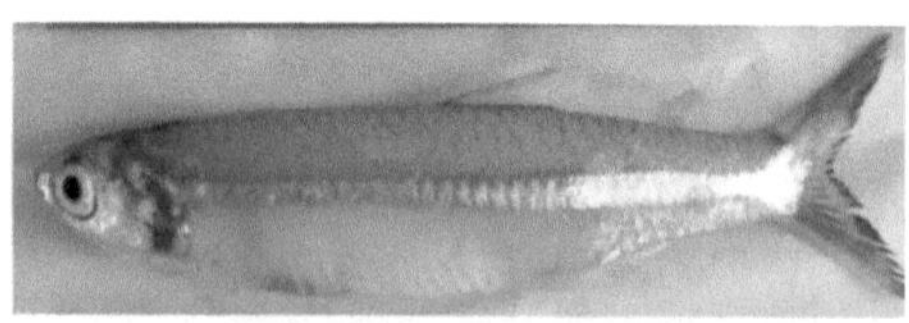

Stolephorus commersonii

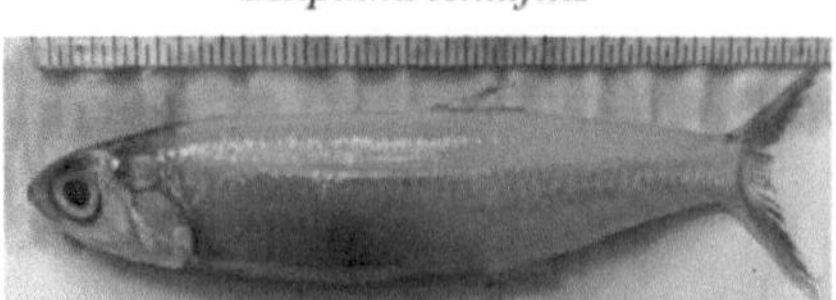

Stolephorus indicus

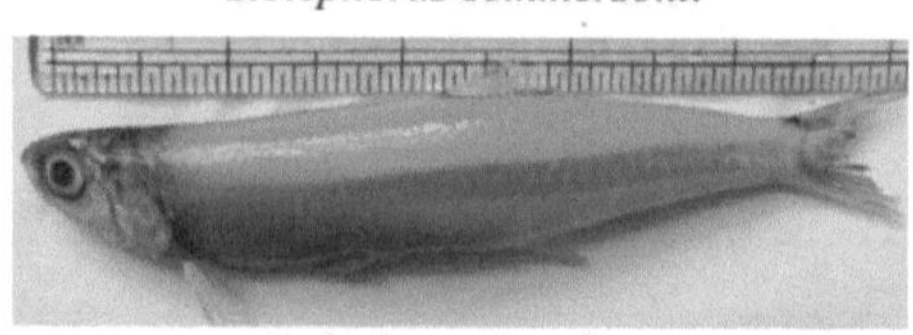

Stolephorus insularis

Thryssa malabarica

Thryssa mystax

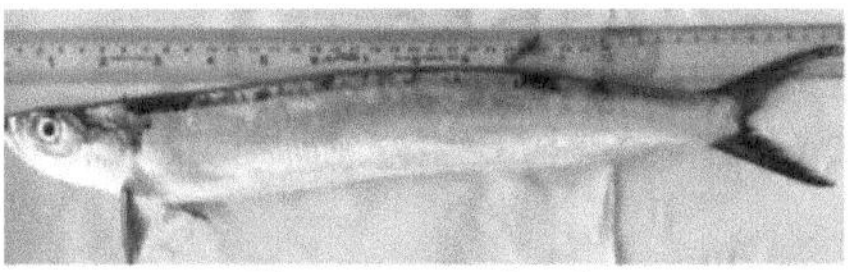

Chirocentrus dorab

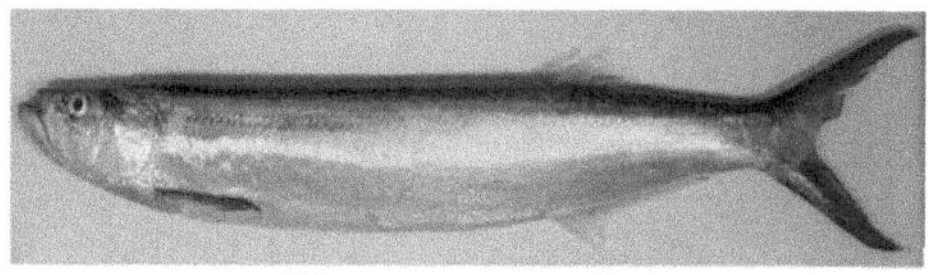

Chirocentrus nudus

Amblygaster sirm

Anodontostoma chacunda

Fig. 5. Clupeiform fishes recorded from Cuddalore coast (See colour version on page 342)

Dussumieria acuta

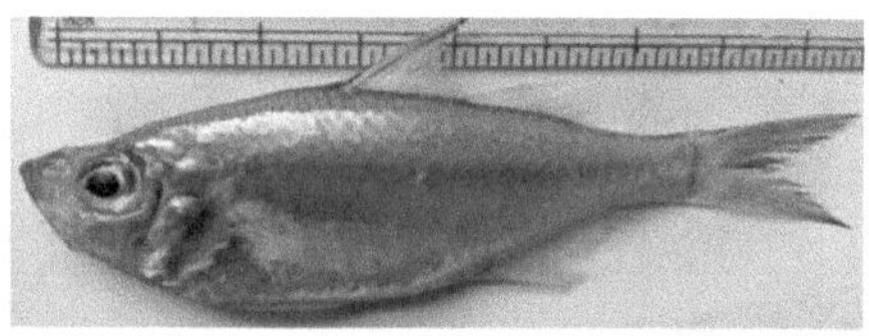

Escualosa thoracata

Sardinella albella

Sardinella fimbriata

Sardinella gibbosa

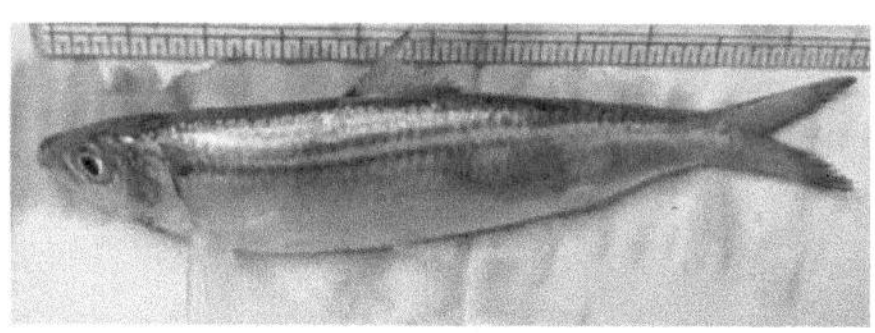

Sardinella longiceps

Fig. 6. Clupeiform fishes recorded from Cuddalore coast (See colour version on page 342)

IUCN Status

In the present study, the IUCN Status of the Clupeiform fishes were evaluated based on the database available in iucn.org. In which, there will be redlist category species are recorded in the present study. Among the 26 species recoded, the Least Concern (LC) are found to be higher (23 species, 88%) than Data Deficient (DD) (2 species, 8%) and Not Evaluated (1 species, 4%) (Table 1 & Fig. 7).

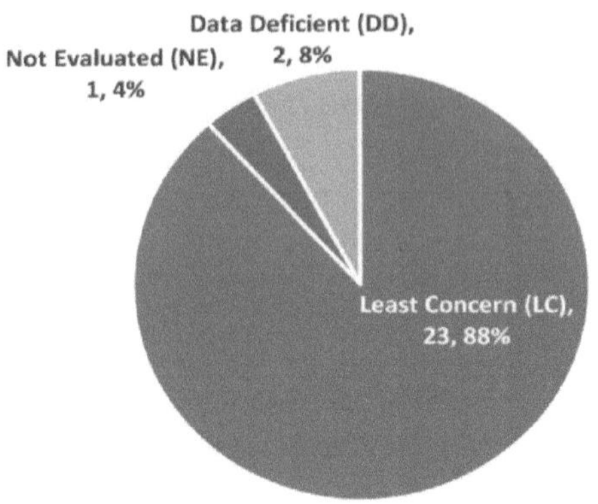

Fig. 7: IUCN Status

DISCUSSION

Biodiversity is important for the future sustainability of marine natural resources that include commercial fisheries. While it is axiomatic that biodiversity is essential for sustainable productive fisheries there is surprisingly little supporting evidence. Fisheries that exploit a range of species or a range of populations may have more stable catches than fisheries that exploit a single species (Dulvy *et al.*, 2000; Hilborn *et al.*, 2003). The main factor that threatens marine fish biodiversity globally is fishing (Dulvy *et al.*, 2003; Garcia *et al.*, 2006). For example, worldwide over 40 local populations of marine fishes have gone extinct as a result of overexploitation (Dulvy *et al.*,2003). These local losses represent an erosion of global biodiversity. Moreover, stresses due to other factors such as climate change, habitat loss, invasive species, eutrophication and pollution can accentuate fishing-induced declines and inhibit or prevent recoveries (Garcia *et al.*, 2006).

The Indian Marine fishing sector has a major impact on the country's economy through the generation of employment, foreign exchange earnings, and above all, the provision of cheap protein food to the people. The current marine fishing scenario in India is characterised by declining fields from the in shore waters, increasing conflicts between different resource users, increasing demand of sea food for the domestic consumption and export, and prospects of large scale formation and coastal Mari Culture, while at the same time the fishing industry has established itself as the largest supplier of animal protein food in India, the largest single sector of rural employment aimed at improving the socio-economic situation of the largest among all animal protein food exporters, and the largest money-generator sharing with the maximum number of rural population and across the country (see Sakthivel (2001)).

The estimated marine fish landings along the coast of the mainland of India during 2022 was 3.49 million tonnes, showing an increase of 14.53% compared to the landings in 2021 (CMFRI, 2023).In Tamil Nadu, the estimated total marine fish landing was 7.22 lakh tonnes in 2022, increasing considerably, by 28.38% from the landings in 2021 (CMFRI, 2023).Among the pelagic fishery resources, there are about 4.2 lakh tonnes of finfishes which were landed in Tamil Nadu in 2022 (CMFRI, 2023). The major contributors of the pelagic finfish landings in the state are sardines, carangids, mackerels, tunas, anchovies, barracudas, ribbonfishes, shads, other clupeids, wolf herrings and seerfishes. In Cuddalore coast, pelagic fishes contributed more, which is attributed to the productive nature of the water of this coast enriched by the fresh water in flow from the rivers Paravanar from the south and Gadilam and Pennayar from the North also the fisherman practice near shore fishing. Oil sardine which constituted the fishery only in west coast earlier. Of late, they contribute substantially along Tamil Nadu coast. In the present study, totally 26 species were recorded in four different Families such as Pristigasteridae, Engraulidae, Chirocentridae and Clupeidae. Of which Family Engraulidae dominated with 12 species in 5 genera, followed by the Family Clupeidae with 8 species in 5 genera.

Species diversity estimates are typically a function of sampling effort and it is important to develop an understanding of taxon-specific sampling requirements (Gotelli and Colwell 2001). Jayaprakash (2002) who described the shift in oil sardine fishery to east coast as noticed in the present study fishing season of oil sardine was projected to be June-December in the present investigation also the peak landing was during June 2003 on the whole the pelagic fishery. Varadharajan *et al.,* (2009) evaluated the fish resources in Mallipattinam coast, Southeast coast of India. Trophic level of fishes associated in the trawl bycatch from Parangipettai and Cuddalore, Southeast coast of India were by Murugesan *et al.,* (2012). The diversity of marine fishes from Nagapattinam coastal waters, south east coast of India are made (Ramu *et al.,* 2015).

Over 419 species of herrings, shads, sardines, anchovies, menhadens, and relatives belong to the Order Clupeiformes and include many of the most important forage fishes. These small, schooling fishes are ecologically, economically, and culturally significant. Immediate conservation priorities include: 1) increasing research and mitigative action directed toward species assessed as threatened or Data Deficient; 2) improving fisheries management regulations for the understudied but heavily exploited species, and 3) promoting local, intensive habitat restoration to reduce pollution and remove dams. These extinction risk assessments and subsequent analyses should be used as an informative tool for fisheries and conservation managers and to monitor conservation progress. Research and conservation prioritization can be difficult in areas of high biodiversity; however, mitigation of anthropogenic stressors in these areas where elevated concern species are distributed is critical. Fishery managers and funding agencies in regions with large proportions of exploited DD species may also consider prioritizing research initiatives to fill gaps in our understanding of these species.

REFERENCES

Alder, J., Campbell, B., Karpouzi, V., Kaschner, K., Pauly, D., 2008. Forage Fish: From Ecosystems to Markets. Annual Review of Environment and Resources, 33, 153-166.

Bloom, D.D. and Egan, J.P. 2018. Systematics of Clupeiformes and testing for ecological limits on species richness in a trans-marine/freshwater clade. Neotropical Ichthyology, 16(3): e180095 (1-14p).

CMFRI 2023. Annual Report 2022. Central Marine Fisheries Research Institute, Kochi. 252 p.

Collette, B.B., Carpenter, K.E., Polidoro, B.A., Juan-Jordá, M.J., Boustany, A., Die, D.J., ... Yáñez, E., 2011. High value and long life-double jeopardy for tunas and billfishes. Science, 333, 291-292.

Comeros-Raynal, M.T., Polidoro, B.A., Broatch, J., Mann, B.Q., Gorman, C., Buxton, C.D., Goodpaster, A.M., Iwatsuki, Y., MacDonald, T.C., Pollard, D., Russell, B., Carpenter, K.E., 2016. Key predictors of extinction risk in sea breams and porgies (Family: Sparidae). Biological Conservation, 202: 88-98.

Compagno, L.J.V., 1997. Stingrays. In: K.E. Carpenter and V.H. Niem (Es.) FAO. Identification Guide for Fishery Purposes. The Western Central Pacific.

Dulvy, N.K., Metcalfe, J.D., Glanville, J., Pawson, M.G., Reynolds, J.D., 2000. Fishery stability, local extinctions and shifts in community structure in skates. Conserv. Biol., 14, 283–293.

Dulvy, N.K., Sadovy, Y., Reynolds, J.D., 2003. Extinction vulnerability in marine populations. Fish Fish.,4, 25–64.

FAO, 2018. The State of World Fisheries and Aquaculture 2018 – Meeting the sustainable development goals. Rome. License: CC BY-NC-SA 3.0 IGO.

FAO, 2020. The State of World Fisheries and Aquaculture 2020. Sustainability in action,Rome.

Garcia, S., Boucher, J., Cury, P., Thebaud, O., Andriantsoa, M., Astudillo, A., Ba, M., Brander,

Gotelli NJ and Colwell RK, 2001. Quantifying biodiversity: Procedures and pitfalls in the measurement and comparison of species richness. Ecology Letters,4:379–391.

Hilborn, R., Quinn, T.P., Schindler, D.E., Rogers, D.E., 2003. Biocomplexity and fisheries sustainability. Ecol. Monogr. 75:3–36.

Jayaprakash, A.A. 2002. Long term trends in rainfall, sea level and solar periodicity: A case study forecast of Malabar sole and oil sardine fishery. J.Mar.Biol.Ass. India,44(1&2):163-175.

K., Charles, A., Dulvy, N., Gauthiez, F., Heip, C., Jennings,S., Joannot, P., McDonald, D., MacKenzie, B., Rice, J., 2006. Workshop 10,Paris Conference: Biodiversity, Science and Governance, January 24–28, 2005 (report of the debates and proposed priority actions). Ministry of Foreign Affairs, Government of France, Paris.

Levin, P.S., Francis, T.B., Taylor, N.G., 2016. Thirty-two essential questions for understanding the social-ecological system of forage fish: the case of Pacific Herring. Ecosystem Health and Sustainability, 2, e01213.

Mohanty, B.P., Ganguly, S., Mahanty, A., Mitra, T., Patra, S., Karunakaran, D., Mathew, S., Chakraborty, K., Paul, B.N., Sarma, D., Dayal, S., Singh, S., Ayyappan, S., 2019. Fish in human health and nutrition. Advances in Fish Research, 7: 189-218.

Mohsin, A.K.M., and M. A. Ambak, 1996. Marine Fishes & Fisheries of Malaysia and Neighbouring Countries. University Pertanian Malaysia Press. Serdang.686pp.

Murugesan P, Purusothaman S, Muthuvelu S. 2012. Trophic level of fishes associated in the Trawl bycatch from Parangipettai and Cuddalore, Southeast coast of India. Journal of Fisheries and Aquatic Sciences, 2012; 7(1): 29-38.

Nelson, J.S., Grande, T.C. and Wilson, M.V.H. (2016) Fishes of the World. 5th Edition, John Wiley and Sons, Hoboken.

Ramu, S. T. Anandaraj, C. Elaiyaraja and A. Panneerselvam, 2015. Check list of marine fish From Nagapattinam coastal waters, southeast coast of India. International Journal of Fisheries and Aquatic Studies, 2(6): 193-197.

Reynolds, J.D., Dulvy, N.K., Goodwin, N.B., Hutchings, J.A., 2005. Biology of extinction risk in marine fishes. Proceedings of the Royal Society B: Biological Sciences 272, 2337-2344.

Roberts, C.M., Hawkins, J.P., 1999. Extinction risk in the sea. Trends in Ecology & Evolution, 14, 241-246.

Sadovy de Mitcheson, Y., Linardich, C., Barreiros, J.P., Ralph, G.M., Aguilar-Perera, A., Afonso, P., Craig, M.T., 2020. Valuable but vulnerable: Over-fishing and under-management continue to threaten groupers so what now? Marine Policy 116, 1-10.

Sakthivel. M.2001.Challenging Tasks for Sustainable Fisheries Development in the Millennium 2000. In: Sustainable Indian Fisheries. Pandian.T.J (Ed.).Academy of Agricultural Sciences. New Delhi, 6-18pp.

Tacon, A.G.J., Metian, M., 2009a. Fishing for feed of fishing for food: increasing global competition for small pelagic forage fish. AMBIO: A Journal of the Human Environment 38, 294-302.

Thornton, T.F., Moss, M.L., Butler, V.L., Hebert, J., Funk, F., 2010. Local and traditional knowledge and the historical ecology of Pacific herring in Alaska. Journal of Ecological Anthropology 14, 81-88.

Varadharajan D, N.Pushparajan and P. Soundarapandian, 2012. Fish Resources in Mallipattinam Coast, South East Coast of India. International Journal of Pharmaceutical & Biological Archives, 3(4): 871-876.

4

Ecology of Coral Reefs: Diversity Interactions and Conservation

Kranthi Kumar Dhande, Praveenjoshi H.S., Hariyala Sirisha, Gora Shiva Prasad and T. Neeraja

ABSTRACT

The coral reefs are distributed in less than one percent of the earth surface yet they are significantly contributing to the nutritional security, ecological services and cultural value. They are hosting a wide variety of fish and other aquatic species for food and shelter. The reefs are formed from calcium carbonate which is excreted by them when free-swimming polyps attach to the submerged rocks or other hard surfaces. Symbiotic relationship between the organisms forms the basis for the presence of diverse organisms in the coral ecosystem. One of such examples is the relationship between hard coral polyps and algae (zooxanthellae). Corals provide protection and components to the zooxanthellae for photosynthesis. Zooxanthellae photosynthesize the organic compounds and then pass on the nutrients to the corals and gives beautiful coloration. The Great Barrier Reef which consists of 1200 miles length in Australia is one of the most popular and largest barrier reefs in the world. Growth of warm water coral reefs dependent on shallow, sunny, warm and alkaline waters. The Scleractinian corals are dominated in shallow and deeper mesophotic coral reefs and they form symbiosis with dinoflagellate protists of the Genus Symbiodinium. Coral reefs suffer from the detrimental effects of shifting water temperatures, ocean acidification, pollution, the invasion of non-native species, alterations in weather patterns, and the physical consequences of ship groundings and storms. For the conservation, strategies like reducing local stressors, creating marine protected areas, using active management techniques like human-assisted development and reef restoration and implementing coordinated governance at various levels are prominent. Restoration plays a crucial role in promoting the revival of ecological characteristics. This includes development of strategies to reduce nuisances and improve the quality in coral habitats, and reduce damages.

Keywords: *Coral polyps, Zooxanthellae, Biodiversity, Community management.*

INTRODUCTION

Coral reefs are fragile and most diverse and valuable ecosystems on earth. The stony corals are formed from the tiny, soft colonial organisms called polyps. They provide enormous ecological, economic services, and cultural value to hundreds of millions of people. They are significantly contributing to the nutrition, economic security and protection from natural disasters. They are often termed as the rain forests of the sea.

Coral reefs were first appeared 485 million years ago during the early Ordovician period. They belong to the class Anthozoa of animal phylum Cnidaria, which includes sea anemones and jelly fish. Coral reef is an under water ecosystem which is characterized by reef-building corals. Unlike sea anemones, colonies of coral polyps secrete a hard exoskeletons called reef which is made up of calcium carbonate to support and protect the corals. These reefs grow best in warm, shallow, clear, sunny and agitated water.

BIODIVERSITY

The term biodiversity was coined by Walter G. Rosen in the year 1985. Biodiversity refers to the variation among living organisms that can be found in a particular area such as terrestrial, marine and desert ecosystems and the ecological complexes to which they are a part. The study of biodiversity helps to understand the ecological stability, economic and ethical importance. Coral reefs are believed to have the highest biodiversity on the planet even more than a tropical rainforest. Though the coral reefs distributed in less than one percent of the ocean floor, they provide habitat for more than 25% marine life with more than 4,000 world's fishes. They act as protective nurseries to many fish species, and shelter small fishes to grow. Coral reefs diversity is so rich that we do not have a firm count on all the species that live within it and every year discover new species." There are about six thousand coral species in the world, in which few are growing in warm shallow waters near coastlines while others in the dark, cold seafloor of the open ocean.

Corals: Coral polyps are tiny little animals which usually forms the coral reefs. They can live individually, or in large colonies and provide the coral structure upon which the entire reef ecosystem depends.

Fish: Coral reefs support a vast array of fishes ranging from small, colourful reef fish like clownfish and damselfish to larger predators like sharks and groupers. Remarkable fish diversity can be observed in coral reef ecosystems.

Invertebrates: Coral reefs host a variety of invertebrates such as sponges, sea anemones, sea stars, sea urchins; and molluscs such as snails, clams, and octopuses.

Crustaceans: abundant planktonic organisms, crabs, shrimp, and lobsters are found in coral reef ecosystem.

Molluscs: Coral reefs harbour a variety of gastropods (snails), bivalves (clams), and cephalopods (octopuses, squid, cuttlefish).

Echinoderms: Species such as sea stars, sea urchins, and sea cucumbers, are ecologically important for the coral reef ecosystem.

Reef building organisms: Besides corals, calcareous algae and various species of sponges contribute for the reef building.

Microorganisms: Organisms such as bacteria and archaea, play crucial roles in nutrient cycling, coral health, and overall ecosystem function within coral reef environments.

Coral reef formation and structure

Most of the coral reefs were formed after the Last Glacial period with the rise of sea level to flood continental shelves due to melting of ice. Most coral reefs are less than 10,000 years old. The corals are formed when free-swimming polyps attach to the submerged rocks or other hard surfaces and sits within a calyx or cup like structure that it excretes. The calyx is the limestone which is also known as calcium carbonate. Over the limestone skeleton, polyps are interconnected to form a mass of living tissue and this is why these corals are called stony corals. As corals grow and expand, reefs would take on one of three major structures viz. fringing, barrier or atoll. These three types of reefs represent the reef developmental stages over time.

1. Fringing Reefs

These reefs are the most common type of reefs which are observed around the islands and continents of coastline. They are considered to be the youngest among 3 types of coral reefs. They usually appeared to be separated from the shore by shallow lagoons. The coral larvae attach itself to rocks or soil near the coasts and they become as polyps. Polyps excrete calcium carbonate to forms their exoskeleton. The secreted calcium carbonate sediments provide a substrate for polyps to get attached. Coral reefs are also found to form where volcanoes have formed due to the shallow sloped walls. As more and more polyps attach, and forms layer over time, they create a coral reef. Calcareous Algae and other organisms with calcareous skeletons add their sediments to the structure after they die and sink. Since fringing reefs are the youngest and grow onshore, they tend to have less diversity of species within that ecosystem. Some locations of these reefs include Kenya, Australia and other parts of Africa shores.

2. Barrier Reefs

This type of reefs found along coastlines with a very wide and deep lagoon separating it from the shore. The coral sticks out of more shallow water and creates a barrier or wall-like structure thus the name *barrier reefs*. One of the most popular and largest barrier reefs is *"The Great Barrier Reef"* in Australia. The reef consists of 1200 miles length and have complex reefs that it made up of. Such reefs are formed when fringing reefs gradually combine with each other to form a borderline along the coast. The calcium carbonate structures attract more polyps to accommodate. It forms a line along the coast and a ring around the island. Large barrier reefs are found rarely in select places on earth such as Belize and parts of the southern Pacific Ocean. Smaller barrier reefs are seen where islands are in an early stage of becoming submerged. The barrier reefs surrounding the Turks and Caicos Islands are formed due to precipitation of $CaCO_3$ on the leftovers of continental debris resulted the separation of North America from South Africa around 200 million years ago. Dissolved minerals in the seawater form small particles called ooids that become cemented together to form oolite rocks of most of the Turks and Caicos Islands and bank.

3. Atoll Reefs

An Atoll reef is a ring-like structure forms around small islands with a lagoon in the middle and are usually located in the middle of the sea. These reefs are formed around a pre-existing structure as an island has sunk down completely due to rise in the sea level. Such islands are usually appeared on the tops of underwater volcanoes. Atoll reefs are found in the Pacific Ocean where the waters are warm and salty. Their size ranges between as small as 1.5 km^2 (Ontong Java reef) to the size of 12.6 km^2 (Great Chagos Bank). Over a million species can live in one square kilometre of coral reef. The sequential order of the three types of reefs from fringing to barrier to atoll represent stages of development of a reef.

4. Patch reefs

Patch reefs are small, isolated reefs found in the bottom of island platform or continental shelf. They usually occur between fringing reefs and barrier reefs. They vary in size and rarely reach the surface of the water.

Symbiotic relationships

Coral reefs are home for a wide variety of living organisms viz. sponges, fishes including large nurse sharks, reef sharks, groupers, clownfish, eels, snappers and parrotfish; jellyfish, anemones, crustaceans, other invertebrates and algae. Majority of these organisms are interdependent on each other for food, shelter and protection from enemies. Symbiotic relationships are quite common in the ocean ecosystem, especially near coral reefs. The symbiotic relationships include mutualism, parasitism, and commensalism. Mimicry is also frequently observed among coral reef organisms. The symbiotic relationship is formed when two different species interact with each other. These interactions are important to create a balance within the ecosystem as at least one of the species are gaining from it.

Types of Mutualism

There are two primary types of mutualism *viz.* obligate mutualism and facultative mutualism. By definition, mutualism refers to the two organisms of different species work together to get benefit from the relationship.

Obligate mutualism states that the organism cannot live without another. The relationship between hard coral and algae (zooxanthellae) is one of the best examples for mutualistic relationship in the coral ecosystem. The microscopic, photosynthetic zooxanthellae reside inside the coral as the later provides protection and components needed to perform photosynthesis. Zooxanthellae photosynthesize the organic compounds and then pass the nutrients in the form of glucose, glycerol and amino acids to the hosting corals and giving essentially the beautiful coloration for them. Approximately 90% of the nutrients produced by zooxanthellae through photosynthesis is transferred to the coral for their use. Zooxanthellae also helps in excretion or removal of waster such as carbon dioxide and nitrogen from the corals. This relationship infers that without algae, corals would starve to death which is known as coral bleaching and in turn the algae would be more vulnerable to several herbivores and other organisms.

The another commonly known mutualistic relationship exists between selected pairs of anemones and clownfish. The Indian and Pacific Oceans warm waters are the habitats of Clownfish. Out of over 1,000 anemone species, only about 10 species coexists with 26 species of tropical clownfish. Sea anemones are predators with stinging polyps attached themselves to rocks, ocean floor and even corals. With the help of poisonous tentacles, anemones entangle the fishes which swim close to them. The toxins paralyze the prey and tentacles direct the prey into the anemone's mouth. However, clownfish are the exception as they live within the anemone's tentacles to get protected from predators, in turns they either letting the anemones to eat on leftovers, or by luring fish over to them. Since the Clownfish are coated with a mucus layer that essentially makes them immune to the deadly sting of the anemone.

Another type of mutualism is the Facultative mutualism which refers to the species benefitted from one another, but do not necessarily fully depend on one another. In the coral ecosystem, the relationship between shrimp or smaller fish and large marine organisms is the perfect example of facultative mutualism. The shrimp or cleaning fish remove materials such as potentially harmful parasites from the body of larger marine organisms, on which they feed on. Smaller fish or cleaner shrimp like Bluehead Wrasse or Spanish Hogfish help the large marine organisms such as fish, sharks and rays by removing potentially harmful parasites and other materials from their body.

Another facultative mutualistic relationship is observed between the root-fouling sponge called *Tedania inis*, and red mangrove (*Rhizophora mangle*). In this relationship, the red mangrove provides carbon to the sponge, and the nitrogen released by the sponge is taken up by the mangrove to enhance growth.

The relationship between spider crabs and algae is another example for mutualism. This relationship benefits both of these species as the greenish-brown algae live on the spiders back, which helps the spider crab blend into the shallow areas of the ocean floor where they live while algae benefitted from a place to live.

The Boxer Crab and anemones form a mutualistic relationship as the former carries two anemones around which can increase their options for finding food whereas the latter assures protection from enemies to the Crab with its poisonous stings.

The Goby (Nes Longus and *Ctenogobius saepepallens*) and a snapping shrimp (*Alpheus floridanus*) are undergo a mutualistic relationship where shrimp prepares a decent sized burrow on the ocean floor and the Goby lives in the entrance of that burrow. When the shrimp comes out of the burrow, it will be in contact with the Goby with the help of antennae. Depending on species of the Goby, it will either signal to the shrimp about predators by darting headfirst back into the burrow or by flicking its caudal tail. In return the Goby gets a free place to live and hide from potential predators.

Commensalism

In commensalism, one species will get benefitted from another species while other species is neither benefitted nor harmed. Such relationship is found between jellyfish and small fish. Small fish will typically hide inside the stinging tentacles of jellyfish which protect the fish from larger predators. This relationship does not affect the jellyfish. The shrimp will blend in with the featherstar to get protection is another similar example.

Mimicry

Mimicry in this case refers to one organism which is harmless evolved to look like another organism which is poisonous. It helps the organisms to get protected against the predators as the latter think that the organisms are poisonous species. Sometimes organisms appear to be larger than they really are. The four-eye butterflyfish uses a large eyespot in order to appear larger to the predators.

The Sabre-tooth Blenny has a mutualistic relationship with larger fish so as to escape from the predators and it evolving to look similar to the Cleaner Wrasse. The Sabre-tooth Blenny bites the large fish's flesh and swims away instead of cleaning the larger fish. This type of mimicry is known as aggressive mimicry. The Bluestreak cleaner wrasse appears to be same like sabre-tooth blenny to mislead the predatory fishes.

Trophic interaction in coral reef ecosystem

Currently, 71% of the earth's surface is covered by the water. Since the Northern Hemisphere has about two thirds of the planet's land area, 61% of it is covered by ocean. The planet's largest aquatic system is the marine ecosystem is made up of estuaries, coral reefs, and oceans. It is quite challenging to handle the system as a whole because it is a single, big, sophisticated system. Because of this, oceanographers have classified the ocean into numerous zones based on physical attributes, namely depth, light, and temperature. In marine there are two important zones according factors of light transparency, depth of the water and intensity of the temperature i.e. pelagic zone and benthos zone. Benthos animals are those that reside on the ocean floor. Invertebrates are animals that, for the most part, do not have backbones. Sea anemones, sponges, corals, sea stars, sea urchins, worms, bivalves, crabs, and many more types of invertebrates are examples of typical benthic organisms. Coral reefs are iconic ecosystems that are renowned for their beauty and biodiversity as well as for the priceless social, economic, and ecological benefits they offer. Here, it is hypothesised that the fundamental reason coral reefs can withstand such adverse conditions is because these three interrelated characteristics enable efficient organic matter recycling and conservation within the system. This theory is supported by small-scale data, such as the symbiotic relationship between coral polyps and their "zooxanthellae" symbionts. It is unknown at this time if this theory holds true for the entire coral reef ecosystem unit. There is also controversy regarding whether or not coral reefs can be thought of as being in a steady-state, that is, whether or not production and respiration balance import and consumption for the system as a whole. Understanding the composition and organization of ecological communities helps to understand the movement of nutrients and energy across food webs. How these procedures help society's use of ecosystem services. Fisheries are one way that many aquatic ecosystems deliver services. Such as nutrition, food security and coastal livelihoods, that can vary regionally in response to interacting human and environmental drivers, and social–cultural contexts. Large-scale comparative studies of community organization along human and environmental gradients have contributed to our understanding of the variation in ecosystem services. Coral reefs are intricate food webs and bio-diverse ecosystems. Even on the relatively species-poor reefs of the tropical Eastern Pacific, nearly 300 interspecific links exist in the sub-web of corals, coral eating animals, and their higher-level predators. However, of all the trophic interactions on tropical reefs, herbivory is the most studied and is widely recognized as critical for maintaining coral dominated reefs. By eating algae that compete with coral for space, spread coral diseases, and contaminate coral tissue, herbivores help coral settle, grow, and survive. Furthermore, in order for corals to recolonize reefs following significant disruptions, herbivory is essential for restricting the growth of macroalgae. Thus, herbivores should be natural allies for restoration practitioners. Coral reef herbivores are likely to be the primary contributor to ecosystem and fisheries services across diverse (and disturbed) benthic habitats, fisheries management strategies and at reefs with varying biomass levels, underlining their importance for tropical food security.

Climate change impact on coral reefs

Coral reefs can be found in a broad variety of habitats, offering a multitude of different creature's food and shelter in addition to a host of other ecological products and services. For example, warm-water coral reefs require shallow, sunny, warm, and alkaline waters to support their rapid growth and calcification, which is essential for the formation and upkeep of their calcium carbonate structures. "Mesophotic" (low light) coral reefs are found in deeper waters (40–150 m). They accumulate calcium carbonate at much slower rates (or not at all in certain circumstances), but they are nevertheless vital as home for a variety of creatures, including some that are vital to fisheries. Eventually, the so-called "cold-water" coral reefs can be discovered even farther down, at least 2,000 meters, in the shadowy depths. Despite their significance, human activities such as pollution, overfishing, physical deterioration, and climate change pose serious threats to coral reefs. Warming temperatures and ocean acidity pose a threat to cold-water corals as well, however it is unclear how directly climate change is affecting them. There is little evidence that coral reefs can adapt quickly enough to keep up with the fast warming and acidification of the ocean, especially considering how slowly corals evolve due to their lengthy lifespans. The calcium carbonate skeletons secreted by warm- and cold-water corals accumulate over time to form a three-dimensional reef matrix that serves as a home for thousands of fish and other species. In many warm-water coral reefs, the production of calcium carbonate, which resembles limestone, is high enough to form carbonate structures. Significant rates of bio-erosion and wave-driven physical erosion can be defeated by high rates of calcification. Barrier reefs and islands, which are vital to tropical coastlines, are supported by these structures. At least 25% of known marine species find home in tropical coral reef ecosystems, while making up less than 0.1% of the ocean floor; many reef species remain undiscovered. Both shallow and deeper mesophotic coral reefs are dominated by Scleractinian corals that form symbiosis with dinoflagellate protists from the genus, Symbiodinium. This symbiosis allows their intracellular symbionts—those that reside within the digestive or gastrodermal tissues of their coral hosts—to photosynthesize and supply the host coral with an abundant supply of lipids, sugars, glycerol, and other organic compounds (Muscatine, 1990). In the clear, warm, and shallow water found along tropical coastlines, this relationship allows corals to grow and calcify at high rates. Depending on the water column's clarity, the number of Scleractinian corals harbouring Symbiodinium drops as depth increases above 20 to 40 metres. In tropical waters, the deepest symbiotic Scleractinian corals with Symbiodinium can be discovered 100 metres or more below the surface. The capacity of corals to absorb and consume plankton and other watery particles (polytrophy) adds to the productivity of this symbiotic relationship. The very productive coral reef ecosystems that adorn many tropical coastlines are made possible by their combined capacity for photosynthesis and feeding. Although certain cold-water corals can be found growing in waters as shallow as 50 m, cold-water coral reefs reach depths of 3,000 m (e.g., Norwegian shelf). There is so little light below 200 m depth that photosynthesis is impossible. Because of this, cold-water corals rely on particle feeding rather than developing a relationship with Symbiodinium. Advances in underwater technologies for surveying and mapping have been the main driving force behind the discovery of the locations and breadth of cold-water reefs.

Coral reef Threats and Stressors

Coral reefs possess immense ecological and economic significance and play a crucial role in sustaining over 25 percent of marine biodiversity. Moreover, these reefs provide a multitude of ecosystem services, catering to the needs of billions of individuals worldwide. These services encompass coastal protection, fisheries production, medicinal resources, recreational advantages, and substantial tourism revenues. However, the coral reefs are encountering many threats which alarming their existence in future. The following are observed to be the threats to the coral reefs.

Warming Temperatures and Coral Bleaching

Corals typically possess a limited capacity to withstand temperature variations. In instances of excessive warmth, corals will expel the symbiotic algae (zooxanthellae) residing within them, resulting in their bleaching (corals become pale and appear bleached in the absence of algae) and deprivation of a crucial food supply. While corals can endure a bleaching event, their susceptibility to diseases increases, and if the prolonged marine heat wave persists, they will ultimately succumb. The anticipated effects of climate change include a rise in sea levels, which may lead to the submergence of numerous coral reefs due to reduced light penetration caused by the greater water volume.

Sea Level Rise

Similar to humans, corals also require sunlight in order to survive. Overgrowth of algae, caused by the rise in nitrogen levels in the water surrounding coral reefs due to pollution, obstructs the sunlight from reaching the corals. The presence of sediments resulting from construction, mining, and agricultural activities in coastal areas, along with inadequately planned coastal development and uncontrolled tourist activities, leads to the turbidity and pollution of seawater. Additionally, this situation hampers the access of sunlight to coral reefs.

Unsustainable Fishing Practices

Fishermen frequently capture a significantly wider range of fish species than is necessary. Accompanying the excessive fishing activities are harmful fishing methods, including purse seining, fine-mesh fishing, cyanide fishing, and blast fishing, which lead to unsustainable damage.

Invasive Species

The intrusion of non-native species into coral reef environments disrupts the intricate web of relationships that maintain the harmony within these ecosystems. Invasive species play a significant role in the significant deterioration of coral reef health worldwide.

Ocean Acidification

Approximately 48% of fossil fuel emissions are taken in by the ocean. The absorption of carbon dioxide (CO_2) by oceans leads to increased acidity, which in turn hampers the growth of reef-building corals and the formation of coral reefs. Corals with weaker skeletons are more susceptible to diseases and damage caused by storms. Studies indicate that corals exposed to elevated CO_2 levels become less productive and face a 50% higher risk of bleaching.

Stronger Storms

With the increase in sea surface temperatures, the intensity of hurricanes, cyclones, and typhoons that harm coral reefs increases. Additionally, the heavy rainfall associated with these storms can lead to coastal erosion and an influx of polluted runoff into the ocean. Certain threats to coral reefs, such as sedimentation and harmful fishing methods, are confined to specific nations. However, other threats facing by coral reefs, such as climate change, sea level rise, ocean acidification, and international trade in coral products, have a global impact.

Stressors

Major stressors to coral reefs fishing, global climate change and decline in water quality.

Conservation strategies for coral reef

Historically, the conservation objectives for coral reefs have primarily cantered on reversing degradation and restoring the ecosystem to its original natural state. However, numerous individuals depend on coral reefs for sustenance and livelihoods and the sustainable utilization of reefs necessitates a detailed assessment of the trade-offs between conflicting management and conservation goals. Additionally, it is crucial to have a complete understanding of the various stakeholders engaged in both direct utilization of reefs and indirect activities that may have an impact on them, such as land-based pollution.

For conservation of coral reefs several Marine Protected Areas (minimalizes local human-induces impacts), were established, where a notable species diversity was observed. Minimizing pollution at these diverse coral sites is essential. Moreover, it is essential to raise awareness to the population about the importance of coral ecosystems for human beings. Equally important is the backing and execution of initiatives that concentrate on tasks like eradicating crown-of-thorns starfish (known coral predators), fostering coral growth, reproducing artificial corals, and other relevant endeavours.

Management now focuses on the entire seascape, considering the complex responses of reef ecosystems to climate change and human pressures. Ecosystem-based management (EBM) takes a comprehensive approach, considering interactions within ecosystems, social-ecological connections, stressors on functional groups, and ecosystem processes, all aimed at preserving ecosystem functioning and services.

To maintain social-ecological system resilience against climate change, we suggest reducing local stressors, creating marine protected areas that address both local and global issues, using active management techniques like human-assisted evolution and reef restoration, and implementing coordinated governance at various levels.

Minimising local stressors

Reducing local human impacts can help coral reefs better withstand climate change and reduce harmful interactions with climate change factors. Improving water quality by reducing pollution and implementing better management practices can have positive effects on ecosystem health in a matter of decades. Eliminating fishing in protected areas can enhance reef resilience by promoting herbivory, reducing predation, and maintaining structural complexity. Enforcement of fishing regulations in marine reserves is crucial for effective protection. Determining the carrying capacity of fish biomass can help balance fisheries objectives with the need to protect reef function.

Marine protected areas

MPAs are designated zones within the ocean that are reserved for the purpose of long-term conservation objectives. MPAs should protect key areas in different environmental scenarios, including climate refugia, ecologically significant areas, and resilient locations. It's crucial to safeguard degraded reefs alongside healthy ones, as they now make up a larger portion of global reef area. Small, isolated MPAs are not sustainable alone, and MPA networks are necessary to manage the surrounding seascape. These networks should replicate diverse habitats, maintain ecological connectivity, and sustain biodiversity and ecosystem functioning. MPA placement should consider current and projected species distributions to aid dispersal. Bottom-up, participatory management and conservation approaches that account for the local context are central to successful management outcomes on many coral reefs, particularly in developing nations. Participatory and climate-focused MPAs are crucial for marine management, but may not distribute benefits equally. MPAs alone cannot solve coral reef conservation issues, as they don't protect against land run-off or work well in degraded reefs. Other management approaches are needed in these cases.

Active management techniques for strategic approaches

To aid in the recovery of degraded reefs, simply removing human pressures may not be sufficient. Active reef restoration, such as introducing artificial complexity and coral transplantation, may be necessary to ensure ecosystem health and resilience to climate change impacts. While these methods may not provide large-scale solutions, they can be effective at a local level. Another strategy is human-assisted evolution, involving the transplantation of genetically modified stress-tolerant corals to enhance their resilience and ability to recover from disturbances. However, there is a risk of reducing overall biodiversity by favoring only a few resilient species.

Integrating various levels of conservation management with spatial planning

Conservation management should not exclusively prioritize biodiversity measures or the preservation of coral reefs in remote areas. The livelihoods of approximately six million fishers from nearly 100 countries are dependent on coral reefs. Therefore, it is crucial for management strategies to promote the resilience of both coral reefs and the communities that rely on them. The impacts of climate change on reef social-ecological systems are reciprocal: the effects of climate change on communities determine their subsequent interactions with the environment.

Large-scale human drivers of coral reef degradation, such as global demand for specific species and changing fishing practices, are often ignored. To tackle these issues, broader management strategies like marine spatial planning are crucial. It is essential to prioritize goals at both national and regional levels, integrate ecosystem services, and engage in thorough conservation planning in order to protect coral reefs. Effective implementation of Marine Spatial Planning (MSP) involves meeting local needs, engaging communities and stakeholders, securing political commitment, and receiving support from NGOs and international allies. Global regulation of markets and the aquarium trade can help reduce stress on coral reefs by encouraging sustainable fishing practices.

World Heritage Convention (WHC)

There are currently more than 878 cultural and natural sites that have been included in the World Heritage List, such as the Great Barrier Reef, Belize's Barrier Reef, and Tubbataha Reef Marine Park in the Philippines. The World Heritage Committee (WHC) mandates member states to ensure the protection and conservation of these listed sites for the benefit of future generations.

Convention on Biological Diversity (CBD)

The CBD mandates member states to implement strategies and establish protected areas in order to conserve biodiversity, with a focus on the conservation of biological diversity, sustainable use of its components, and fair and equitable sharing of the benefits of genetic resources.

The collective implementation of these measures will aid in the restoration of coral reef populations globally, benefiting future generations and promoting the sustainable development of coral reefs.

Community based management of coral reefs

Corals are two-layered invertebrates that live in groups (*i.e.* they are colonial) and are related to jellyfish and sea anemones. Coral reef is an important economic asset; the marine tourism development is largely depending on coral reefs. Coral reefs provide clear waters, white sandy beach and wonderful underwater world. Therefore, coral reef management is desperately needed everywhere, but it's especially critical in areas where maritime tourism is significant. Corals are made up of tiny individuals called polyps. Each polyp is like a fluid filled bag with a ring of tentacles surrounding its mouth, and looks like a tiny anemone. Polyps within a colony are linked by living tissues and can share their food. In some corals, the polyp extracts calcium carbonate from the sea and secretes it as a cup of calcium carbonate from the bottom half of its body. These cups provide anchorage for the polyps but when threatened, the polyp can retreat into the safety of the hard cup. When the calcium carbonate cups of many billions of these polyps fuse together, they form coral reefs. The coral reef ecosystem performs a range of functions. It provides a habitat to fish and protects coastlines from erosion and natural calamities. Corals are home to a variety of species of plants and animals thus, constituting an important link in the marine biodiversity. Millions of people depend on coral reefs for their sustenance and livelihood, yet these vital resources are in great danger today. 27 per cent of the world's coral reefs have been lost and it is estimated that another fourteen per cent will be destroyed in the next ten to twenty years. Since ancient times, local and Indigenous communities have been managing and protecting their natural resources through a variety of customs and behaviours that are deeply ingrained in their social and cultural contexts. According to Christie *et al.* (2005), the transition of authority over the use and management of natural resources from local communities to the state has frequently resulted in the deterioration of coral reefs and other ecosystems. Trade liberalisation policies and international supply chains, along with centralised command and control governance systems that weaken local communities, have the potential to spark local unrest, reduce the motivation for group action, and jeopardise the stability of coral reef governance systems. When communities can overcome these challenges, often in collaboration with partners in government, non-governmental organizations, and/or private enterprises, renewed systems of local environmental governance can be strengthened and maintained. Although the social and ecological outcomes of these efforts can vary due to local conditions, one recent study found that "bright spots" in coral reef conservation - reefs with unexpectedly high levels of reef fish biomass given ecological conditions - tended to be found where there were high levels of local engagement in management processes, with rules restricting access and/or use of resources rooted in local cultures or social practices. There is no consensus on the definition of 'community'. The term is used broadly in this paper to describe geographically distinct groups of people, who live in close proximity to reefs (not necessarily small in areas occupied). Communities are supposed to be heterogeneous groupings of people with multiple interests, variable influences on decision-making, and influence from both internal and external institutions; however, some degree of social identity and coherence is implied.

Protection Measures of Coral Reefs in India

Since the coral reefs are one of the fragile ecosystems on the earth surface need to be given priority to protect and conserve so as to enjoy the ecological services. For this, the India has included corals in Schedule-I list of the Wild Life Protection Act, 1972 of Ministry of Environment and Forest vide its notification dated on 11 July 2001 has included all the hard coral in the Schedule List of WLPA of 1972.

Environment Protection Act, 1986 (EPA) was brought into force to confers exclusive jurisdiction to the Central Government to preserve and protect the marine environment and to prevent and control marine pollution.

Coastal Regulation Zone Notification (CRZ) 1991 under the EPA is the only law that explicitly outlaws coral mining in India.

Marine Protected Areas (MPAs): to preserve the certain areas of the nation's waters, including some areas with coral reefs.

There are a five-coral reef MPAs in India

Gulf of Mannar Biosphere Reserve *viz.* Gulf of Mannar Marine National Park, Gulf of Kutch Marine National Park, Mahatma Gandhi Marine National Park in Andamans, Great Nicobar Biosphere Reserve and Rani Jhansi Marine National (Richies Archipelago).

No take zones (NTZs) are the places selected from a number of world's MPAs, where all forms of exploitation are prohibited and restricts human activities. These no take zones are extended to cover either entire MPA or a specific portion.

Global Protection Measures

Aichi Target 10 working to reduce the pressures on coral reefs and other vulnerable ecosystems impacted by climate change or ocean acidification. The Global Coral Reef Monitoring Network (GCRMN) is an operational network of International Coral Reef Initiative (ICRI) with a purpose of reporting on the condition of the world's coral reefs.

Coral reef restoration and rehabilitation efforts

Restoration plays a crucial role in promoting the revival of ecological characteristics like vegetation cover, complexity, and biodiversity, thereby mitigating the decline of natural ecosystems. Across the globe, restoration efforts are gaining momentum to neutralize the persistent stressors that are adversely affecting the extent and resilience of coral reefs. However, the effectiveness of restoration in facilitating reef ecological recovery is a subject of debate, with some arguing that it is an expensive diversion from addressing the underlying causes of reef degradation. Various initiatives and agreements at the international level have played a crucial role in accelerating the restoration of reefs and enhancing the overall well-being of natural ecosystems. Coral reef restoration initiatives like employing methods such as underwater coral cultivation (Growing and Planting Healthy Corals) and reattaching fragmented coral pieces, these endeavours relocate and revive numerous coral colonies on impaired reef locations. Also the UN Decade for Ecosystem Restoration, the Kunming-Montreal Global Biodiversity Framework, the Coral Reef Breakthrough, and the implementation of diverse financial mechanisms such as parametric insurance. These commitments and instruments have provided the necessary force to restore and preserve the vitality of reef ecosystems on a significant scale.

Other restoration strategical efforts

Improving habitat quality for corals

This includes assisting in the research and development of strategies to reduce the presence of nuisance and invasive species that compete with corals for their habitat.

Preventing loss of corals and their habitat

Reducing damage to coral reefs involves various aspects, including the identification of high-risk areas, the provision of emergency response support, and the recovery of damages caused by physical events like vessel groundings.

Enhancing coral population resilience

The advancement of new methods through research and development will enhance the ability of coral larvae to withstand challenges and decrease mortality rates.

Improving coral health and survival

Enhancing methods to manage the spread of coral diseases and minimize the effects of organisms that prey on corals will enhance the survival rates of corals in crucial reef locations.

Future perspective for coral reef ecology and conservation

Millions of people depend on coral reefs, which are shallow sub-tidal ecosystems of tropical oceans that are created at the border between land and sea and offer a wealth of resources. Their geologic component—the calcium carbonate deposited by algae, foraminfera, mollusks, and corals—distinguishes them from other marine ecosystems. Evolutionary change on coral reefs is better documented compared to many other ecosystems, and ancient reefs have been the subject of countless studies of past global change. The *Scleractinia*, or modern stony corals, have, however, undergone a major radiation since the Cretaceous 65 million years BP with their species numbers increasing until the present. Over the previous 1.8 million years, there have likely been 22 ice ages, which has given rise to repeated cycles of global climate change. Changes in the Earth's eccentricity and obliquity, or Milankovitch frequencies, have caused glacial cycles in the past, with atmospheric CO_2 concentrations trailing temperature increases by about 600 years. Sea level and water temperature fluctuations that have accompanied glacial and other cycles have not destroyed coral reefs. However, over these cycles, species extinction has happened and has been linked to climate change and synergistic habitat reductions. The present atmosphere is one of the most CO_2-rich in recent geologic history with the current level being greater than at any time in the last 420 000 years. The level projected for the year 2100 is greater than that seen in the last 20 million years. Understanding the fundamental processes of ecosystems—such as carbonate accretion, herbivory, and nutrient cycling in the case of coral reefs—is the goal of functional ecology. As coral reefs change, so too does our use of them. As society changes, technological innovations may create novel opportunities for ecosystem services to be drawn from coral reefs. These changes to the ecosystem and society will not benefit all user groups equally, and winners and losers will emerge. Thus, going forward, thoughtful consideration and a strong integration of social science into ecosystem management are necessary for how we choose and manage for a specific functional goal.

The larvae of *scleractinian* corals may encounter a variety of suitable habitats for settlement on tropical reefs. An optimal habitat for settlement would promote the survival of spat (early single polyp recruits) through consolidation (e.g. reef matrix that won't easily be dislodged). Protection from grazing, predation and sediment smothering, and adequate irradiance, requirements which could differ amongst coral species. Importantly, substrata that have developed a microbial bio-film community can often induce metamorphosis. In addition to microbial biofilms, colonisers such as coralline algae (phylum Rhodophyta, subclass Corallinophycidae) have been demonstrated as important biochemical inducers for coral larval settlement. The potential for coralline algae to facilitate the recovery of coral populations highlights their importance for the resilience of coral reefs, which are in a state of global decline. The ways and rates at which interactions between coral reefs and human society are altering are unprecedented. The inherent biophysical context of a reef is complemented by a network of cross-scale human socioeconomic and cultural forces that increasingly control the dynamics of Anthropocene reefs. The primary structural factor of coral reefs has been human-induced global climate change, which has overtaken their local surroundings and has led to the adoption of innovative configurations. Due to these changes, historical ecosystem services are disappearing in some areas while, in others, as human society adjusts, new ecosystem services and functions are emerging. Future reef configurations and society's ongoing adaptation will probably lead to roles that are not yet understood or valued.

Case studies of successful coral reef conservation

Coral reefs support many coastal societies and economies throughout the tropics and have been calculated to generate USD 375 billion annually (Bryant *et al.*, 1998). According to the Global Coral Reef Monitoring Network 20% of reefs have been destroyed and a further 24% of reefs have lost 50–90% of their corals (Wilkinson, 2004). It has long been believed that reef restoration is crucial to halting the deterioration of coral reefs (Precht, 1998). Coral bleaching is becoming more frequent as the oceans warm. It is often associated with strong El Niño events (Hoegh-Guldberg, 1999). Coral bleaching in 1998 and 2010 caused large-scale coral deaths in reef systems around the globe, with the 1998 event heavily impacting Palau, in the Western North Pacific sub-region, and Palmyra Atoll in the Central North Pacific sub-region (Turgeon, *et al.*, 2002). In Palau, nearly one-half (48%) of 946 surveyed colonies were totally bleached, and a further 15% were partially bleached (Bruno, 2001). Coral bleaching has also been observed elsewhere in the Micronesian, Marianas, Samoan, and Hawaiian archipelagos. The Reefs at Risk Revisited (2011) report predicts that by 2050 many of the reefs in the Pacific will bleach every year. This frequency of bleaching is worrying because it allows little time for corals to recover. The stress of high temperatures is compounded by the ocean's acidity, which is brought on by rising air carbon dioxide levels that seawater absorbs. Less carbonate in the form required for coral reefs to construct their calcium carbonate skeletons is one effect of ocean acidification. The Palmyra Atoll National Wildlife Refuge encompasses a robust tropical Pacific coral reef. The USFWS photo credit goes to J. Maragos. These tiny coral polyps' skeletons are an essential component of coral reef ecosystems. In the Pacific, the average coral cover is predicted to drop to 15% by 2035 based on the pace of coral loss documented over the last 20 years as well as the anticipated effects of more frequent coral bleaching and ocean acidification. Since it is impossible to cool down significant amounts of saltwater, coral reef managers have limited options for stopping or minimizing coral bleaching. Reducing human influences like overfishing, sediment and pollution discharge, and invasive species can help increase the potential resilience of reefs. Furthermore, it is now feasible to determine which reefs are maybe more resistant to bleaching and have a greater chance of recovering thanks to early-warning systems that forecast coral bleaching and track its effects on reef ecosystems. Nurfadli (2012) studied on the role of habitat creation in coral reef conservation: a case study from Aceh, Indonesia and they conclude Nonetheless, Pak Dodent's project demonstrates that artificial reefs can play an important role within an integrated marine conservation strategy by raising community awareness of reef conservation and providing economic benefits by increasing employment. Edwards and Clark (1998) studied and carried the harvesting of coral colonies or fragments from a donor site and the transplantation of these to a restoration site without the intermediate step of growing coral fragments in a nursery. Chamberland *et al.*, in 2017 conducted research regarding larval enhancement in coral reef: flow-through facilities are used to aid fertilization and competent coral larvae are then seeded to a reef restoration site. The reared coral larvae can be either free swimming or attached to substrate.

REFERENCES

Bambridge T. (Ed). 2016. The Rahui: Legal pluralism in Polynesian traditional management of resources and territories. Australian National University Press. 269 pp.

Christie, P., Lowry, K., White, A.T., Oracion, E.G., Sievanen, L., Pomeroy, R.S., Pollnac, R.B., Patlis, J.M., Eisma, R.L.V. 2005. Key findings from a multidisciplinary examination of integrated coastal management process sustainability. Ocean & Coastal Management, 48, 468-483.

Cinner, J., Huchery, C., MacNeil, M. *et al.* 2016. Bright spots among the world's coral reefs. Nature, 535, 416–419. https://doi. org/10.1038/nature18607

Cinner, J.E., Barnes, M.L., Gurney, G.G., Lockie, S., Rojas, C. 2021. Markets and the crowding out of conservation-relevant behavior. Conservation Biology, 35, 816-823.

Clive, W., 'Status of the Coral Reefs of the World', in Report of the Global Coral Reef Monitoring Network (GCRMN) (2004).

Hicks, C.C. *et al.* 2019 Harnessing global fisheries to tackle micronutrient deficiencies. Nature 574, 95–98. (doi:10.1038/s41586-019-1592-6)

Jupiter, S.D., Cohen, P.J., Weeks, E., Tawake, A. and Govan, H. 2014. Locally-managed marine areas: multiple objectives and diverse strategies. Pacific Conservation Biology, 20, 165-179.

Kumar, C. 2005. Revisiting 'community' in community-based natural resource management. Community Development Journal, 40: 275–285.

Levinton, Jeffrey S. 1995. Marine Biology: Function, Biodiversity, Ecology. Oxford University Press, Oxford, England.

Micheli, F. *et al.* 2014 High vulnerability of ecosystem function and services to diversity loss in Caribbean coral reefs. Biol. Conserv.171: 186-194. (doi:10. 1016/j.biocon.2013.12.029)

5

Marine Protected Areas in India

V.B. Sakhare

Increasing pressures exerted on the environment by humans make protection of natural areas crucial for the preservation of biological diversity. Protected areas are one of the most effective tools available for conservation of biodiversity.

The IUCN defines a Protected Area as 'an area of land and/or sea specially dedicated to the protection and maintenance of biological diversity, and of natural and associated cultured resources, and managed through legal or other effective means.

Since 1986, the IUCN commission on National Parks and Protected Areas (CNPPA) has been promoting the establishment and management of a global representative system of Marine Protected Areas (MPAs) which play a critical role in the conservation of biodiversity. MPAs are growing in importance globally as practical and potentially effective options for the management of fisheries, the protection of biodiversity and the generation of income from eco-tourism. The effective management of MPAs to ensure that they meet their declared objectives poses many challenges.

The IUCN has defined an Marine Protected Area (MPA) as 'Any area of the intertidal or subtidal terrain, together with its overlying water and associated flora, fauna, historical and cultural features, which has been reserved by law or other effective means to protect part or all the enclosed environment (Kelleher,1999).

In the United States of America, the term 'marine protected areas' is defined by a Presidential Executive order as 'any area of the marine environment that has been reserved by federal, state, tribal, territorial, or local laws or regulations to provide lasting protection for part or all of the natural and cultural resources there in'.

At present, only about 6% of the ocean is under marine protected areas. Of this, 2.4 per cent are fully and highly protected and 3.6 per cent are highly protected. The remaining 0.8 per cent are designated and 2 per cent have been proposed and committed. According to experts roughly 10 million square kilometres of the ocean must be annually brought under Marine Protected Areas (MPAs) to protect 30 per cent of the world's ocean by 2030. Fully protected areas prevent extractive and destructive activities, while highly protected MPAs allow light extractive activities. Taking the global coverage as of 2010, there are approximately 6800 MPAs. The New Caledonia MPA, created in 2014, is currently the largest protected area in the world at 1,300,000 km^2.

Reasons for establishment of MPAs: MPAs are generally designated with biodiversity conservation objectives, to protect fishery resource species or habitat, or with a broader ecosystem purpose within the framework of EAF (Ecosystem Approach to Fisheries).Within this context, MPAs tend to be conceived and implemented to achieve a subset of a wide variety of potential objectives. A list of examples of objectives for establishment of MPAs is provided below (FAO, 2011):

1. Rebuilding fish stocks;
2. Ensuring sustainability of fish stocks and fisheries;
3. Protection of marine biodiversity and critical habitats;
4. Support local and traditional sustainable marine-based lifestyles and communities;
5. Increase resilience to climate and other environmental changes;
6. Facilitate the resolution of multiple stakeholder conflicts;
7. Facilitate scientific research, education and recreation;
8. To protect cultural and archaeological sites.

Benefits from MPAs: Some of the benefits from MPAs for fisheries management have been supported by empirical studies. There is strong evidence about the protection of spawning biomass as a result of MPAs in Australia, Philippines, and elsewhere (Russ and Alcala 1996; Mapstone *et al.*,1997).Since the fishing effort is eliminated in MPAs, adults keep on reproducing while juveniles grow into maturity and become part of the spawning biomass (Alcala,1988;Alcala and Russ 1990).Diversity, density and biomass of target species of groupers, snappers and emperors increased when the community protected the coral reef to Apo and Sumilon Islands, Philippines (Rus,1985;Rus 1989;Alcala and Russ 1990;Rus and Alcala,1996). Fish catches from adjacent marine areas increased after protection and declined when the protection was lifted (Alcala,1988; Alcala and Russ,1990; Robertson,1999a; Robertson 1999b).

MPAs are favourable areas for prompt fisheries management. There is no need for a thorough understanding of the biology of the stock. Estimates of the density of target species would be useful at the start as a preference point for evaluating the benefit of the closure at some point in time. It is also considered easy to enforce; no quotas are set, because fishing is completely prohibited in the MPA.

Categories of MPAs

Category I: National parks and sanctuaries have an entire area of mangroves, coral reefs, creeks, and sea grass beds, estuaries, and lagoons.

Category II: Includes islands, which have major parts in the marine ecosystem and some parts in the terrestrial ecosystem.

Category III (A): Includes sandy beaches beyond the intertidal line but occasionally interacting with seawater.

Category III (B): Includes evergreen or semi-evergreen forests of the islands.

Underfunded MPAs: A well-managed and sufficiently funded MPA can restore good health to vulnerable ecosystems, help achieve global biodiversity conservation goals. A lack of funding poses challenges to the benefits that marine protected areas (MPAs), areas managed for long-term conservation, can potentially provide, according to experts. As many as 70 per cent of MPAs are underfunded. A well-managed and sufficiently funded MPA can restore good health to vulnerable ecosystems. MPAs can generate sustainable revenues for their own management. Revenue can be generated from statutory and non-statutory MPA fees for tourism programmes, blue carbon credits generated from mangrove conservation and avoided deforestation as well as seaweed farming and sustainable coastal fisheries.

MPAs in India: India has a coastline of 8.118 km, with an Exclusive Economic Zone (EEZ) of 2.02 million sq.km and a continental shelf area of 372,424 km^2, spread across 9 maritime states and seven Union Territories, including the island of Andaman and Nicobar, and Lakshadweep.

In India, PAs that fall in whole or in part within swath of 500 meters from the high tide line and to marine environment are included in the Marine Protected Areas in India. Out of these, there are four Marine National Parks, Sixty-seven marine sanctuaries, National parks and Wild Life Sanctuaries, three Marine Biosphere Reserves, three Ramsar Coastal Wetlands, one Tiger Reserve (Sunderbans), one National Mangrove Genetic Resource Centre and Gene Centre and the coral reefs of Lakshadweep (Seventeen),thirty two mangrove notified forests, four national parks (having area 130 km^2) and 16 wildlife sanctuaries (185 km^2) have been identified for conservation measures. Apart from this,17 parks and 28 wildlife sanctuaries have been proposed/existing on the island territories of India. It is necessary to monitor these areas to assess impact of conservation measures, status of habitats within the protected areas, impact of development on the quality and quantity of the resources etc. No such database exists for the Indian Marine Protected Areas (MPAs). The repetitive nature of satellite data helps in monitoring vital ecosystem areas as well as in assessing the impact of conservation measures.

Currently, there are about 31 MPAs along India's coastline that have been officially declared for conserving and protecting coastal and marine biodiversity. Most of the MPAs were designated during the 1980s and early 1990s.They were notified as either 'national parks' or ' wildlife sanctuaries', under the wild Life (Protection) Act (WLPA) 1972,where,in most cases ,no extractive activity is allowed.

Indian Marine Protected Areas cover a total area of 6271.2 km^2 with an average size of 202.1 km2.East coast and Andaman and Nicobar Islands have adequate areas in the MPAs whereas west coast and Lakshadweep Islands have poor representation. Also, another 100 Pas (10 in main Indian coast and 90 islands Pas in Andaman and Nicobars have terrestrial or fresh water ecosystems which constitute boundaries with sea water or partly contain marine environment, but they are not listed as MPAs as per the criteria. Although, India has a very long coastline and varied coastal habitats, contribution of the MPAs is only 4% to the total area of the protected Areas and 1.3% of the continental shelf area of the country (Singh,2003).

The Andaman and Nicobar groups of islands located in the southeast of the Bay of Bengal consists of 350 islands. The deeply idented coastline results in innumerable creeks, bays and estuaries and facilitates the development of mangroves. There are 60 species of mangroves making it the second most diversity rich area in the Indian sub-continent. Coastal reefs are the important habitats with diverse marine fauna, particularly in the western side of the Andaman Islands.

The Gulf of Kachchh in Gujarat has an assemblage of ecologically sensitive ecosystems, viz., coral reefs, mangroves, sea grasses, algae/seaweeds etc. It is the largest inlet of the Arabian sea, about 60 km wide at its widest and 170 km long. It is designated as both a wildlife sanctuary and a national park, to protect the coral reefs and mangroves. It was first designated as a sanctuary in 1980, and later in the same year, to provide complete protection to the islands and intertidal area, it was declared as a national park (Rajagopalan,2008).

The Malvan Marine Sanctuary, Sindhudurg forms a part of the Western Ghats is considered as one of the biologically rich regions along the Maharashtra coast. It was designated on 13th April 1987 by a notification by the Forest Department, Government of Maharashtra. It covers an area of 29.12 sq.km, with a core zone of 3.18 sq.km. The core zone includes the Sindhudurg fort,

Padamgad island and other submerged rocky structure. The core zone is used for anchoring fishing vessels, and for fishing by a small number of hook-and-line fishermen. The buffer zone has seven villages, with a population of over 7000 that depends on fishing for a livelihood. Though, the sanctuary has been designated, it exists mainly on paper, as the regulations have not been implemented. Sindhudurg is a low fortified island on the coastal reef, which is joined to the mainland by a reef. At Ratnagiri and Malvan, sandy beaches are interspersed with rock formation extending over a considerable distance into the sea and forming small bays. During low tide, the exposed area of the bay has large rock pools with rocks partially exposed and the corals are generally seen attached to the rocks. These corals are legally protected under the Malvan Marine Sanctuary. This sanctuary, together with some adjoining areas also harbours mangrove vegetation (Laxmilatha *et al.,* 2015).

Sunderbans Biosphere Reserve, West Bengal along with three other wildlife sanctuaries covers a total area of 1,73,645 sq.km of coastal wetlands. Suderbans Biosphere Reserve is a majestic natural region in the world. The Sunderbans is a vital coastal asset with an annual economic value of Rs.2114 crores, far greater than any other coastal and marine ecosystem. It is a well-protected wetland, managed with legal systems, such as Reserve Forest, Protected Areas, National Park and Wildlife Sanctuary. The Sunderbans is an 'endangered ecosystem' under the IUCN Red List. It is the most stressed environment with rise in sea surface, temperature, sea-level rise, salinity, erosion, accretion and natural disasters.

The Gulf of Mannar Biosphere Reserve (Tamil Nadu) lies in the south of the Palk Bay harbours 21 islands and is globally significant because of its unique biological diversity. The Gulf has a number of offshore platform reefs, patch reefs and coral pinnacles, which lie from south of the Pamban Pass to north-east of Tuticorin. The Gulf of Mannar is considered as 'Biologists' paradise' as it has 3600 species of flora and fauna. Coral reefs comprise of 94 coral species under 37 genera,10 species of sea grasses and high density of macro-algal species. Ninety species of sponges,119 species of annelid fauna,450 species of molluscs, and 22 species of bivalves have been reported from this region. This region harbours mangrove ecosystems. The region was declared as marine National Park in 1980 comprising an area of 6.23 sq.km and was given the status of Biosphere Reserve in 1989 covering an area of 10,500 sq.km. Rough estimates suggest that there are 125 fishing villages and 35,000 active fishers who depend on the resources in Gulf of Mannar area, especially on fishing, and collection of seaweed and other marine resources. There are approximately 5000 fisherwomen who depend on seaweed collection in and around the 21 islands, and 25,000 fishermen who dive to collect sea cucumbers (Rajagopalan, 2008).

Bhitarkania is a mangrove forest located in Kendrapara district of Orissa. It was declared as a sanctuary in 1975 and National Park in 1988.It comprises of Bhitarkanika, Kalibhanjadian and Gahirmatha mangrove areas and is fed by the Dharma River, the Manipura River and the Bhitarkanika River. The floral diversity includes a total of more than 300 plant species with both mangrove and non-mangrove species. The diverse habitat ranging from mangrove forests to agricultural fields, rivers, streams, freshwater and brackish water wetlands, riverine islands, off-shore islands, muddy and sandy coastlines etc provides home to a large number of animal species which includes longest estuarine crocodile of the world, partial albino reptilian species, salt water crocodiles, three species of Indian monitor lizards, the python and various migratory birds (Laxmilatha *et al.,* 2015).

Table 1: List of Marine and Coastal Protected Areas in India

S.No.	Name	State	Year of establishment	Total Area (Sq.km)
1	Gulf of Mannar	Tamil Nadu	1986	560.00
2	Sunderbans	West Bengal	1984	1330.10
3	Bhitarkanika	Orissa	1988	145
4	Gulf of Kachchh	Gujarat	1980	162.89
5	Rani Jhansi	Andaman & Nicobar	1996	256.14
6	Mahatma Gandhi	Andaman & Nicobar	1983	281.50
7	North Button	Andaman & Nicobar	1987	0.44
8	Middle Button	Andaman & Nicobar	1987	0.64
9	South Button	Andaman & Nicobar	1987	0.03
10	Malvan	Maharashtra	1987	29.12
11	Gulf of Kachchh	Gujarat	1980	295.03
12	Khijadiya	Gujarat	1981	6.05
13	Sajnekhali	West Bengal	1976	362.04
14	Lothian	West Bengal	1976	38
15	Haliday	West Bengal	1976	5.95
16	Bhitarkanika	Orissa	1975	672
17	Gahirmatha	Orissa	1997	1435
18	Chilika	Orissa	1987	15.53
19	Coringa	Andhra Pradesh	1978	235.70
20	Pulicat	Andhra Pradesh	1976	500.00
21	Krishna	Andhra Pradesh	1999	194.81

22	Point Calimere	Tamil Nadu	1967	17.26
23	Pulicat	Tamil Nadu	1980	153.67
24	Lohabarrack	Andaman & Nicobar	1987	100
25	North Reef Island	Andaman & Nicobar	1987	3.48
26	South Reef Island	Andaman & Nicobar	1987	1.17
27	Cuthberr Bay	Andaman & Nicobar	1987	5.82
28	Cingue	Andaman & Nicobar	1987	9.51
29	Galathea	Nicobar	1997	11.44
30	Parkinson Island	Andaman & Nicobar	1987	0.34
31	Mangroves Island	Andaman & Nicobar	1987	0.39
32	Blister Island	Andaman & Nicobar	1987	0.26
33	Sandy Island	Andaman & Nicobar	1987	0.26
34	Pitti	Lakshadweep	2000	0.01
35	Sunderbans (Tiger Reserve)	West Bengal	1973	2,585
36	Gulf of Mannar (Biosphere Reserve)	Tamil Nadu	1989	10,500

REFERENCES

Alcala, A.C.1988.Effects of protective management of marine reserves on fish abundances and fish yields in the Philippines, Ambio, 17:194-199.

Alcala, A.C. and Russ, G.R.1990.A direct test of the effects of protective management on abundance and yield of tropical marine resources.J.Const.Int.Explor.Mer., 46:40-47.

FAO.2011. Fisheries Management,4. Marine Protected Areas and Fisheries, FAO Technical Guidelines for Responsible Fisheries,198.

Kelleher, G.1999.Guidelines for marine protected areas. Best Practice Protected Area Guidelines Series No.3, Gland, Switzerland, International Union for Conservation of Nature (IUCN), and Cordiff, Wales, UK, Cordiff University.

Laxmilatha, P., T.S. Sruthy and M.S. Varsha. 2015.Marine Protected Areas, Summer School on Recent Advances in Marine Biodiversity Conservation and Management,16 February-8 March 2015, Central Marine Fisheries Research Institute, Kochi, Kerala,199-206.

Madhuri, Ramesh and Shimul, Bijoor. 2021. Management of Marine Protected Areas in India. Economic and Political Weekly, Vol. LVI (25):35-41.

Mapstone, B.D., Davies, C.R., Higgs, J.B. and Wlech, D.J.1996.The effect of re-opening Bramble Reef to bottom fishing on fishing behavior and catch rates of commercial and recreational line fishers. Proceedings, 2nd World Fisheries Congress,1:24.

Rajagopalan, Ramya. 2008.Marine Protected Areas in India. International Collective Support of Fishworkers, Samudra Monograph, Chennai, pp 49.

Robertson, J.1999a.Reef closures-do they really protect reef communities? pp 109-120. In: Proc. APEC Workshop on Impacts of Destructive fishing Practices on the Marine Environment,16-18 December 1997, Hongkong,315pp.

Robertson, J.1999b.Effectivness of temporary reef closures to replenish reef fish stocks in the Great Barrier Reef. Pp 145-154. In: Dight, I.R. Kenchington and J. Baldwin, International Tropical Marine Ecosystems Management Symposium, November 1998, Townsville, Australia,451 pp.

Russ, G.R. and Alcala, A.C.1996.Do marine reserves export adult fish biomass? Evidence from Apo Island, Central Philippines.Mar.Ecol.Progr.Ser.,132:1-9.

Singh, H.S.2003.Marine Protected Areas in India. Indian Journal of Marine Sciences, 32(3):226-233

6

Indian Marine Copepods: A Review

V.B. Sakhare and A.I. Shaikh

Study done in India on various aspects of marine copepods are those of Krishnaswamy (1952), Kartha (1959), Ummerkutty (1960, 1966), Ganapathi *et al.* (1961), Kasturirangan (1963), Saraswathy (1966), Abraham (1970), Silas and Pillai (1973), Sarladevi (1979), Sarkar *et al.* (1985), Madhupratap and Haridas (1986), Wells and Rao (1987), Gajbhiye *et al.* (1991), Madhupratap (1999), Kesarkar and Anil (2010), Ramanibai and Shanthi (2011), Srichandan *et al.* (2014). Jayabarathi *et al.* (2015), Nishida *et al.* (2015), Arunpandi *et al.* (2017), Loka *et al.* (2017), Deepika *et al.* (2019), Kaviyarasan *et al.* (2019), Khandagale *et al.* (2022), Nawaz *et al.* (2023), Thangaraj and Vijayanand (2023).

Copepods are planktonic crustaceans, which occur in marine, fresh as well as brackish waters. They represent about 80% of the zooplankton in the ocean. There are more than 210 families, 2400 genera and 24000 species identified in this group. Planktonic copepods are considered to be the most abundant metazoans on earth.

The name 'copepod' is derived from the Greek words meaning 'animals with oar shaped foot' i.e., 'kope' means oar and 'podos' means foot (Stottrup,2003). Copepods are important secondary producers and primary consumers and ultimately contribute significantly to the food chain in large ecosystems.

Copepods are very hardly planktonic forms which can withstand most of the unfavourable conditions and can produce diapause eggs and resting eggs to survive in these conditions. Most of the copepods can adjust wide range of salinity and temperature levels. Due to its hardy nature, copepods can be easily introduced into all types of water bodies.

Copepods forms important food for many aquatic animals especially fishes. Certain fishes and fish larvae were evolutionary adapted to feed on copepods. Copepods have been found to be nutritionally superior to almost all other live feeds.

Classification

Copepods are classified under the phylum Arthropoda, subphylum crustacea, class Hexanauplia and subclass Copepoda. Most of the free-living planktonic copepods feed on phytoplankton and many benthic copepods feed on other animals, and many copepods are parasitic in nature. Basically there are nine orders in the subclass copepods i.e., Calanoida, Cyclopoida, Harpacticoida, Platycopioida, Mormonilloida, Misophrioida, Siphonostomatoida, Monstrilloida and Gelyelloida.

The subclass Poecilostomatoida is mostly considered as a group under cyclopoida but phylogenetically separated from it. Members of the subclass Calanoida, Cyclopoida and Harpacticoida form important components of marine plankton.

Body Structure

The body structure comprises a large cephalothorax formed by the fusion of head and thoracic segments and a small segmented abdomen called urosome. The thorax has basically six segments. All segments possess a pair of legs or pleopods which are meant for swimming. The 5th and 6th legs are considered to be taxonomically very important. Cephalic region has a rostrum, a pair of median eyes, a pair of antennule, antennae, mandible, maxilla and maxilliped. Most of the copepod appendages except the antennule are generally biramous. The abdominal segments are reduced and without any limbs except for the caudal furca which form a tail fan with long setae. The sixth thoracic segment unites with first abdominal segment to form a genital double somite. The genital double somite together with abdominal somites form a tail like portion called urosome.

Male copepods are distinguished from females by their smaller body size and modified antennule. Antennules of males are often modified for holding females during mating. Mouth parts in copepods are modified as cutting, piercing or sucking types. In parasitic forms, males are mostly free living and retain their original basic crustacean feathers throughout their life.

Some copepods carry egg sacs and some are without egg sacs. Many calanoid copepods broadcast their eggs directly in water. In some genera like Acartia, eggs have an additional protective covering with minute spines. Some other have simple and smooth shell. Many of them have a pair of egg sacs. Most of the cyclopoids have a pair of egg-sac while calanoids and harpacticoids generally have a single egg sac (Santosh *et al.,* 2018).

Culture and Composition in cultured copepods: The copepods constitute an important component of the food chain in aquatic system and they are known to be the preferred natural prey for most marine finfish larvae and form the dominant component of their diets in wild.Culture of copepods was carried out with different types of food. Algae and diatoms were found to be excellent diet for the culture of copepods. A calanoid *Gladioferan pactinatus* has produced multiple generations in unialgal diet of *Isochrysis galbana,* but with *Tetraselmia chuii* and *Phaeodactylum tricornatum*. The development was slow (Arnott *et al.,* 1986).

Main problem in copepod culture is the growth of undesirable organisms in culture tanks. Overfeeding, accumulated debris and faecal pellets result the emergence of ciliates and other organisms in culture system. Stock and mass culture methods for different species developed at Vizhinjam Research Centre of CMFRI are ideal to meet the requirements of large hatcheries where marine food fish larval rearing has been taken up. The methods are applicable for small scale ornamental fish hatcheries also. This method is ideal for most of the calanoid copepods and cycloid copepods. But among harpacticoids studied, only *E.acutifrons* can be cultured using thus method (Santosh *et al*., 2018).

The most commonly used species in mariculture are free-living copepods belonging to orders Calanoida, Harpactocoida and Cyclopoida.

The calanoid copepods are predominantly pelagic and are selective feeders,75% of the known species are marine,and some are benthopelagic or commensal. They feed mainly on small phytoplankton cells by filtration, or preying upon a variety of animal prey including copepod eggs. They are distinguished by their long antennules, as long as the body itself or even longer, with up to 27 segments and biramous antennule used as accessory locomotory appendages (Ranjan *et al*., 2023).

The harpacticoids include more than 50% of the copepod species, are primarily marine, free living, benthic organisms. They are distinguished by their short antennules, fewer than 10 segments and biramous antennae (Ranjan *et al.,* 2023).

The cyclopoids include pelagic, epibenthic, benthic and parasitic species, which inhabit both freshwater and marine environments. In the marine environment cyclopoida belonging to the family cyclopinidae are predominantly benthic and those of the Oithonidae are planktonic. Cyclopods have uniramous antennae used to help catching food. The antennules in cyclopoids are shorter than in calanoids, rarely reaching beyond the cephalothorax (Ranjan *et al*., 2023).The cyclopoids mainly feed on detritus, many are predacious and few are parasitic.

Information on biochemical composition of copepod is of great importance in understanding the nutritive value and energy pathways in the pelagic food web. The marine copepods are considered to be 'nutritionally superior' as food for marine fish and prawn larvae as they are a valuable source of proteins.

Lipids, carbohydrates, amino acids, fatty acids and enzymes, all of which play an important role in larval digestion and enhance metamorphosis of larvae (Fluchter and Rembold 1986; Munilla-Moran *et al*., 1990; Millamena *et al*., 1990 and Stottrup 2000). Most of the fish and crustacean species mainly depend on copepods throughout their life stages and some species even feed exclusively on copepods during their entire life. Perumal *et al.,* (2009) studied biochemical composition of wild copepods, *Acartia spinicauda* and *Oithona similis*, from Parangipettai coastal waters in relation to environmental parameters. Of the principal biochemical constituents, protein formed the major component followed by lipid and carbohydrate. Totally 16 amino acids were observed in these wild copepods, with threonine, glutamic acid, alanine, aspartic acid, serine, valine and methionine as the dominant one. Ashok Prabu and Rajkumar (2007) studied biochemical composition of cultured copepod (*Acartia spinicauda*, Giesbrecht) and Ashok Prabu *et al.,* (2005) also studied biochemical composition of some marine copepods. On similar lines Rajkumar *et al*., (2008) also studied biochemical composition of wild copepods, *Ascartia erythraea* Giesbrecht and *Oithona brevicornis* Giesbrecht from Coleroon coastal waters of south east coast of India.

The live-feed organisms like copepods are called 'living capsules of nutrition' for fish and prawn larvae. According to Rajkumar *et al.,* (2008) the two species (*Acartia erythraea* Giesbrecht and *Oithona brevicornis* Giesbrecht) are nutritionally equal to any other food fish and they could be exploited successfully for feed purpose and for preparation of various by-products similar to fish and prawn by-products or selected animals and may be even used as live feed for aqua farming purpose.

The copepods are rich in proteins, lipids, essential amino acids and essential fatty acids (Watanabe *et al*., 1983; Altaff and Chandran 1989; Safiullah,2001).

Use of Copepods in Marine fin Fish Larval Rearing

Copepods are the best option as initial feed for larvae of a variety of marine food fishes and ornamental fishes. But due to the difficulties in rearing these are being used as feed for a critical period when larvae are unable to feed or survive with artemia and rotifers. The main demerit of copepods as a commercial larval feed is that, these cannot be cultured in high densities.

Copepods are used in larval rearing of many ornamental fishes in CMFRI trials on seed production of damsel fishes were successful only after using copepods as initial feed (Santosh *et al*., 2018). Gopakumar and Santoshi (2009) reported use of copepods as live feed larviculture of damselfishes. On similar lines and Santosh *et al.,* (2016) also documented potential species of copepods for marine finfish hatchery.

Marine Copepod Diversity in India

Distribution of calanoid copepods in the Arabian sea and Bay of Bengal was reported by Stephen (1984). The inshore offshore zonation of a neritic, shelf and oceanic communities was traced based on the species composition. Numerical abundance was high in the shelf stations. Secondary production considered in terms of copepod density showed the ratio 2:1 between the Arabian sea and Bay of Bengal. The water column from the top layer of thermocline to surface was dominated by *Acrocalanus spp, Paracalanus spp, Eucalanus spp, Undinula vulgaris* and *Euchaeta spp*. In the discontinuity larger *Eucalanus attenuates, Scolecithricella ctenopus, Haloptilus longicornis* and *Lucicutia flavicornis* were encountered. In the deeper collections from the Bay of Bengal bathypelagic copepods viz., *Heterostylilis longicornis, Lucicutia clausi, Amallothrix indica* and *Euaugaplilus nudus* have been recorded here for the first time.

Biodiversity of copepods of Chennai coast (India) was reported by Altaff *et al.,* (2004). They recorded 33 species, out of which calanoids, cyclopoids and Harpacticoids constituted 14,13 and 6 species respectively. In general, higher female ratio was observed in all copepods.

Robin *et al.,* (2009) studied zooplankton distribution, density and diversity along the southern Kerala coast. Among different groups of zooplankton copepods contributed maximum numerical abundance contributed up to 31.93%.26 species of copepods constituted by 20 species of calanoida,3 species of harpatocoida and 3 species of cyclopida. The predominant species of copepods were *Acartia spinicuda, Calanus finmarchcus, Paracalanus parvus, Acrocalanus gracilis* and *Euterpina acutiferons*.

Studies on copepods from Chennai coast (Cooum and Adyar),Bay of Bengal were made by Shanthi and Ramanibai (2011).Altogether 35 copepod species were identified, out of these 31 species belong to calanoids and 4 to cyclopoids. The low occurrence of copepods in the near shore stations compared to that of offshore stations may be due to the inflow of effluents, sewage and domestic waste discharged from the Cooum and Adyar river.

Biraja *et al.*, (2012) reported copepoda as the major component of zooplankton with various stages of life cycles like nauplii larvae and copepod eggs. Copepoda comprised 78.07% to 86.59% of the total groups found in the sample. The distribution of copepods density and zooplankton biomass shows almost identical patterns. The copepod population ranged from 9379 to 16203 nos/100m^3.

Santhanam *et al.,* (2013) studied diversity and abundance of marine copepods in Muthupet mangrove waters (southeast coast of India) and recorded 75 species of copepods comprising 51 species of calanoida,14 cyclopoida, and 10 Harpacticoida. The calanoid copepods were dominant with 68% followed by cyclopoida (18.66%) and Harpacticoida (13.33%). High species evenness was observed during the monsoon season indicating that the species were equally distributed.

Sivaleela and Venkataraman (2014) studied diversity and distribution of harpacticoid copepods from Tamil Nadu coast.Harpacticoid copepods increased in coarser sand.A lower density of harpactoicoid copepods was observed at Nagapattinam and Tuticorin.It is situated near the harbour, which was considerably organically enriched due to harbour activities.

Srichandan *et al.,* (2014) studied summer distribution of zooplankton in coastal waters of Odhisa and reported copepod as dominant among all groups/ taxa, and well distributed all along the odhisa coast. The population density of copepod showed well marked spatial variation and it ranged from 301 org.m^{-3} to 5027 org.m^{-3}.

Srichandan *et al.,* (2015) studied zooplankton distribution in coastal water of the North-Western Bay of Bengal, off Rushikulya estuary, east coast of India. They reported calanoida as the most dominant order among the copepods. Copepods not only remained as the most dominant group but also appeared in large numbers. They were represented by 112 species, belonging to 26 families and 4 orders. Calanoida emerged as the most dominant group comprising 68 species. The dominance of calanoida over others could be due to their continuous breeding, quick larval development and adaptation to wide range of environmental conditions.

Kavitha *et al.,* (2018) reported copepod abundance and diversity from offshore region of Tuticorin, South East Coast of India and recorded 56 copepod species belonging to 20 families under 4 orders. The percentage composition of different groups of copepod species was composed of Calanoida (35 numbers), 62.5%, Cyclopoida (4 numbers), 7.14%, Harpacticoida (8 numbers)14.3% and Poecilostomatoida (9 numbers) 16.1%.

In India,44 species of pontellid copepod belonging to 6 genera have been reported (Roy,2018). Kasturirangan (1963) described 14 species of pontellid copepods in his manual. Seven species of pontellid have been reported from the offshore region of Tuticorin (Kavitha *et al.,* 2018). Nawaz *et al.,* (2023) reported 11 species of pontellid copepods from the Covelong (Kovalam) coast.

An inventory in the form of a catalogue of Indian calanoid copepod has been prepared by Dev Roy (2018). It comprises 287 species and subspecies belonging to 82 genera and 25 families. Among these 99 are marine and 55 freshwater forms, remaining 143 species occur in estuarine and diversified water conditions. Maximum species have been recorded in Kerala (114 species) while the minimum is recorded from the states of Delhi, Manipur and Uttarakhand being represented by one species each.

Copepods developed at Vizhinjam Research Centre of CMFRI gave promising results in trials conducted in larviculture orange spotted grouper, *Epinephelus coioides* and Indian Pompano, *Trachinotus mookalee* were successfully bred and reared at Visakhapatnam Regional Centre of CMFRI (Ranjan *et al.,* 2018).

Copepods parasitizing marine fish along the Cochin and Malabar coast (Kerala) of the Indian ocean were identified by Nikhila *et al*., (2019). One hundred species of copepods from 31 genera in eight families were found living on 85 fish species. Copepod of the order Siphonostomatoida were the dominant parasites.

Sivakumar *et al.,* (2021) studied population composition of calanoid copepods of the Chennai coast (Tamil Nadu) and calanoid copepod population variations between the monthly sampling. The changes of calanoid copepod densities were due to the influence of physico-chemical parameters.

Arunpandi *et al.,* (2022) identified 48 copepod species, with 26 belonging to the calanoida order,119 to the cyclopoida order and three to the harpacticoida order. Nearshore copepod density ranged from 700-2580 No.m^{-3}, while offshore copepod density ranged from 112 to 260 No.m^{-3}. Calanoid copepods were the most abundant in the nearshore (35-65%),while cyclopoid copepods dominated the offshore (52.5-72.6%).Harpacticoid copepod contribution was less than 8% of overall copepod density

Observations of copepod diversity in Bay of Bengal coast during the cruise expedition were made by Michael Beaven *et al.,* (2024). They identified copepods were 24 species belonging to 17 families and three orders which is represented in the sequence Calanoida (83.8%), Cyclopoida (13.5%) and Harpacticoida (2.7%).

REFERENCES

Abraham, S.1970.On the occurrence and seasonal distribution of Acartia plumosa T. Scott (Copepoda: Calanoida)-a new record for the west coast of India.Curr.Sci.,39(5):115-116.

Altaff, K. and Chandran, M.R.1989.Sex-related biochemical investigation of the diaptomid Heliodiaptomus viduus (Gurney) (Crustacea: Copepoda). Proceedings of the Indian National Science Academy. (Animal Science),98:175-179.

Altaff, K., Dilshad Begum I., Gopikrishna, M. and Siva Kumar, K.2004. Biodiversity of copepods of Chennai coast-India.Convergence,6(1-4):44-60.

Arnott, G.H., Brand, G.W. and Kos, L.C.1986. Effects of food quality and quantity on the survival development and egg production of Gladioferens pectinatus (Brady) (Copepoda: Calanoida). Aust. J.Mar.Freshw.Res.,37:467-473.

Arunpandi, N., Jyothibabu, R., Jagadeesan, L., Gireshkumar, T.R., Karan, C. and Naqvi, S.A. 2017. Noctiluca and copepods grazing on the phytoplankton community in a nutrient -enriched nearshore environment along the southwest coast of India. Environmental Monitoring and Assessment,189(7):351.

Arunpandi, N., Jyothibabu, R., Jagadeesan, L., Lekshmi, S., Surya, S. and Biju, A.2022. Copepod carcasses in the western Bay of Bengal and associated ecology. Environmental Monitoring and Assessment,194 (2):1-29.

Ashok Prabu, V., Perumal, P. and Rajkumar, M.2005.Biochemical composition of some marine copepods. Res.J.Chem.Environ.,9:36-41.

Ashok Prabu, V. and Rajkumar, M.2007.Biochemical composition of cultured copepod, Acartia spinicauda, Giesbrecht.Aquacult.,8:219-224.

Biraja Ku. Sahu, S.K. Baliarsingh, S. Srichandran and K.C. Sahu. 2012. Zooplankton abundance and composition in surf zone of Gopalpur Part, east coast of India-A case study. Marine Science,2(6): 120-124.DOI:10.5923/j.ms.20120206.02.

Deepika, S., Srinivasan, A., Padmavathy, P. and Jawahar, P.2019.Diversity of zooplankton in seagrass ecosystem of Mandapam coast in Gulf of Mannar. International Journal of Current Microbiology and Applied Sciences,8(7):2034-2042.

Dev Roy, M.K.2018.Catalogue on copepod fauna of India-Part I. Calanoida. J. Environ. & Sociobiol.15(1):1-73.

Fluchter, J. and Rembold, H. 1986.Soluble factor essential for metamorphosis of coreconid larvae has been partially purified from Artemia salina. Arch.Hydrobiology,22:197-202.

Gajbhiye, S.N., Rosamma Stephan, Vijayalakshmi R. Nair and Desai, B.N.1991.Copepods of the near shore waters of Bombay.J.Mar.Sci.,20:187-194.

Ganapathi, PN. and Shanthakumari, K.1961.The systematics and distribution of planktonic copepods in the Lawson's Bay, Waltar. J. Mar. Biol. Assoc.India,3:6-18.

Gopakumar, G. and Santoshi, I.2009.Use of copepods as live feed for larviculture of damselfishes. Asian Fisheries Science,22:1-6.

Jagadeesan, L. and Jyothibabu, R.2016.Tumour like anomaly of copepods-an evaluation of the possible causes in Indian marine waters. Environmental Monitoring and Assessment,188(4):244.

Jayabarathi, R., Padmavati, G. and Anandavelu, I. 2015.Spatial heterogeneity of benthic copepods: A comparative aspect of composition, abundance, and correlation. Zoological Studies,54,51.

Kartha, K.N. Krishna. 1959. A study of the copepods of the inshore waters of Palk Bay and Gulf of Mannar. Indian Journal of Fisheries,6(2):256-267.

Kasturirangan, L.R.1963.A key for the identification of the more common planktonic copepoda of the Indian coastal waters. CSIR, New Delhi, India,1-87.

Kavitha, M., Padmavathy, P., Srinivasan, A., Jawahar, P., Ranjith, L. and Linga Prabu, D.2018.Copepod abundance and diversity from offshore region of Tuticorin, South Erast Coast of India. International Journal of Current Microbiology and Applied Sciences,7(4):2767-2792.

Kaviyarasan, M., Santhanam, P., Balakrishnan, S., Kumar, S.D., Ananth, S., and Kandan, S.2019. Seasonal variations in copepods diversity along the Vellar coastal waters, South Eastern India. Environment and Ecology,37(2):471-480.

Kesarkar, K.S. and Anil, A.C.2010.New species of Paracalanidae along the west coast of India: Paracalanus arabiensis.J.Mar.Biol.Assoc.UK,90(2):399-408.

Khandagale, P.A., Mahatre, V.D. and Thomas, S.2022.Seasonal and spatial variability of zooplankton diversity in North Eastern Arabian Sea along the Maharashtra coast. J. Mar. Biol. Ass.India,64(1):25-32.

Krishnaswamy, S.1952.Some new species of copepods from Madras coast. Records of the Zoological Survey of India.49 (3-4):321-336.

Loka, Jayasree, Philipose K.K., Sonali S.M., Santosh B., Anzeer F Muhammed, Saha Purubali, Sukumaran Sandhya, Kumar Rahul G. and Gopalakrishnan, A.2017. Apocyclops cmfri sp. nov. (Cyclopoda: Cyclopodia: Cyclopidae) a new copepod species from Arabian sea off Karwar, Karnataka, India. Indian Journal of Fisheries, 64(2):1-9.

Madhupratap, M.1999.Free-living copepods of the Arabian sea: Distributions and research perspectives. Indian Journal of Marine Sciences,28:146-149.

Madhupratap, M. and Haridas, P.1986.Epipelagic calanoid copepods of the northern Indian ocean. Oceanologica Acta,9(2):105-117.

Michael Beaven, M., G. Ananthan and M. Thangaraj. 2024. Observations of copepod diversity in Bay of Bengal coast during the cruise expedition. International Journal of Zoology and Applied Biosciences,9(2):5-12.

Millamena, O.M., Penaflorida, V.D. and Subosa, P.F.1990.The macronutrient composition of natural food organisms mass cultured as larval food for fish and prawns. Israeli J.Aqua. Bamidgeh,42:77-83.

Munilla-Moran, R., Stark, J.R. and Barbour, A.1990.The role of exogenous enzymes in digestion in cultured turbot larvae (Scophthalmus maximus L.). Aquaculture,88:337-350.

Nawaz, M.A., Siva Kumar, K., and Baskar, G.2023.Seasonal dynamics of body size in calanoid copepods (Calanoida: copepoda) from the stressed tropical coast of India, Chennai, Bay of Bengal, Aquatic Ecology, https://doi.org/10.1007/s10452-023-10075-1.

Nawaz, M.A., Siva Kumar, K., and Baskar, G. and Radha, V.2023.Diversity rhythm in pontellid copepods (Pontellidae: Copepoda) from the Covelong past pre-and post-COVID-19 lockdown, Bay of Bengal. Turkish Journal of Zoology,47:71-80.

Nikhila, R.M.V., Drisya, O.K., Rijin K., Mumthaz, T.M.V., Helna, A.K., Ivanenko, V.N., Kappalli, S.2019.Copepod crustaceans parasitizing marine fish of the Kerala coast, India//Arthropoda Selecta,28(4): 529-544.DOI:10.15298/arthsel.28.4.05.

Nishida, S., Anandavelu, I. and Padmavati, G.2015.Two new species Tortanus (Atortus) (Copepoda, Calanoida, Tortanidae) from the Andaman Islands.Crustaceana,88:216-230.

Perumal,P., Rajkumar, M. and Santhanam,P.2009.Biochemical composition of wild copepods, Acartia spinicauda and Oithna similis, from Parangipettai coastal waters in relation to environmental parameters.J.Environ.Biol.,30(6):995-1005.

Rajkumar, M., Kumaraguru Vasagam, K.P. and Perumal, P.2008.Biochemical composition of wild copepods, Ascartia erythraea Giesbrecht and Oithona brevicornis Giesbrecht from Coleroon coastal waters, Southeast coast of India, pp 1-20. In:Advances in Aquatic Ecology (Volume 2) Ed. V.B. Sakhare, Daya Publishing House, New Delhi.

Ramaiah, N. and Nair, V.1997.Distribution and abundance of copepods in the pollution gradient zones of Bombay harbour-Thane creek-Basin creek, west coast of India. Indian J. Mar. Sci., 26:20-25.

Ramanibai, Ravichandran and Shanthi, Jordi. 2011. Seasonal profiles of copepod abundance and diversity in relation to water masses in the coastal ecosystem of Chennai, Bay of Bengal, India. Geophysical Research Abstracts, Vol.13.EUG2011-13825, EGU General Assembly 2011.

Ranjan,R., Avadhanula Ravi, K. and Shiva ,Ponnaganti.2023.Copepod culture techniques for marine finfish larval rearing. In: Training Manual on Livefeed Culture Techniques for Mariculture Applications, CMFRI Training Manual Series No.36/2023.ICAR-Central Marine Fisheries Research Institute, Visakhapatnam, pp 65-78.

Ranjan, R. Sekar Megarajan, Biji Xavier, Shubhadeep Ghosh, Santosh, B. and Gopalakrishnan, A.2018.Broodstock development induced breeding and larval rearing of Indian Pompano, Trachinotus mookalee (Cuvier,1832)-A new candidate species for aquaculture,495:550-557.

Robin, R.S., Srinivasan, M. and Chandrasekar, K.2009.Distribution of zooplankton from Arabian sea, along southern Kerala (Southwest coast of India) during the cruise. Current Research Journal of Biological Sciences, 1(3):155-159.

Roy, D.M.K.2018.Catalogue on copepod fauna of India, Part I-Calanoida. Journal of Environment and Sociobiology,15(1):1-75.

Safiullah, A.2001.Biochemical and nutritional evaluation and culture of freshwater live food organisms for aquahatcheries, Ph.D. thesis, University of Madras,103 pp.

Santhanam, P., Jeyaraj, N., J. Sivakumar, K. Jothiraj, T. Jayalakshmi, S. Ananth, R. Nandakumar, A. Shenbaga Devi, B. Balaji Prasath and S. Dinesh Kumar.2013.Diversity and abundance of marine copepods in Muthupet mangrove waters, southeast coast of India. In: Venkataraman *et al.*, (eds.), Ecology and Conservation of Tropical Marine Faunal Communities, DOI:10.1007/978-3-642-38200-0_8, Springer-Verlag Berlin Heidelberg 2013, pp 117-136.

Santhanam, P. and Perumal, P.2003.Diversity of zooplankton in Parangipettai coastal waters, southeast coast of India. J. Mar. Biol. Assoc.India,45:144-151.

Santosh, B., Anil, M.K., Ritesh Ranjan, Muhammed Anzeer ,F., Aneesh, K.S., Mijo V. Abraham, Dasrsana S., Unnikrushnan, C., Jose Kingsley H., Udayakumar, A. and Greever Yoyak.2018.General methods for stock and mass production of copepods as live feed,pp.19-35.In:Culture Techniques of Marine Copepods (Eds. Santosh, B., Anil, M.K., Muhammed Anzeer, F., Aneesh, K.S., Mijo V. Abraharr, Gopakumar, G., Rani Mary George, Gopalkrishnan, A. and Unnikrishnan,C.2018.ICAR-Central Marine Fisheries Research Institute, Kochi, Kerala.

Santosh, B., Muhammed Anzeer, F., Unnikrishnan, C. and Anil, M.K.2016.Potential species of copepods for marine finfish hatchery. In: Course Manual Winter School on Technological Advances in Mariculture for production Enhancement and Sustainability. Imelda Joseph and Boby lgnatius (Eds.), Central Marine Fisheries Research Institute, Kochi,170-176.

Santosh, B., Saleela K.N., Muhammed Anzeer F., Aneesh K.S., Mijo V. Abraham, Unnikrishnan, C., Jose Kingsley H., Uadayakumar A. and Greever Yoyak.2018.Basic biological aspects of copepods relevant to culture, pp 9-18.In:Culture techniques of marine copepods, (Eds. Santhosh B., Anil M.K., Muhammed Anzeer, F., Aneesh ,K.S., Mijo V. Abraham, Gopakumar, G., Rani Mary George, Gopalakrishnan, A. and Unnikrishnan), ICAR-Central Marine Fisheries Research Institute, Kochi, Kerala, India,144 pp.

Saraswathy, M.1966.Pelagic copepods from the inshore waters off Trivandrum coast, Proc. Symp. Crustacea. J. Mar. Biol. Assoc.India,1:74-106.

Sarkar, S.K., Singh, B.N. and Choudhury, A.1985.Copepod components of inshore zooplankton of the Bay of Bengal off Sagar island, West Bengal, India. Current Science,54(23):1217-1220.

Sarladevi, K., Stephen, R. and Rao, T.S.S.1979.Distribution of Haloptilus (Copepoda: Calanoida) in the Indian ocean, Indian Journal of Marine Sciences,8:159-165.

Shanti, M. and Ramanibai, R.2011. Studies on copepods from Chennai coast (Cooum and Adyar), Bay of Bengal-During the Cruise. Current Research Journal of Biological Sciences,3(1):132-136.

Silas, E.G. and Pillai, P.P.1973.The calanoid copepod family Pontellidae from the Indian ocean. J. Mar. Biol. Assoc.India,15:771-858.

Sivakumar, K., Nawaz, Md. A., and Saboor, A.2021.Population composition of calanoid copepods of the Chennai coast, Tamil Nadu, Indian Journal of Geo Marine Sciences,50(9):693-700.

Sivaleela,G. and Venkataraman,K.2014.Diversity and distribution of Harpacticoid copepods from Tamil Nadu coast,India.Rec.Zool.Surv.India,114(Part 1):1-11.

Srichandan, S., Biraja K. Sahu, R. Panda, S.K. Baliarsingh, K.C. Sahu and R.C. Panigrahy. 2015. Zooplankton distribution in coastal water of the North-western Bay of Bengal, off Rushikulya estuary, east coast of India. Indian Journal of Geo-Marine Sciences,44(4):546-561.

Srichandan, S., Panda, C.R. and Rout, N.C.2014.Summer distribution of zooplankton in coastal waters of Odhisa, East coast of India. International Journal of Oceanography and Marine Ecological System,3(1):9-25.

Stephen, Rosamma. 1984. Distribution of calanoid copepods in the Arabian sea and Bay of Bengal. Mahasagar-Bulletin of the National Institute of Oceanography,17 (3):161-171.

Stottrup, J.G.2000.The elusive copepods: Their production and suitability in marine aquaculture. Aqua.Res., 31:703-711.

Stottrup, J.G.2003.Production and nutritional value of copepods. In: Live feeds in marine aquaculture, Stottrup, J.G. and McEvoy, L.A. (Eds.), Blackwell Publishing Company, Oxford, U.K.p.145-205.

Thangaraj, M. and Vijayanand, P.2023.Diversity and distribution pattern of copepods in Palk Bay coastal waters of India. Manga Scientia Advanced Research and Reviews,7(1):42-48.

Ummerkutty, A.N.P.1960.Studies of Indian copepods I. Paralepeopsyllus mannarensis, a new genus and species of cyclopoid copepod from the Gulf of Mannar. Journal of the Marine Biological Association of India,2(1):105-114.

Ummerkutty,A.N.P.1966.Studies on Indian copepods-16. On some rare and interesting copepods from south east coast of India. J. Mar. Biol. Ass.India,8 (2):302-319.

Watanabe, T., Kitajima, C. and Fujita, S.1983.Nutritional value of live organisms used in Japan for mass propagation of fish: a review, Aquaculture, 34:115-143.

Wells, J.B.J. and Rao, G.C.1987.Littoral Harpacticoida (Crustacea: Copepoda) from Andaman and Nicobar Islands. Memoirs of the Zoological Survey of India,16:1-385.

7

A Review on Mariculture in India

Ketan Makwana, H.L. Parmar, Hardik Sikotariya, Sayan Biswas and Mayur Bhadarka

ABSTRACT

According to the FAO's "Global Aquaculture Production" database, the annual global production of aquaculture has increased at an average rate of around 8% over the past three decades, outpacing any other major animal food production sector. This rapid global increase in production is occurring alongside a slow decline in marine fisheries catches. India, with its extensive coastline and abundant marine resources, has immense potential for the development of its fisheries sector. With a record fish production of 175.45 lakh tons in FY 2022-23, India has become the third-largest fish-producing country in the world. Mariculture, including open sea cage farming of finfish and shellfish, assumes great importance in this context. Mariculture involves the cultivation of marine organisms for food and other products in the open ocean, tanks, ponds, or raceways filled with seawater. Species such as sea bass, cobia, and pompano have been identified as suitable candidates for mariculture due to their high commercial value and market demand.

***Keywords**: Mariculture, Fisheries, Fish, Resources.*

INTRODUCTION

The Food and Agriculture Organization (FAO) of the United Nations defines aquaculture as the cultivation of aquatic organisms such as fish, mollusks, crustaceans, and aquatic plants by individuals or corporations. In aquaculture, interventions are made in the rearing process to enhance production. While aquaculture is a practice with origins dating back more than 3000 years, Aquaculture products have only recently been recognized by the United Nations Statistical Commission as distinct global commodities, despite the practice having origins dating back more than 3000 years. According to the FAO's "Global Aquaculture Production" database, the annual global production of aquaculture has increased at an average rate of around 8% over the past three decades (since 1980). Despite a recent slowing in global trends, growth remains higher than any other major animal food production sector. This rapid global increase in production is occurring alongside a slow decline in marine fisheries catches. There is a growing global focus, largely driven by demand from Western countries, on increasing the intensive production of omnivorous and carnivorous species farmed and harvested in maritime and brackish coastal environments, a practice known as mariculture. The farming of these species, such as salmon, groupers, seabasses, and prawns, typically relies heavily on inputs such as wild post-larvae and juveniles, fishmeal, fish oils, as well as significant amounts of water, land, and energy. As a result, mariculture has faced widespread criticism due to its often negative impacts on marine and coastal ecosystem health. Negative associations have been drawn between mariculture and its effects on marine capture fisheries and food security.

Resources

India, with approximately 8,118 kilometers of coastline, nearly 2 million square kilometers of Exclusive Economic Zone (EEZ), and half a million square kilometers of Continental Shelf, possesses abundant marine resources. From these resources, India has an estimated fisheries potential of 4.41 million tonnes. In this context, the optimum utilization of resources becomes pivotal to achieve the targeted production. It is against this backdrop that we aim to harness all possibilities for the intensive and integrated development of the fisheries sector.

Marine Capture Fisheries

Marine capture fisheries are undergoing significant changes due to increasing fishing effort. The catch of almost all commercially important marine finfish and shellfish is on a declining trend, leading to severe resource depletion and unemployment. The decline in marine capture fishery also affects the availability of affordable protein for the public and impacts the GDP growth of the country. It is in this context that mariculture, including open sea cage farming of finfish and shellfish, assumes great importance[9]. With a record fish production of 175.45 lakh tons in FY 2022-23, India has become the third-largest fish-producing country in the world, accounting for 8% of global production. The fisheries sector contributes about 1.09% to the country's Gross Value Added (GVA) and over 6.724% to the agricultural GVA. The sector has immense potential for growth, which calls for focused attention through policy and financial support to ensure sustainable, responsible, inclusive, and equitable development .The overall marine fish landings in the country included 18.72 lakh tonnes of pelagic fishes, 8.74 lakh tonnes of demersal fishes, 2.23 lakh tonnes of molluscs, and 3.73 lakh tonnes of crustaceans. These landings are made by various fishing crafts at 1269 designated locations, which include fifty fishing harbors across the coastal regions of the country.

Mariculture

Mariculture is a specialized branch of aquaculture involving the cultivation of marine organisms for food and other products in the open ocean, or in tanks, ponds, or raceways filled with seawater. Examples include the farming of marine fish such as cobia, pompano, and sea bass, as well as shellfish like lobster and oysters, along with seaweed, all in saltwater environments. Non-food products produced by mariculture include fish meal, nutrient agar, jewelry (e.g., cultured pearls), and cosmetics. Fish raised through mariculture practices are often perceived to be of higher quality than those raised in ponds or tanks, and mariculture offers a more diverse choice of species.

Open Sea Cage Culture

Open sea cage farming is eco-friendly and typically conducted in areas of the open sea where wave action is minimal. The fishes cultured in these cages are high-value species, leading to a significant export demand for cage-cultured fishes.A cage, which can be of any shape or size, is an enclosure used for the culture of biotic organisms such as finfish and shellfish in captivity, with a specified objective. The cages are enclosed on all sides, except for an opening at the top for providing feed.The size of a sea cage can vary from 6 to 12 meters in diameter for fish farming in the open sea, where the wave and tidal impact is suitable for farming. A series of cages are typically spaced in a battery for better operation and management.

Marine Finfish Culture

[1] Cobia (*Rachycentron canadum*)

Cobia is one of the suitable species for open sea cage farming. It can reach a body weight of about 3-4 kg in one year and 8-10 kg in two years.Brood stock development of cobia was initiated by CMFRI at Mandapam during 2007-08. The technology for cobia seed production and farming was standardized by CMFRI and has been successfully adopted by fish farmers.The successful sea cage farming of cobia by the 'Cobia Farming Association' in the Palk Bay region, Rameswaram, and its subsequent adoption by many fishermen groups in Tamil Nadu, Kerala, Karnataka, and Goa, indicate that the technology is commercially viable.Currently, in the Mandapam-Rameswaram area, more than 50 cages are deployed for cobia farming by fishermen groups. There is a significant demand for seed to deploy further cages for cobia farming.

Cobia were reared in a three-phase system in nearshore cages for 12 months according to the following protocol:

Phase 1: Nursery Phase (Months 1-3)

- Cobia juveniles were stocked in nursery cages.
- Feeding with commercial pellets containing 45% crude protein, three times a day.
- Regular monitoring of water quality parameters.

Phase 2: Grow-out Phase (Months 4-9)

- Cobia transferred to larger grow-out cages.
- Continued feeding with commercial pellets, adjusted for growth.
- Regular monitoring of water quality parameters and health status.

Phase 3: Final Grow-out Phase (Months 10-12)

- Continued feeding and monitoring.
- Fish monitored for optimal growth and health.
- Harvesting of mature cobia at the end of the 12-month period.

[2] Sea bass (*Lates calcarifer*)

Sea bass is one of the most important candidate species for open sea cage culture and pond culture due to its high commercial value. Seed production technology has been developed and standardized by ICAR Institutes. Protocols for nursery and rearing of sea bass have been developed by CMFRI and CIBA. Several successful cage and pond farming demonstrations have been conducted.CMFRI has perfected cage culture of sea bass at Karwar in Karnataka and has successfully achieved production of about 3 tonnes of fish from a 6-meter diameter cage.Sea bass seeds are produced in the hatcheries of Rajiv Gandhi Centre for Aquaculture (RGCA) and CIBA. However, hatchery production meets less than 40% of the demand. Sea bass is successfully farmed along the coasts of Maharashtra, Goa, Karnataka, Tamil Nadu, Andhra Pradesh, and Kerala.

Seabass fingerlings (5.5 cm each) were procured from the Rajiv Gandhi Centre for Aquaculture (RGCA), Tamil Nadu. Each oxygen-filled polythene bag contained 100 juveniles and was transported during October-November. Farmers typically formed procurement groups to share transportation expenses and associated costs. Seed bags were transported in air-conditioned vans for distances up to 1,000 km, incurring an average transportation cost of US$ 671-745 per trip to bring 10,000 juveniles.During transportation, approximately 15-20 percent of the fish died due to stress factors and the acclimation process. Farmers stocked 5,000-6,000 fingerlings per cage, but only 1,000-1,200 survived, largely due to cannibalism during the juvenile stages.Typically, farmers cultured seabass for 18-20 months to achieve a weight of 3-4 kg. Some farmers experimented with stocking stunted fingerlings procured from private hatcheries in Vijayawada. However, due to disease problems, the survival rate was very poor, and one farmer incurred a loss of US$ 7,455 on seed and transportation.Maintaining the quality of seabass fingerlings is crucial to maximize survival during transportation. Over the years, the demand for seabass fingerling production has been increasing, with production increasing from 436,000 fingerlings in 2009 to 1 million fingerlings in 2013-14. During the first and second month, farmers use shrimp feed at 200 g per 1000 juveniles of 5-6 cm. Feeding is done three times daily. Weaning fingerlings to natural food is carried out within a week.After two months, fish are fed a fish paste mainly consisting of sardine. Each box of 30 kg fish waste costs around US$ 11.93.

[3] Silver pompano (*Trachinotus blotchii*)

Silver pompano is also a suitable species for marine aquaculture due to its wide acceptability and medium size, making it suitable for small families. It has the ability to accustom to pellet feeds, wide tolerance to water quality, and high market demand.The Central Marine Fisheries Research Institute (CMFRI) has achieved successful brood stock development, induction of spawning, and larval production of silver pompano. The technology for silver pompano seed production and farming, developed and standardized by CMFRI during 2011-15, has been successfully adopted by fish farmers.The techno-economic viability of coastal pond farming of silver pompano was demonstrated by CMFRI in Andhra Pradesh during 2012. Silver pompano is found to grow faster in low salinities (10-25 ppt), is less cannibalistic, and more resistant to a wide range of diseases. It commands a good price in both domestic and international markets.Pompano seeds have been supplied to many farmers in Tamil Nadu, Andhra Pradesh, Kerala, Karnataka, Goa, Gujarat, and West Bengal for commercial culture.

Weight of the fish	Feed Size	Crude Protein %	Crude Fat %	% to be fed as per the biomass	Feeding/day
> 1 gram	800 - 1000 µ	50	10	30	4
1-10 gram	1.0 - 1.5 mm	40	8	20	4
10-100 gram	1.8 mm	35	8	8	3
100-250 gram	3.5 mm	30	6	5	3
250-500 gram	4.5 mm	30	6	3	3

Source (Jayakumar *et al.*, 2014)

Some way to commercialization of mariculture of sea bass, cobia and pompano are:

a) Establishment of brood-banks
b) Establishment of hatcheries
c) Establishment of land based nurseries
d) Feed production Technology
e) Construction of Cage farms
f) Establishment of land based support facilities including cold chain and processing
g) Marketing strategies

CONCLUSION

The Practice Of Aquaculture, As Defined By The Food And Agriculture Organization (FAO) Of The United Nations, Involves The Cultivation Of Aquatic Organisms Such As Fish, Mollusks, Crustaceans, And Aquatic Plants By Individuals Or Corporations, With Interventions Made In The Rearing Process To Enhance Production. Mariculture, which involves the intensive production of omnivorous and carnivorous species farmed and harvested in maritime and brackish coastal environments, has been criticized for its negative impacts on marine and coastal ecosystem health, as well as its implications for marine capture fisheries and food security. Mariculture, as a specialized branch of aquaculture, offers opportunities for the cultivation of marine organisms for food and other products in the open ocean or in tanks, ponds, or raceways filled with seawater. Species such as sea bass, cobia, and pompano have been identified as suitable candidates for mariculture due to their high commercial value and market demand.

REFERENCES

Ail, S. K.S., and Bhatta, R. 2016. Small-scale Cage Culture of Asian Seabass in Kundapur Region, Karnataka, India. World Aquaculture, 47.

Campbell, B. and Pauly, D. 2013. Mariculture: a global analysis of production trends since 1950. Marine Policy, 39, 94-100.

de Bezerra, T.R.Q., Domingues, E.C., Maia Filho, L.F.A., Rombenso, A.N., Hamilton, S., and Cavalli, R. O. 2016. Economic analysis of cobia (Rachycentron canadum) cage culture in large-and small-scale production systems in Brazil. Aquaculture international, 24, 609-622.

FAO. 2008. Strategy and outline plan for improving information on status and trends of aquaculture. Rome: FAO Fisheries and Aquaculture Department,73 p.

FAO. 2009. FAO Fisheries Glossary. Url: /www.fao.org/fi/glossary/default.aspS; [accessed 28.04.09].

FAO. 2012. The state of world fisheries and aquaculture 2012. Rome: FAO Fisheries and Aquaculture Department, 209 p.

Goldburg, R.2005. Naylor R. Future seascapes, fishing, and fish farming. Front. Ecol. Environ., 3(1):21–8.

Hanson, J. A.1974. Open Sea Mariculture Perspectives, Problems And Prospects. In Offshore Technology Conference (pp. OTC-2085). OTC.

Jayakumar, R., Nazar, A. K. A., Tamilmani, G., Sakthivel, M. and Gopakumar, G. 2014. Farming of Silver Pompano Trachinotus blochii in coastal aquaculture ponds.

Jayakumar, R., Nazar, A. K. A., Tamilmani, G., Sakthivel, M., Ramesh Kumar, P., Anikuttan, K. K., and Krishnaveni, N. 2020. Farming of Cobia and Silver Pompano.

Johnson, B., Nazar, A. A., Jayakumar, R., Tamilmani, G., Sakthivel, M., Ramesh Kumar, P., ... & Zacharia, P. U. 2019. Adoption of sea cage farming of cobia (Rachycentron canadum) by fishermen self-help groups as a diversified livelihood option: A success story from Ramanathapuram District, Tamil Nadu, India. Indian Journal of Fisheries, 66(3): 118-124.

Ling S.W.1977. Aquaculture in Southeast Asia-A Historical Overview. Seattle,WA: University of Washington Press; 1977 108 p.

Marine Fisheries Census. 2016. India. ICAR-Central Marine Fisheries Research Institute, Indian Council of Agricultural Research, Ministry of Agriculture and Farmers Welfare; Fishery Survey of India and Department of Fisheries, Ministry of Fisheries, Animal Husbandry and Dairying, Government of India).

Pullin RSV, Rosenthal H, Maclean JL. Eds.1992. Environment and aquaculture in developing countries. Summary Report of the Bellagio conference on environment and aquaculture in developing countries. Bellagio, 17–22 September, 1990.International Center for Living Aquatic Resources Management Conf. Proc. 36: 16.

Sathianandan, T. V., Jayasankar, J., Kuriakose, S., Mini, K. G. and Mathew, W. T. 2011. Indian marine fishery resources: optimistic present, challenging future. Indian Journal of Fisheries, 58(4), 1-15.

Sathianandan, T. V.2013. Status of marine fisheries resources in India–an overview.

Trujillo, P.2007. A global analysis of the sustainability of marine aquaculture. Master of science, resource management & environmental sciences dissertation. Vancouver, BC: University of British Columbia; 127 p.

8

Dynamics of pH and Its Regulation in The Bio-FLOC Technology (BFT) System: A Review

Mohammed A.A. Abo-Taleb, Safwat Gobran and Mostafa A.M. Soliman

ABSTRACT

Aquaculture is one of the most important pillars that contribute significantly to achieve food security, especially in developing countries. In order to combat the scarcity of land, water, and feed supplies, sustainable aquaculture aims to maximize the use of existing resources. Bio-floc Technology (BFT) is an innovative approach of sustainable aquaculture that has the ability to overcome these challenges and increase revenue. Along with preserving water quality, the primary function of bio-floc is to transform fish wastes into microbial protein that fish could consume, which led to increase the profits, decrease costs and strengthening concepts of sustainability. Numerous variables influence bio-floc aggregation and proliferation, which in turn impacts water quality and fish farming success. The pH is one of the most crucial of these factors since it can alter the water's quality and have an impact on fish raised in farms. Fluctuations in pH are critical because they impact the balance between carbonate and bicarbonate ions, which in turn affect the buffering ability of water culture system. Moreover, the pH range influences the prevalence of nitrifying bacteria. Nitrosomonas, for instance, prefer a pH range of 7.8–8.8, whereas Nitrobacter prefers a pH range of 7.2–9. Thus, this review will focus on the most significant pH-related parameters in the BFT system and how pH fluctuation is controlled in such conditions.

***Keywords**: Biofloc technology, pH dynamics, Nitrification process, Alkalinity, Stocking density, Carbon source, C/N ratio.*

INTRODUCTION

The aquaculture industry is regarded as one of the most advanced food production industries globally. From 2010 to 2020, it grew at an average annual rate of 4.6%, reaching 122.6 million tonnes of live weight (FAO, 2022). Sustainable aquaculture proposes numerous profits to attain economic progress through well usage of natural resources (FAO 2017). The expansions in the sustainable aquaculture business are challenging due to the inadequate number of available natural resources and industrialized effects that released to the surround environments (Costa-Pierce *et al.* 2012; Verdegem 2013). Particularly in intensive system, increased higher organic matter concentrations raise several future issues, including acute harmful, chronic impacts (Piedrahita 2003). Recently, Innovative biofloc technology system (BFT) is one of the most popular methods to combat these problems.

Biofloc is a climate-smart technology that relies on mass synthesis of in situ microorganisms (Ogello *et al.*, 2021). The BFT permits high stocking densities and higher fish productivity (Ogello *et al.*, 2021) while requiring limited or zero water exchange in to sustain the flocs (Emerenciano *et al.*, 2017). The power of BFT aligns with the "cradle to cradle" concept that outlined by McDonough and Braungart (2002), which omits the phrase "waste”. In BFT, nitrogen wastes that generated from feces of aquatic animals and residual feeds are transformed into the microbial protein that becomes feed for the same creatures (Minaz and Kubilay 2021). Numerous microorganisms, including bacteria, microalgae, protozoa, and zooplankton, are present in biofloc particles (Ray *et al.* 2010a). Hence, biofloc can incorporate the diet of aquatic species, which utilises the natural productivity in the culture system as a feed source (Wasielesky *et al.* 2006; Emerenciano *et al.* 2012a). This microorganisms are recognised for their contributions to (i) preserving high water quality (Emerenciano *et al.*, 2017), (ii) enhancing culture efficiency by lowering feed conversion ratio (FCR) and costs of feed (Hargreaves, 2006). (iii) biosecurity (Defoirdt *et al.*, 2004), and (iv) conserving greenhouse gases (GHG) (Manan *et al.*, 2016). These four biological roles of microorganisms in BFT units are factors of high fish production, profitability and environmental protection (Ogello *et al.*, 2021).

Regarding the applications, in the past years, BFT has been used in grow-out phase for tilapia (Avnimelech, 2007 and Crab *et al.*, 2009) and marine shrimp (Burford *et al.*, 2003 and Burford *et al.,* 2004), nursery phase (Samocha *et al.,* 2007, Ballester *et al.*,2010 and Emerenciano *et al.*, 2012a), freshwater prawn culture (Crab *et al.*, 2010 and Pérez-Fuentes *et al.*, 2013), broodstock formation and maturation in fish (Ekasari *et al.*, 2015) and shrimp (Emerenciano *0k* 2012b – Emerenciano *et al.*, 2013), and as aquafeed ingredient also called as “biofloc meal” (Kuhn *et al.,* 2009 , Bauer *et al.,* 2012 and Neto *et al.*, 2015). In addition, recently BFT also has been applied in carp culture (Najdegerami *et al.*, 2016), catfish culture (Yusuf *et al.,* 2015).

Beyond oxygen levels, the primary concerns for the BFT system are an overabundance of particulate organic matter and hazardous nitrogen compounds (linked to pH) (Emerenciano *et al.,* 2021). The BFT system's characteristics might mediate pH effects. The ideal pH is necessary for heterotrophic systems to support bacterial activity (Martinez-Porchas *et al.,* 2020). In contrast, pH affects both autotrophic algae and heterotrophic bacteria in mixotrophic systems, which include complicated relationships between both (Pinho *et al.*, 2022).

The pH scale determines the concentration of hydrogen ions in a solution and describes the balance between acids and bases in water. The pH influences almost all chemical reactions and phenomena that occur in the water (Lemonnier *et al.*, 2004). So, the current review will focus on pH as one of the most crucial elements that affects biofloc aggregation and, consequently, the efficiency of farming process.

1. The role of pH on nitrification process in biofloc technology system

Nitrification: the procedure by which bacteria belonging to the *Nitrobacteraceae* family oxidise nitrogen in ammonia and organic substrates to nitrites and nitrates. Thus, nitrification is important because it keeps the water body's oxygen supply from being depleted by nitrogenous oxygen demand (NOD) (Amatya *et al.*, 2011). According to Ebeling *et al,* (2006) the primary variables influencing nitrification rate include ammonia and nitrite concentrations, the C:N ratio, dissolved oxygen, pH, temperature, and alkalinity.

Boyd (2019) interpreted this process as the follows:

$NH_4^+ + 1½\, O_2 \rightarrow NO_2^- + 2H^+ + H_2O$ Equation (1)

$NO_2^- + 0.5\, O_2 \rightarrow NO_3^-$ Equation (2)

The bacteria *Nitrosomonas* and *Nitrobacter*, respectively, are responsible for the first and second oxidations. Typically, these two bacterial species are found together in the environment. As nitrite is changed into nitrate in the second reaction, nitrite seldom ever accumulates as a result of the first reaction. One can combine the two reactions to obtain the overall nitrification equation.

$NH_4^+ + 2\, O_2 \rightarrow NO_3^- + 2H^+ + H_2O$ Equation (3)

At pH levels between 7 and 8, and at temperatures between 25 and 35 C, nitrification proceeds most quickly. However, it might proceed slowly at lower temperatures and pH levels of 3 or 4. One essential component of nitrification is a large amount of dissolved oxygen. Ammonia can be anaerobically oxidised by certain bacteria. Van der Graaf *et al.* (1995) interpreted this process by the following equation.

$5\, NH_4^+ + 3\, NO_3^- \rightarrow 4N_2 + 9\, H_2O + 2H^+$ Equation (4)

The nitrification process adds to acidity because it generates hydrogen ions through both aerobic and anaerobic processes. In waters that receive considerable inputs of ammonia nitrogen, aerobic nitrification also significantly increases oxygen consumption and necessitates a plentiful supply of molecular oxygen.

The conversion of hazardous ammonia to nitrate makes nitrification a crucial process in BFT systems. Before a BFT system is started and run, nitrification is a procedure that needs to be carefully considered (Abakari *et al.,* 2021). Numerous variables affect the rate of nitrification, such as alkalinity, pH, temperature, dissolved oxygen (DO), ammonia-nitrogen concentration, C/N ratio, and ammonia-nitrogen concentration (Koops and Pommerening- Röser, 2001; Timmons *et al.,* 2002). To be more precise, the nitrification process depends on the pH of the biofloc water because it influences the nitrifying bacteria's activities (Abakari *et al.*, 2021). Lower pH values result in a slower rate of nitrification; Odegaard (1992) found that when pH dropped from neutral (7) to 6, the rate of nitrification decreased by 90%. According to Henze and Harremoës (1990), the ideal pH range is between 8 and 9. This is because it promotes the nitrification process.

The concentration of dissolved oxygen (DO) in the BFT system culture media is one of the most important elements influencing the nitrification process. Since nitrifying bacteria are obligate aerobes, the presence of oxygen makes them more active, which raises the nitrification rate. Consequently, it's crucial to maintain an oxygen concentration high enough to support these bacteria's activity (Lekang, 2007 , Abakari *et al.,* 2021). As demonstrated by Huang and McCarty (1971), *Nitrosomonas* activity decreases at low oxygen levels (< 4 mg/L), but *Nitrobacter* activity is also impacted at low oxygen levels (< 2 mg/L).

Numerous studies have shown that there is a wide range of pH values (7.0–9.0) that are ideal for nitrifying bacterial development (Zhang *et al.,* 2017). The optimal pH range for nitrification can be ascertained, according to Villaverde *et al.* (1997), based on three distinct effects of pH on nitrifying bacteria: the nitrifying bacteria's activation and deactivation; nutritional effects associated with alkalinity (inorganic carbon); and the inhibition of bacteria through free ammonia and free nitrous acid. For example, *Penaeidae* growth can be inhibited in environments with pH values over 9.5 or below 6.5. Furthermore, heterotrophic and nitrifying bacteria prefer a pH range of 7.0 to 9.0 (Chen *et al.*, 2006; Van Wyk and Scarpa, 1999).

To summarize, the ideal pH range for the nitrification process varies depending on the genera of the nitrifying bacteria and the conditions necessary for their growth. For instance, the ideal pH range for *Nitrosomonas*, which oxidise ammonia to nitrite (AOB), is 7.8–8.8, but the optimal pH range for *Nitrobacter*, which oxidise nitrite to nitrate (NOB), is 7.2–9.0 (Chen *et al.* 2006). Moreover, studies by Chen *et al.* (2006) and Ebeling *et al.* (2006) have shown that ammonia can be converted to nitrate through the consumption of alkalinity. Finally, it is significant to remember that the nitrification process causes a pH decrease, thus it is always necessary to take the appropriate actions to bring the pH back into balance in order to increase the activity of the nitrifying bacteria (Ebeling *et al.,* 2006).

2. Alkalinity and pH

In actuality, nitrifying bacteria use alkalinity in the form of carbonate and bicarbonate from the culture media. In addition, alkalinity offers the buffering ability required to prevent pH shifts brought on by the nitrification process's creation of acid. Thus, the relationship between alkalinity and pH affects the rate of nitrification (Zhang *et al.*, 2017). In the farming environment, pH and alkalinity are directly correlated; an increase in alkalinity will result in a fall in pH (Van Wyk and Scarpa 1999). Alkalinity is consumed by heterotrophic, chemosynthetic, and autotrophic bacteria in the biofloc technology system, lowering the alkalinity and pH (Dauda 2020). In the process of assimilating ammonia, heterotrophic and nitrifying bacteria consume alkalinity Chen *et al.* (2006) and Ebeling *et al.* (2006). Therefore, a decrease in alkalinity could happen as a result of heterotrophic bacteria consuming alkalinity often and there being no water exchange (Ebeling *et al.* 2006; Furtado, *et al.,* 2011). Thus, systems with or limited water exchange alkalinity suggested to be remained between 100 and 150 mg $CaCO_3$ L^{-1} (Ebeling *et al.,* 2006). The conversion of 1 g of ammonium-N into 8.07 g of microbial biomass and 9.65 g of carbon dioxide as a by-product requires heterotrophic bacteria to require 15.17 g of carbohydrates, 4.71 g of dissolved oxygen, and 3.57 g of alkalinity (Ebeling *et al.*, 2006). A lower or variable pH will result from the reduction to alkalinity because it will change the equilibrium of carbonate and bicarbonate ions in the culture system and could reduce the system's buffering ability (Furtado *et al.* 2011).

Natural productivity in *Litopenaeus vannamei* juveniles was examined by favourable impact on growth, food consumption, weight gain, feed conversion rates, and shrimp survival. Therefore, to produce a better particle aggregation and high quality of the bioflocs, the influence of the management factors such as mixing intensity, dissolved oxygen, pH, temperature, and organic carbon sources is crucial (De Schryver *et al.* 2008). According to studies by Ebeling *et al.* (2006) and Furtado *et al.* (2015), the growth of shrimp and the consumption of inorganic carbon by nitrifying and heterotrophic bacteria in BFT systems result in a drop in both pH and alkalinity, two indicators that were previously tightly associated. Lara *et al.* (2017) found that the aggregation of biofloc was affected by the use of three different aerators in the *Litopenaeus vannamei* biofloc culture system: (1) PR—propeller aspirator pumps aerator; (2) VP—vertical pump aerator; and (3) BL—diffused air blower. The lower alkalinity levels in the BL treatment indicate there were an increase in heterotrophic and nitrifying bacterial growth and, consequently, greater alkalinity consumption in this treatment *L. vannamei* requires an alkalinity greater than 100 mg $CaCO_3$ L^{-1} for optimal development, according to Van Wyk and Scarpa (1999). According to the same writers, a high alkalinity decreases the pH's diurnal fluctuations. The physical and chemical parameters of the water quality must be constantly monitored when rearing *L. vannamei* using BFT because, during the biofloc formation process, three key events take place: (1) an increase in the concentration of total ammonia nitrogen (TAN); (2) a reduction in TAN and an increase in nitrite concentration due to the action of ammonia-oxidizing autotrophic bacteria (AOB); and (3) a reduction in nitrite concentration and an increase in nitrate concentration due to the action of nitrite-oxidizing autotrophic bacteria (NOB). Increases in total suspended solids (TSSs) and decreases in alkalinity and pH have also been confirmed to occur during the rearing process (Furtado *et al.* 2011; Gaona *et al.* 2011).

However, the effects of pH and alkalinity on water quality and shrimp performance in BFT systems have not been extensively investigated (El-Sayed, 2021). In juvenile *L. vannamei* tanks, adding sodium bicarbonate greatly raised the water alkalinity (Zhang *et al.* 2017). Better growth rates, immunological response, and water quality measures were obtained with pH adjusted to 7.6–8.1 than with non-adjusted pH treatment. According to this study, optimal shrimp development, biofloc stability, and the formation of nitrifying bacteria in *L. vannamei* raised in BFT require a pH above 8.1 and an alkalinity >100 mg $CaCO_3$ L^{-1}. The similar finding was drawn by Furtado *et al.* (2015), who discovered that in a BFT system, low alkalinity (75 mg $CaCO_3$ L^{-1}) produced the greatest levels of ammonia and nitrite and the lowest growth rates of juvenile white shrimp (*L. vannamei*). The data showed a noteworthy enhancement upon increasing the alkalinity to 150 and 300 mg $CaCO_3$ L^{-1}. In contrast to the above results, Pierri *et al.* (2015) discovered that juvenile white shrimp (*L. vannamei*) (5.6 g) in the BFT system can be successfully grown at reduced alkalinity. When the sheimp were raised at varying alkalinities (40, 80, 120, and 160 mg $CaCO_3$ L^{-1}), survival and growth rate were not significantly impacted, while pH and TSS fell with lower alkalinity. The authors came to the conclusion that *L. vannamei* can be cultured in biofloc with success at 40–160 mg $CaCO_3$ L^{-1}. Although the pH in this study was lower at 40 mg $CaCO_3$ L^{-1} than at other alkalinity treatments, it was still very near to 7 (neutral, not acidic), indicating that additional alkalinity may not have been required. It should be noted that, as shown by Furtado *et al.* (2011), the type of alkalinizing compounds utilised may have an impact on maintaining pH and alkalinity in BFT-based shrimp production.

3. Stocking density and pH

An intense aquaculture system is distinguished by a high stocking density, which is followed by the requirement for both high-quality and quantity of artificial diets (Piedrahita, 2003) and Avnimelech *et al.* (2008). Fish production and profitability in aquaculture practices are significantly influenced by stocking density, which is considered one of the major factors (Refaey *et al.,* 2018). Avnimelech and Ritvo (2003) declared that approximately 20–25% of the protein in feed is assimilated by fish, with the remainder released as ammonia and organic nitrogen in faeces and residual feeds. In parallel, the bacteria that break down organic matter in faeces and uneaten feeds further mineralize the organic nitrogen, producing ammonia. Since fish pellets typically include at least 25% protein, large feed input in intensive aquaculture systems results in a high ammonia excretion (Brune *et al.* 2003), which is extremely harmful to aquatic creatures (Stickney 2005). On the other hand, system water may become acidic at high stocking densities and high feeding rates as a result of higher nitrification. The process of nitrification lowers the pH of water and consumes a substantial amount of alkalinity (El-Sayed, 2021). McIntosh (2001) declared that a high stocking density lowers pH by raising CO_2 levels due to high respiration rate. Consequently, fish health may be harmed by the on-going stress that high density provides (Oliveira *et al.,* 2022).

As a matter of fact, there are numerous factors such as growth and survival, size variation, mortality, and water quality measurements, which are directly impacted by stocking density (North *et al.,* 2006; Hussain *et al.,* 2014) which, in turn affect the ecology of farmed fish. So, it's important to figure out the stocking density for every species and life stage in various production systems because of the harmful consequences of inadequate storage (Oliveira *et al.,* 2022). Consequently, inappropriate stocking density may limit the advantages of the biofloc system. Fish performance was negatively impacted by unchecked high stocking density. It lowers feed utilisation efficiency and growth performance by reducing water quality and crowding, which cause physiological changes (Azim & Little, 2008; Crab *et al.*, 2012; Minabi *et al.*, 2020; Minaz and Kubilay, 2021). As a result, adequate aeration is needed to sustain a high particle density at high stocking densities. Additionally, the supply of organic carbon is required to maintain an adequate C/N ratio, which in turn optimises bacterial growth and the generation of microbial flocs (El-Sayed 2021). As a result of increased CO_2 production from animal respiration, stocking density predominantly affects DO and subsequently causes a pH reduction. Heterotrophic conversion and a decrease in alkalinity are linked to elevated total dissolved solids TDS (Majhi *et al.*, 2023).

4. Carbon source, C/N ratio and its relations to pH

BFT system operated with zero or minimal water exchange, which promotes the development of a primarily heterotrophic and aerobic biota. This is achieved by continuous aeration and fertilization of the water column via organic carbon sources (Wasielesky *et al.,* 2006; Azim and Little, 2008; Avnimelech, 2009). In BFT system, the increase in the amounts of nitrite and total ammonium nitrogen can be avoided by adding organic carbon sources, such as sugar cane molasses (Samocha *et al.*, 2007). The carbon source, which often varies in carbon content and degradability, is one of the primary elements influencing the properties of biofloc. As a result, several sources were identified as excellent sources for producing biofloc, which encourages quick ammonia removal and promotes biofloc aggregation (Khanjani and Sharifinia (2020); (El-Sayed 2021), and could affect the pH of fish bond.

Heterotrophic bacteria compete with autotrophic bacteria for dissolved oxygen and available space when the C:N ratio is high. Nevertheless, autotrophic organisms are more common in low C:N ratios (Michaud *et al.,* 2006). The synthesis and recycling of nutrients can be enhanced by an ideal C:N ratio. The ideal C:N ratio, according to Schneider *et al.* (2005), is roughly 15:1, however Burford *et al.* (2003) and Pérez-Fuentes *et al.* (2016) state that the ratio must be greater than 10:1. While, the ideal C:N ratios for microbial floc production were noted by Wasielesky *et al.* (2006) to be between 14:1 to 30:1, with heterotrophic bacteria predominating.

The pressing necessity to identify the optimal carbon source for aquaculture systems utilising biofloc technology has spurred research (Wei *et al.*, 2016; Zhang *et al.,* 2016). Numerous carbon sources, including grain, sugar, molasses, sugarcane bagasse, chopped hay, and tapioca, have been investigated in BFT systems. The cost and accessibility of the sources must be taken into account while selecting one of them (Hargreaves, 2013; Fugimura *et al.*, 2015; Serra *et al.,* 2015; Ahmad *et al.,* 2016). Molasses were employed for tilapia culture by Widanarni *et al* (2012) (C:N ratio of 15:1) and Lima *et al.* (2015) (C:N ratio of 6:1).

Silva *et al.* (2017) used three carbon sources: sugar, molasses, and cassava starch, as well as two carbon to nitrogen (C:N) ratios (10:1 and 20:1) for each carbon source to culture Nile Tilapia under biofloc system conditions. Alkalinity, settleable solids (SS), turbidity, and total suspended solids (TSS) were all impacted by the C:N ratio and carbon source; the variables with the highest values were those with a C:N ratio of 20:1 ($P < 0.05$). Molasses and sugar, at C:N ratios of 10:1 and 20:1, were the most effective carbon sources for microbial floc production.

Dauda *et al.* (2018) cultivated African catfish (*Clarias gariepinus*) in glycerol-based biofloc systems with varying C/N ratios of 0 (control), 10, 15, or 20. They found that the pH remained consistent throughout the trial weeks, with the exception of week 3, when the C/N 20 group's pH significantly dropped ($P < 0.05$) compared to the other groups. For the entire six-week period, the DO was considerably ($P < 0.05$) lower in the C/N 20 group than in any other group, with the exception of the C/N 15 treatment in weeks two and five. In all weeks, the DO was highest in the control group but did not differ significantly ($P > 0.05$) from the C/N 10 treatment group. This is an explanation of how the C/N ratio affects DO and, in turn, the nitrification process.

El-Husseiny *et al.* (2018) investigate how various carbon sources affect tilapia performance and biofloc conditions. In the presence of a control (clear water), five distinct organic carbon simple sources (glucose, molasses) and complex sources (starch, wheat bran, cellulose) were subjected to treatments with biofloc. A C:N ratio of 10:1 was maintained in various experimental tanks in order to determine the ideal biofloc environment. The results showed that no significant variations ($p > 0.05$) were found in the values of water temperature, pH, NH^{4+}, NO_2, and NO_3 among the various carbon sources. There were notable variations in DO values ($p < 0.05$) between the various treatments.

In order to study the potential effects of varying the C:N ratio on growth performance, physico-chemical parameters, microbial dynamics, feed utilisation, and immunological parameters, Panigrahi *et al.* (2019) conducted a 16-week indoor culture trial. A control group with 100 nos/m3 stocking density in 500 L tanks with continuous aeration was also included in the experiment, along with three iterations of each of the five biofloc treatment groups (with varying C:N ratios of 5:1, 10:1, 15:1, and 20:1). Since molasses is an organic carbon source, the treatments' C:N ratios were changed, but the control group received no carbon additions at all. The observed pH levels were substantially lower in BFT treatments than the control due to the variations of C/N ratios.

Using sugar cane molasses, Saha *et al.* (2022) cultivated stinging catfish, *Heteropneustes fossilis*, in a biofloc-based culture system to evaluate four levels of C/N ratio (12, 15, 18, 21). They discovered that constant aeration in the tanks led to a higher value of DO concentration in all experiment tanks. As the trial progressed, the pH declined with the rising C/N ratio presumably due to introducing molasses (pH 5.3) and the proliferation of nitrifying bacteria in the system (Goldman *et al.,* 1987).

According to Saha *et al.* (2022), increasing the C/N ratio will result in an increase in total dissolved solids (TDS), which relates to the decreased water exchange and a significant amount of organic materials in biofloc treatments. The biofloc system's elevated TSS levels as a result of the rising C/N ratio Because TSS concentration and water quality are related, it is important to monitor BFT tank TSS concentrations closely (Ray *et al.*, 2010b). Avnimelech (2009) and Martins *et al.* (2017) indicate that TSS is connected to the optimal microbial population and food reserves. The morpho-structure, microbial community, and nutritional makeup of biofloc can all be impacted by varying carbon sources (Wei *et al.*, 2016). Nevertheless, high TSS levels can block fish gills, thus they should be avoided as they increase oxygen demand. Moreover, solid particles may form and build, creating anaerobic layers, if the water mixture is not carefully managed or if the TSS concentration is higher than the system's capability (Avnimelech, 2009).

The effects of the carbon-to-nitrogen ratio (C:N) on prawn growth performance and water quality are examined by Huang *et al.* (2022) as they grow-out culture *Litopenaeus vannamei* in a biofloc system with low salinity. Three different biofloc treatments were created to stock shrimp juveniles (0.8 g $shrimp^{-1}$) at a density of 270 individuals m^{-3}, for a 63-day culture experiment at a salinity of approximately 5%. The C:N (contained in feed and exogenous carbon source with the assumption that 75% of the feed nitrogen is excreted) of 8:1 (CN8), 16:1 (CN16), and 24:1 (CN24), respectively. The results indicated a significant ($p<0.05$) decrease in pH, carbonate alkalinity, biofloc volume, and TSS in the CN8 treatment. Conversely, TAN, nitrite, and nitrate levels were considerably greater ($p<0.05$) than those in the other two treatments.

BFT typically exhibit fluctuations in pH, DO, and amounts of nitrogenous waste (Azim and Little, 2008; Luo *et al.*, 2014). As indicated by (Solanki *et al.,* 2023) at a greater stocking density (Burford *et al.*, 2003; Yuniasari, D., & Ekasari, J. (2010)), increasing the weight of culture species and increasing the bacterial population frequently lowers DO of the BFT system. The effect of stocking density (3, 4, and 5 spawn L−1) and C/N ratio (10, 15, and 20) on *Gibelion catla* spawn nursery rearing in the indoor biofoc system is examined by Solanki *et al.* (2023). The carbon source utilised to adjust the C/N ratios was rice bran. Average length (6.7 ± 0.4 mm) and weight (1.6 ± 0.2 mg) of catla spawn were used to stock each treatment. They observed that increasing stocking densities significantly ($p < 0.05$) decreased the amount of DO in water. Significant differences ($p < 0.05$) were found in pH between various C/N ratios, with a lower pH at higher C/N ratios. However, no significant difference ($p > 0.05$) was found in stocking densities. As the C/N ratio increased, total alkalinity decreased significantly ($p < 0.05$), but there was no significant difference ($p > 0.05$) in C/N 15 and 20. Total alkalinity was significantly ($p < 0.05$) lower in areas with higher stocking densities. A significant ($p < 0.05$) decrease in TAN and NO2-N was seen with an increase in the C/N ratio.

For more instance, Souza *et al.* (2019) reported that the pH value in the BFT system was lower than in the control group, probably because the carbon dioxide concentration in the biofloc-treated water increased as a result of heterotrophic organism respiration (Wasielesky *et al.*, 2006, Furtado *et al.*, 2011). According to Kumar *et al.* (2018), the predominance of heterotrophic bacteria over autotrophic bacteria was the reason for the lower pH in the BFT system. This is due to the fact that carbon dioxide is produced as a by-product of heterotrophic bacteria consuming organic substances in the water. The pH may then drop as a result of the water's excess carbon dioxide. Furthermore, carbon dioxide is excreted by microorganisms during their respiration process, which can help lower pH levels even more Ahmed Alkhamis *et al.* (2023). Consequently, a lower pH in the BFT system may arise from the interaction of microbial respiration and heterotrophic dominance. In general, pH has a negative effect on BFT, according to the meta-analysis carried out by Ahmed Alkhamis *et al.* (2023).

5. Regulation of pH in biofloc culture system

The water's ability to act as a buffer determines the availability of free hydrogen ions, which are linked to lower pH levels (Lekang, 2007). To keep the conditions favourable for the development of biofloc, alkalinity supplements are required (Ogello *et al.,* 2021). As nitrification proceeds in biofloc systems, liming agents such calcium carbonates can be utilised to raise pH levels when nitrification has led to a pH decrease (De Oliveira Alves *et al.,* 2017; Azim and Little, 2008). In contrast, Whangchai *et al.* (2004) revealed that high purity sodium bicarbonate and calcium hydroxide were effective in regulating the alkalinity of a shrimp culture treated with ozone, but calcium carbonate was ineffective in raising the alkalinity. Contrary to sodium carbonate, calcium hydroxide and sodium bicarbonate were effective in raising the alkalinity in BFT cultures when used as a correction for alkalinity (Furtado *et al.*, 2011). According to Martins *et al.* (2017), the utilisation of sodium bicarbonate proved to be effective than calcium carbonate in preserving alkalinity and pH in the BFT for *Oreochromis niloticus*.

REFERENCES

Abakari, G., Luo, G., & Kombat, E. O. 2021. Dynamics of nitrogenous compounds and their control in biofloc technology (BFT) systems: A review. Aquaculture and Fisheries, 6(5), 441-447.

Ahmad, I., Verma, A. K., Rani, A. B., Rathore, G., Saharan, N., and Gora, A. H. 2016. Growth, non-specific immunity and disease resistance of Labeo rohita against Aeromonas hydrophila in biofloc systems using different carbon sources. Aquaculture, 457, 61-67.

Ahmed Alkhamis, Y., Sultana, A., Tareq Arafat, S., Abdur Rouf, M., Mustafizur Rahman, S., Thomas Mathew, R., ... & Rahman, M. 2023. The Impact of Biofloc Technology on Water Quality in Aquaculture: A Systematic Meta-Analysis. Aquaculture Nutrition, 2023.

Amatya, I. M., Kansakar, B. R., Tare, V., and Fiksdal, L. 2011. Role of pH on biological Nitrification Process. Journal of the Institute of Engineering, 8(1-2), 119-125

Avnimelech Y, Ritvo G. 2003. Shrimp and fish pond soils: processes and management. Aquaculture 220:549-567. http://dx.doi. org/10.1016/S0044-8486(02)00641-5.

Avnimelech Y. Feeding with microbial flocs by tilapia in minimal discharge bioflocs technology ponds. Aquaculture. 2007;264:140–147.

Avnimelech Y., Verdegem, M.C.J., Kurup, M. and Keshavanath, P. 2008. Sustainable land-based aquaculture: rational utilization of water, land and feed resources. Mediterr Aquac. J. 1:45-55.

Avnimelech, Y.2009. Biofloc Technology: a practical guide book. 2. ed. Baton Rouge, US: The World Aquaculture Society,182 p.

Azim, M.E, and Little, D.C.2008. The biofloc technology (BFT) in indoor tanks: water quality, biofloc composition, and growth and welfare of Nile tilapia (Oreochromis niloticus). Aquaculture, 283: 29–35. https://doi.org/10.1016/j.aquaculture.2008.06.036.

Ballester, E.L.C., Abreu, P.C., Cavalli, R.O., Emerenciano, M., Abreu, L. and Wasielesky, W.2010. Effect of practical diets with different protein levels on the performance of Farfantepenaeus paulensis juveniles nursed in a zero exchange suspended microbial flocs intensive system. Aquaculture Nutrition.16:163–172.

Bauer W, Prentice-Hernandez C, Tesser MB, Wasielesky W, Poersch LHS. Substitution of fishmeal with microbial floc meal and soy protein concentrate in diets for the Pacific white shrimp Litopenaeus vannamei. Aquaculture. 2012;342–343:112–116.

Boyd, C. E. 2019. Water quality: an introduction. Springer Nature.

Brune, D.E., Schwartz, G., Eversole, A.G., Collier, J.A., Schwedler, T.E. 2003. Intensification of pond aquaculture and high rate photosynthetic systems. Aquac. Eng., 28:65-86. http://dx.doi. org/10.1016/S0144-8609(03)00025-6

Burford, M.A., Thompson, P.J., McIntosh, R.P., Bauman, R.H., Pearson, D.C.2003. Nutrient and microbial dynamics in high-intensity, zero-exchange shrimp ponds in Belize. Aquaculture,219:393–411.

Burford, M.A., Thompson, P.J., McIntosh, R.P., Bauman, R.H. and Pearson, D.C.2004. The contribution of flocculated material to shrimp (Litopenaeus vannamei) nutrition in a high-intensity, zero exchange system. Aquaculture,232:525–537.

Chen, S., Ling, J. and Blancheton, J.P. 2006. Nitrification kinetics of biofilm as affected by water quality factors. Aquaculture Engineering, 34, 179–197.

Costa-Pierce, B.A., Bartley, D.M., Hasan, M., Yusoff, F., Kaushik, S.J.,Rana, K et al .2012. Responsible use of resources for sustainable aquaculture. In: Subasinghe RP, Arthur JR, Bartley DM, De Silva SS, Halwart M, Hishamunda N, Mohan CV, Sorgeloos P (eds) Farming the waters for people and food. Proceedings of the global conference on aquaculture 2010. Phuket, Thailand, pp 113–147 http://www.fao.org/3/i2734e/i2734e03a.pdf.

Crab R, Kochva M, Verstraete W, Avnimelech Y.2009. Bio-flocs technology application in over-wintering of tilapia. Aquacultural Engineering,40:105–112.

Crab R, Chielens B, Wille M, Bossier P, Verstraete W.2010. The effect of different carbon sources on the nutritional value of bioflocs, a feed for Macrobrachium rosenbergii postlarvae. Aquaculture Research,41:559–567

Crab, R., Defoirdt, T., Bossier, P. and Verstraete, W. 2012. Biofloc technology in aquaculture: beneficial effects and future challenges. Aquaculture, 356: 351–356. https: // doi. org/ 10.1016/j.aquaculture.2012.04.046.

Dauda, A. B., Romano, N., Ebrahimi, M., Teh, J. C., Ajadi, A., Chong, C. M. and Kamarudin, M. S. 2018. Influence of carbon/nitrogen ratios on biofloc production and biochemical composition and subsequent effects on the growth, physiological status and disease resistance of African catfish (Clarias gariepinus) cultured in glycerol-based biofloc systems. Aquaculture, 483, 120-130.

Dauda, A. B. 2020. Biofloc technology: a review on the microbial interactions, operational parameters and implications to disease and health management of cultured aquatic animals. Reviews in Aquaculture, 12(2), 1193-1210.

De Oliveira Alves, G. F., Fernandes, A. F. A., de Alvarenga, É. R., Turra, E. M., de Sousa, A. B., and de Alencar Teixeira, E. 2017. Effect of the transfer at different moments of juvenile Nile tilapia (Oreochromis niloticus) to the biofloc system in formation. Aquaculture, 479, 564-570.

De Schryver P., Crab, R., Defroidt, T., Boon, N.2008. The basics of bio-flocs technology: the added value for aquaculture. Aquaculture, 277:125–137

Defoirdt, T., Boon, N., Bossier, P. and Verstraete, W. 2004. Disruption of bacterial quorum sensing: an unexplored strategy to fight infections in aquaculture. Aquaculture, 240 (1-4), 69-88.

Ebeling, J.M., Timmons, M.B., Bisogni, J.J. 2006. Engineering analysis of the stoichiometry of photoau- totrophic, autotrophic, and heterotrophic control of ammonia-nitrogen in aquaculture production sys- tems. Aquaculture, 257:346–358

Ekasari J, Zairin M, Putri DU, Sari NP, Surawidjaja EH, Bossier P. Biofloc-based reproductive performance of Nile tilapia Oreochromis niloticus L. broodstock. Aquaculture Research, 2015;46:509–512.

El-Husseiny, O. M., Goda A, M. A.-S., Rania S. M. and Soaudy, M.2018. Complexity of carbon sources and the impact on biofloc integrity and quality in tilapia (Oreochromis niloticus) tanks. AACL Bioflux, 11(3).

El-Sayed, A. F. M. 2021. Use of biofloc technology in shrimp aquaculture: a comprehensive review, with emphasis on the last decade. Reviews in Aquaculture, 13(1): 676-705.

Emerenciano M, Ballester ELC, Cavalli RO, Wasielesky W.2012a. Biofloc technology application as a food source in a limited water exchange nursery system for pink shrimp Farfantepenaeus brasiliensis (Latreille, 1817). Aquaculture Research,43:447–457.

Emerenciano, M., Cuzon, G., Goguenheim, J., Gaxiola, G., Aquacop.2012b. Floc contribution on spawning performance of blue shrimp Litopenaeus stylirostris. Aquaculture Research,44:75–85.

Emerenciano M, Cuzon G, Arevalo M, Mascaró M, Gaxiola G.2013. Effect of short-term fresh food supplementation on reproductive performance, biochemical composition, and fatty acid profile of Litopenaeus vannamei (Boone) reared under biofloc conditions. Aquaculture International,21:987–1007.

Emerenciano, M. G. C., Martínez-Córdova, L. R., Martínez-Porchas, M., and Miranda-Baeza, A.2017. Biofloc technology (BFT): a tool for water quality management in aquaculture. Water quality, 5, 92-109.

Emerenciano, M. G., Fitzsimmons, K., Rombenso, A., Martins, G., Lazzari, R., Fimbres-Acedo, Y. and Pinho, S. 2021. Biofloc technology (BFT) in tilapia culture. Biology and Aquaculture of Tilapia, 258-293.

FAO. 2017. FAO and the SDGs. Indicators: measuring up to the 2030 Agenda for Sustainable Development Fleckenstein LJ, Tierney TW, Ray AJ (2018) Comparing biofloc, clear-water, and hybrid recirculating nursery systems (part II): tilapia (Oreochromis niloticus) production and water quality dynamics. Aquac. Eng.,82:80– 85. https://doi.org/10.1016/j.aquaeng.2018.06.006

FAO - Food and Agriculture Organization of the United Nations. 2022. The State of World Fisheries and Aquaculture 2022: Towards Blue Transformation. FAO, Rome. 266p.

Fugimura, M. M. S., dos Reis Flor, H., de Melo, E. P., da Costa, T. V., Wasielesky, W. and Oshiro, L. M. Y. 2015. Brewery residues as a source of organic carbon in Litopenaeus schmitti white shrimp farms with BFT systems. Aquaculture International, 23, 509-522]

Furtado, P.S., Poersch L.H. and Wasielesky W. Jr.2011. Effect of calcium hydroxide, carbonate and sodium bicarbonate on water quality and zootechnical performance of shrimp Litopenaeus vannamei reared in bio-flocs technology (BFT) systems. Aquaculture, 321, 130–135.

Furtado, P.S., Poersch, L.H., Wasielesky, W. Jr. 2015. The effect of different alkalinity levels on Litopenaeus vannamei reared with biofloc technology (BFT). Aquac. Int., 23:345–358

Gaona, C.A.P., Poersch, L.H., Krummenauer, D., Foes, G.K. and Wasielesky, W.J.2011.The effect of solids removal on water quality, growth and survival of Litopenaeus vannamei in a biofloc technology culture system. IJRA, 12:54–73

Goldman, J. C., Caron, D. A., and Dennett, M. R. 1987. Regulation of gross growth efficiency and ammonium regeneration in bacteria by substrate C: N ratio 1. Limnology and Oceanography, 32(6): 1239-1252]

Hargreaves,J.A.2006. Photosynthetic suspended-growth systems in aquaculture. Aquacultural engineering, 34(3):344-363.

Hargreaves, J. A. 2013. Biofloc production systems for aquaculture,4503, pp1-11). Stoneville, MS: Southern Regional Aquaculture Center]

Henze, M., and Harremoës, P. 1990. Chemical-biological nutrient removal — the HYPRO concept. In H. H. Hahn, and R. Klute (Eds.). Chemical water and wastewater treatment. Berlin, Heidelberg: Springer

Huag, R. T. and McCarty, P. L. 1971. Nitrification with submerged filters. Water Pollution Control Federation, 44, 2086–2102.

Huang, H. H., Li, C. Y., Liang, T., Lei, Y. J., Yang, P. H. and Wu, M. X. 2022. Effects of carbon-to-nitrogen ratio (C: N) on water quality and growth performance of Litopenaeus vannamei (Boone, 1931) in the biofloc system with a salinity of 5‰. Aquaculture Research, 53(15):5287-5299.

Hussain, T., Verma, A.K., Tiwari, V.K., Prakash, C., Rathore, G., Shete, A.P. and Nuwansi, K.K.T.2014. Optimizing koi carp, Cyprinus carpio var. koi (Linnaeus, 1758), stocking density and nutrient recycling with spinach in an aquaponic system. Journal of the World Aquaculture Society, 45 (6): 652–661. https://doi.org/10.1111/jwas.12159.

Khanjani, M. H. and Sharifinia, M. 2020. Biofloc technology as a promising tool to improve aquaculture production. Reviews in aquaculture, 12(3), 1836-1850.

Koops, H.-P. and Pommerening- Röser, A. 2001. Distribution and ecophysiology of the nitrifying bacteria emphasizing cultured species. FEMS Microbiology Ecology, 37(1), 1–9. https://doi.org/10.1111/j.1574-6941.2001.tb00847.x.

Kuhn, D., Boardman, G., Lawrence, A., Marsh, L. and Flick, G.2009. Microbial floc meal as a replacement ingredient for fish meal and soybean protein in shrimp feed. Aquaculture,296:51–57.

Kumar, V. S., Pandey, P. K., Anand, T., Bhuvaneswari, G. R., Dhinakaran, A. and Kumar, S. 2018. Biofloc improves water, effluent quality and growth parameters of Penaeus vannamei in an intensive culture system. Journal of Environmental Management, 215, 206-215.

Lara, G., Krummenauer, D., Abreu, P. C., Poersch, L. H. and Wasielesky, W. 2017. The use of different aerators on Litopenaeus vannamei biofloc culture system: effects on water quality, shrimp growth and biofloc composition. Aquaculture International, 25, 147-162.

Lekang, O. 2007. Aquaculture engineering (pp. 121–130). Blackwell Publishing.

Lemonnier, H., Bernard, E., Boglio, E., Goarant, C., Cochard, J., 2004. Influence of sediment characteristics on shrimp physiology: pH as principal effect. Aquaculture, 240, 297–312.

Lima, E. C. R. D., Souza, R. L. D., Wambach, X. F., Silva, U. L. and Correia, E. D. S. 2015. Cultivo da tilápia do Nilo Oreochromis niloticus em sistema de bioflocos com diferentes densidades de estocagem. Revista Brasileira de Saúde e Produção Animal, 16:948-957.

Luo, G., Gao, Q., Wang, C., Liu, W., Sun, D., Li, L. and Tan, H. 2014. Growth, digestive activity, welfare, and partial cost-effectiveness of genetically improved farmed tilapia (Oreochromis niloticus) cultured in a recirculating aquaculture system and an indoor biofloc system. Aquaculture, 422, 1-7.

Majhi, S. S., Singh, S. K., Biswas, P., Debbarma, R., Parhi, J., Ngasotter, S., ... & Patel, A. B. 2023. Effect of Stocking Density on Growth, Water Quality Changes and Cost Efficiency of Butter Catfish (Ompok bimaculatus) during Seed Rearing in a Biofloc System. Fishes, 8(2):61.

Manan, H., Kasan, N. A., Suratman, S. and Ikhwanuddin, M. 2016. Study on carbon sinks by classified biofloc phytoplankton from marine shrimp pond water.

Martinez-Porchas, M., Ezquerra-Brauer, M., Mendoza-Cano, F., Higuera, J. E. C., Vargas-Albores, F. and Martinez-Cordova, L. R. 2020. Effect of supplementing heterotrophic and photoautotrophic biofloc, on the production response, physiological condition and post-harvest quality of the whiteleg shrimp, Litopenaeus vannamei. Aquaculture Reports, 16, 100257.

Martins, G. B., Tarouco, F., Rosa, C. E. and Robaldo, R. B. 2017. The utilization of sodium bicarbonate, calcium carbonate or hydroxide in biofloc system: water quality, growth performance and oxidative stress of Nile tilapia (Oreochromis niloticus). Aquaculture, 468, 10-17.

McDonough W, Braungart M .2002. Cradle to cradle: remaking the way we make things, North Point Press. https://mcdonough.com/writings/cradle-cradle-remaking-way-make-things

Mcintosh, B.J., 2001. Changing paradigms in shrimp farming: establishment of heterotrophic bacterial communities. Global Aquaculture Advocate 53–58 February.

Michaud, L., Blancheton, J. P., Bruni, V. and Piedrahita, R. 2006. Effect of particulate organic carbon on heterotrophic bacterial populations and nitrification efficiency in biological filters. Aquacultural engineering, 34(3), 224-233.

Minabi, K., Sourinejad, I., Alizadeh, M., Rajabzadeh Ghatrami, E. and Khanjani, M.H. 2020. Effects of different carbon to nitrogen ratios in the biofloc system on water quality, growth, and body composition of common carp (Cyprinus carpio L.) fingerlings. Aquaculture International, 28:1883–1898 https://doi.org/10.1007/s10499-020-00564-7

Minaz, M. and Kubilay, A. 2021. Operating parameters affecting biofloc technology: carbon source, carbon/nitrogen ratio, feeding regime, stocking density, salinity, aeration, and microbial community manipulation. Aquaculture International, 29(3), 1121-1140.

Najdegerami, E.H., Bakhshi, F., Lakani, F.B.2016. Effects of biofloc on growth performance, digestive enzyme activities and liver histology of common carp (*Cyprinus carpio* L.) fingerlings in zero-water exchange system. Fish Physiology Biochemistry, 42(2):457–465.

Neto, H.S., Santaella, S.T., Nunes, A.J.P.2015. Bioavailability of crude protein and lipid from biofloc meals produced in an activated sludge system for white shrimp, Litopenaeus vannamei. Brazilian Journal of Animal Science,44(8):269–275.

North, B.P., Turnbull, J.F., Ellis, T., Porter, M.J., Migaud, H., Bron, J. and Bromage, N.R. 2006. The impact of stocking density on the welfare of rainbow trout (Oncorhynchus mykiss). Aquaculture, 255 (1–4): 466–479. https://doi.org/10.1016/j.aquaculture. 2006. 01.004.

Odegaard, H. 1992. Norwegian experiences with chemical treatment of raw wastewater. Water Science and Technology, 25(12), 255–264.

Ogello, E. O., Outa, N. O., Obiero, K. O., Kyule, D. N. and Munguti, J. M. 2021. The prospects of biofloc technology (BFT) for sustainable aquaculture development. Scientific African, 14, e01053.

Oliveira, L. K., Wasielesky Jr, W. and Tesser, M. B. 2022. Fish culture in biofloc technology (BFT): Insights on stocking density carbon sources, C/N ratio, fish nutrition and health. Aquaculture and Fisheries.

Panigrahi, A., Sundaram, M., Chakrapani, S., Rajasekar, S., Syama Dayal, J. and Chavali, G. 2019. Effect of carbon and nitrogen ratio (C: N) manipulation on the production performance and immunity of Pacific white shrimp Litopenaeus vannamei (Boone, 1931) in a biofloc-based rearing system. Aquaculture Research, 50(1), 29-41.

Pérez-Fuentes J.A., Pérez-Rostro C.I., Hernández-Vergara, M.P. Pond-reared Malaysian prawn Macrobrachium rosenbergii with the biofloc system. Aquaculture, 2013;400–401:105–110.

Pérez-Fuentes, J. A., Hernández-Vergara, M. P., Pérez-Rostro, C. I. and Fogel, I. 2016. C: N ratios affect nitrogen removal and production of Nile tilapia Oreochromis niloticus raised in a biofloc system under high density cultivation. Aquaculture, 452, 247-251.

Piedrahita, R.H. 2003. Reducing the potential environmental impact of tank aquaculture effluents through intensification and recirculation. Aquaculture, 226:35-44. http://dx.doi.org/10. 1016/S0044-8486(03)00465-4.

Piérri, V., Valter-Severino, D., Goulart-de-Oliveira, K., Manoel-do-Espírito-Santo, C., Nascimento-Vieira, F. and Quadros-Seiffert, W. 2015. Cultivation of marine shrimp in biofloc technology (BFT) system under different water alkalinities. Brazilian Journal of Biology, 75, 558-564.

Pinho, S. M., de Lima, J. P., David, L. H., Emerenciano, M. G., Goddek, S., Verdegem, M. C., ... & Portella, M. C.2022. FLOCponics: The integration of biofloc technology with plant production. Reviews in Aquaculture, 14(2):647-675.

Ray, A.J., Seaborn, G., Leffler, J.W., Wilde, S.B., Lawson, A. and Browdy, C.L. 2010a. Characterization of microbial communities in minimal-exchange, intensive aquaculture systems and the effects of suspended solids management. Aquaculture, 310:130–138

Ray, A. J., Lewis, B. L., Browdy, C. L. and Leffler, J. W. 2010b. Suspended solids removal to improve shrimp (Litopenaeus vannamei) production and an evaluation of a plant-based feed in minimal-exchange, super intensive culture systems. Aquaculture, 299 (1-4): 89-98.

Refaey, M. M., Li, D., Tian, X., Zhang, Z., Zhang, X., Li, L. and Tang, R. 2018. High stocking density alters growth performance, blood biochemistry, intestinal histology, and muscle quality of channel catfish Ictalurus punctatus. Aquaculture, 492, 73–81.

Saha, J., Hossain, M. A., Mamun, M. A., Islam, M. R. and Alam, M. S. 2022. Effects of carbon-nitrogen ratio manipulation on the growth performance, body composition and immunity of stinging catfish Heteropneustes fossilis in a biofloc-based culture system. Aquaculture Reports, 25, 101274.

Samocha, T.M., Patnaik, S., Speed, M., Ali, A.M., Burger, J.M., Almeida, R.V., Ayub, Z., Harisanto, M., Horowitz, A., Brook, D.L., 2007. Use of molasses as carbon source in limited discharge nursery and grow-out systems for Litopenaeus vannamei. Aquacultural Engineering, 36:184–191.

Schneider, O., Sereti, V., Eding, E. H., & Verreth, J. A. J. 2005. Analysis of nutrient flows in integrated intensive aquaculture systems. Aquacultural Engineering, 32(3-4):379-401.

Serra, F. P., Gaona, C. A., Furtado, P. S., Poersch, L. H. and Wasielesky, W.2015. Use of different carbon sources for the biofloc system adopted during the nursery and grow-out culture of Litopenaeus vannamei. Aquaculture International, 23: 1325-1339.

Silva, U. L., Falcon, D. R., Pessôa, M. N. D. C. and Correia, E. D. S. 2017. Carbon sources and C: N ratios on water quality for Nile tilapia farming in biofloc system. Revista Caatinga, 30:1017-1027.

Solanki, S., Meshram, S. J., Dhamagaye, H. B., Naik, S. D., Shingare, P. E. and Yadav, B. M. 2023. Effect of C/N Ratio Levels and Stocking Density of Catla Spawn (Gibelion catla) on Water Quality, Growth Performance, and Biofloc Nutritional Composition in an Indoor Biofloc System. Aquaculture Research, 2023.

Souza, R. L. D., Lima, E. C. R. D., Melo, F. P. D., Ferreira, M. G. P. and Correia, E. D. S. 2019. The culture of Nile tilapia at different salinities using a biofloc system. Revista Ciência Agronômica, 50, 267-275.

Stickney, R.R. 2005. Aquaculture : an introductory text. Cambridge (Mass.): CABI publishing

Timmons, M. B., Ebeling, J. M., Wheaton, F. W., Summerfelt, S. T. and Vinci, B. J. 2002. Recirculating aquaculture systems (2nd ed.). New York: Cayuga Aqua Ventures769.

Van der Graaf, A.A., Mulder, A., de Bruijin, P., Jetten, M.S.M., Robertson, L.A. and Kuenen J.G. 1995. Anaerobic oxidation of ammonium is a biologically mediated process. App Env Microb., 61:1246–1251

Van Wyk P. and Scarpa J. 1999. Water quality and management. In: Farming Marine Shrimp in Recirculating Freshwater Systems (ed. by P. Van Wyk & J. Scarpa), pp. 128–138. Florida Department of Agriculture and Consumer Services, Tallahassee, FL, USA

Verdegem, M.C.J. 2013.Nutrient discharge from aquaculture operations in function of system design and production environment. Rev Aquac 5:158–171. https://doi.org/10.1111/raq.12011

Villaverde, S., Garcia-encina, P.A. and Fdz-Polanco F (1997) Influence of pH over nitrifying biofilm activity in submerged biofilters. Water Res 31:1180–1186

Wasielesky, W., Atwood, H., Kegl, R., Bruce, J., Stokes, A., Browdy, C.L., 2006. Efeito do phna sobrevivência e crescimento do camarão branco Litopenaeus vannamei em cultivos superintensivos. Aquaciência 2006. Anais do congresso.

Wei, Y., Liao, S. A. and Wang, A. L.2016. The effect of different carbon sources on the nutritional composition, microbial community and structure of bioflocs. Aquaculture, 465, 88-93.

Whangchai, N., Veronica, P., Migo, P., Alfafara, C.G., Young, K.H., Nomura, N. and Matsumura, M., 2004. Strategies for alkalinity and pH control for ozonated shrimp pond water. Aquacultural Engineering, 30, 1–13.

Widanarni , Ekasari, J. and Maryam, S. 2012. Evaluation of biofloc technology application on water quality and production performance of red tilapia Oreochromis sp. cultured at different stocking densities. Hayati journal of Biosciences, 19(2):73-80.

Yuniasari, D., & Ekasari, J. (2010). Nursery culture performance of Litopenaeus vannamei with probiotics addition and different C/N Ratio under laboratory condition. Hayati Journal of biosciences, 17(3), 115-119.

Yusuf, M.W., Utomo, N.B.P., Yuhana, M., Widanarni.2015. Growth performance of catfish (Clarias gariepinus) in biofloc-based super intensive culture added with Bacillus sp. Journal of Fisheries and Aquatic Science, 10(6):523–532.

Zhang, N., Luo, G., Tan, H., Liu, W. and Hou, Z. 2016. Growth, digestive enzyme activity and welfare of tilapia (Oreochromis niloticus) reared in a biofloc-based system with poly-β-hydroxybutyric as a carbon source. Aquaculture, 464, 710-717.

Zhang, K., Pan, L., Chen, W. and Wang, C. 2017. Effect of using sodium bicarbonate to adjust the pH to different levels on water quality, the growth and the immune response of shrimp Litopenaeus vannamei reared in zero-water exchange biofloc-based culture tanks. Aquaculture Research, 48(3), 1194-1208.

9

Use of Some Non Conventional Animal Protein Sources in Culture of Fresh Water Cat Fish *Clarias Magur*

S.M. Hussain, S. Borthakur, B.N. Hazarika and Kardivel Govindasamy

ABSTRACT

The aquaculture activities are expanding very fast in the world which leads to the unavailability of main fish feed ingredient fishmeal. Due to its high cost and high demand, as it is used in fisheries as well as in animal husbandry the research on other alternative for non conventional protein sources to replace fish meal in fish feed becomes of great importance for fish farming activities. Among the catfishes, Clarias magur, an air breathing species and popularly known as 'magur' or 'walking cat fish' is a high valued fish in the Indian subcontinent and fetches higher price than many other fishes including major carps. Very sparse information is available regarding the use of non conventional animal protein sources in practical diet formulation for fresh water catfish Clarias magur. The outcomes of the present research will definitely help the farming community of India and North Eastern Region in particular. It will also save lots of edible fishes (suitable for human consumption) from being converted into fish meal and contribute significantly towards food security. The experiments were conducted in triplicates using twelve equal size (6m × 4m × 1.5m) outdoor cement cisterns as T-I (03 no's), T-II (03 no's) , T-III (03 no's) & T-IV (03 no's) and the study was conducted for a period of 180 days. The juveniles of magur were stocked @ 4 no's/m^2. In the study juvenile specimens of Clarias magur were reared by feeding with four iso-nitrogenous experimental diets containing 35% of protein level which were prepared using ingredients such as Fish meal, Vermi meal, Chicken viscera, rice polish, wheat flour, vitamin & mineral mixture. Each group having both the sexes of Clarias magur were fed @ 10-5% body weight in two split doses daily once in morning and next in evening. For first 90 days they were fed with 10 % body weight and after that fishes were fed with 5% of their body weight till harvesting. Findings of the present study indicated that non conventional animal protein like vermi meal and chicken viscera *meal is an acceptable ingredient for the replacement of fish meal protein in practical diets of fishes but still a lot of improvement is required in strategy and technology of culturing of Clarias magur to make it more successful and adoptable at farmers level.*

Keywords: *Clarias magur, Fish meal, Non conventional protein sources, Vermi meal, Chicken viscera.*

INTRODUCTION

Aquaculture is a technology driven industry which relies a lot on research to develop species diversification in culture system as well as in the appropriate technology for commercial production. The rapid development of aquaculture in the world depends largely on the availability of fish feed; which are traditionally based on fish meal as the main protein source (Hardy & Tacon, 2002; Krishnankutty, 2005; Yigit *et al.*, 2006) but the major challenge is its availability, quality and cost (Ghazala *et al.*, 2011; Tabinda and Butt, 2012). In the aqua feed formulation, protein is the main but expensive ingredient and its quality and quantity in fish feeds formulation plays a vital role in promoting fish growth (Pandian, 2001). Dietary protein provides essential and non-essential amino acids to synthesize body protein and energy for maintenance. According to fish species, fish size, dietary protein sources and environmental conditions protein requirements ranges from 300 to 550 g /kg (NRC 1993).

In India there is a significant development in fresh water aquaculture in last few decades contributing greatly to overall fish production of the country. The fresh water aquaculture is crossing the traditional boundary of concentration on farming of major carp and entered the arena of diversification to the culture of other species like catfishes. So far in India majority of research works were concentrated on development of commercial fish feed mainly for carps and shrimps. To fulfill the objectives of large scale culture of cat fishes and development as an economic proposition, it is necessary to take up induced breeding and larval rearing of commercially important catfishes in controlled conditions and develop easily adoptable package of practices of the same to cater to the needs of interested fish entrepreneurs. Among the catfishes, *Clarias magur,* an air breathing species and popularly known as 'magur' or 'walking catfish' is a high valued fish in the Indian subcontinent and fetches higher price than many other fishes including major carps and it is highly revered for its nutritious flesh which is therapeutic in nature, hence, better prospects are there for developing its culture (Hussain *et al.*, 2016). Vishwanath (2010) reported that this species is very much popular in Bangladesh due to use as important part of the diet for children and lactating mothers and also prescribed as diet for the convalescent of the patients. This fish is highly regarded for food due to its high protein (15.0%), low fat (1.0%) and high iron content (710mg/100g tissue). Paul *et al.* (2015) in their studies on Indian catfish namely magur (*Clarias batrachus*) and singhi (*Heteropneustes fossilis*) for proximate composition and mineral content concluded that both the fishes are rich in protein, fat and minerals content irrespective of their size and season.

However, the production of large quantities of magur seed and desired feed still remains a major bottleneck in popularization of its extensive and intensive culture. Khedkar *et al.* (2014) warned that the stock of magur has declined drastically from natural habitats in India during the last decade. The occurrence of this precious and economically important species in natural water bodies has greatly been reduced due to over exploitation and disease outbreaks (Ahmad *et al.* 2012; Binoy, 2010). The *C. batrachus* available in India have recently been renamed as *Clarias magur*. As per the IUCN Red list of threatened species (www.iucnredlist.org; Version 2014.3; downloaded on 20 November, 2014), this species has been listed under endangered (A3cde + 4acde) categories.

It is well known that formulation of balanced diets and their adequate feeding are important for successful aquaculture. Dietary nutrients needs of a species are of fundamental importance in aquaculture, since adequate dietary nutrients significantly influence growth, survival of the fish and feed cost. Nahar *et al.* (2000), Giri *et al.* (2003) and Nyina *et al.* (2010) has opined that to achieve a good growth, catfish requires complete artificial feed of 35% to 45% protein level. The utilization of dietary protein for growth depends on the quantity as well as quality of feed proteins (Giri *et al.*, 2003). Because the catfish requires relatively high levels of dietary animal protein for rapid growth (Mishra & Mukhopadhyay, 1996) fish meal has traditionally been used as a major protein source due to its balanced essential amino acids profile, fatty acids, vitamins and minerals (Kaushik, 1998). However, due to its high cost and high demand as it is used in fisheries as well as in animal husbandry and the unavailability of this ingredient locally being a finite resource, the research on other alternative for non conventional protein sources to replace fishmeal in fish feed becomes of great importance for the catfish farming industry in India especially in *Clarias magur.*

Very scanty information is available regarding the use of non conventional animal protein sources in practical diet formulation for fresh water cat fish *Clarias magur*. Giri *et al.* (2010) have showed that dried chicken viscera can be incorporated up to 30% in the diet for juveniles of *Clarias batrachus* without affecting nutrients digestibility and can be used as a replacer of expensive fishmeal in the diet. The dietary protein requirements of *Clarias batrachus* are about 30 to 40%, their energy requirements range between 13 and 17 kJ/g

(Van Weerd, 1995). Researchers have used animal sources such as poultry by- product (Tucker *et al.* 2005; Shapawi *et al.* 2007), meat and bone meal (Gimenez *et al.,* 2009) with favorable results in some cases. The earthworm, *Eisenia fetida* is among the non conventional protein sources, which have been tested with relatively promising results, thanks to its high protein levels, proper amino acid profile, high reproduction rate, low mortalities, fast growth and ease of production. The use of earthworms has also been documented by different researchers. Earthworms are unique as fish food as they contain many essential amino acids along with hemoglobin in their blood serum, which provides the required iron for developing fish (Chakrabarty *et al.*, 2009). Mohanta *et al.* (2016) opined that Earthworm (*Eisenia foetida*), a non-conventional protein source with 520-530 g protein per kg on dry matter basis and can be used as a dietary protein source for rohu (*Labeo rohita*) advanced fry. The protein quality of earthworm meal was reported to be comparable to that of fishmeal with high levels of essential amino acids for fish (Dong *et al.*, 2010). Sobana and Jegadeesan (2016) reported that earthworm fed *Catla catla* had good nutritional impact as compared to the fish with conventional diet. Using earthworms in aquaculture has potential positive ecological and economic effects which solves the challenge of organic waste recycling (Boaru *et al.*, 2017). Chicken intestine is easily available at very low price throughout the year and is protein rich poultry byproduct but not commonly been utilized as animal protein source in aqua feed. Giri *et al.* (2000a) have showed that dried chicken viscera can be incorporated up to 30% in the diet for juveniles of *Clarias batrachus* without affecting nutrients digestibility and can be used as a replacer of expensive fishmeal in the diet. The utilization of dried chicken viscera and vermi meal as potential animal protein sources in fish feeds has not been fully investigated. Considering above facts the present research is aimed to standardize and develop adoptable practical diet of this valuable fish.

MATERIAL AND METHODS

The experiments were conducted in triplicates using twelve equal size (6m × 4m × 1.5m) outdoor cement cisterns. Cisterns were initially dried and cleaned. The study was conducted for a period of 180 days. The experimental tanks were provided with six inch of soil bed and water level was maintained 50 ± 5 cm. Water from experimental tanks were analyzed for temperature, pH, DO, CO_2, alkalinity etc. on the day of stocking and thereafter at 30 days interval, by following the standard methods of APHA, 2017. Lime application was done @ 200 kg/ha in a 30 days interval for maintaining the water pH between 6.5 to 8.0 ranges.

The juveniles of magur were stocked @ 4 no's/m^2. The tanks were covered with net to avoid the prey of fishes by predatory birds. In the study juvenile specimens of *Clarias magur* were reared by feeding with four different compositions of feeds in tanks. The four iso-nitrogenous experimental diets containing 35% of protein level were prepared using ingredients such as Fish Meal, Vermi meal, Chicken viscera, rice polish, wheat flour, vitamin and mineral mixture. Proximate analysis of the selected ingredients was carried out by standard methods (AOAC, 2005). The Diet 1 (D_1) is formulated by 100% replacement of Fish meal with Vermi meal; Diet 2 (D_2) is formulated by 100% replacement of Fish meal with Chicken Viscera meal; Diet 3 (D_3) is formulated by 100% replacement of Fish meal with mixture of Vermi meal and Chicken Viscera meal and Diet 4 (D_4) is Fish meal based reference diet. The analysis for Amino acid profile of prepared diets were done at Biochemistry and Nutrition Division, ICAR-Central Institute of Fisheries Technology, Cochin 682029, India, the total amino acid composition was determined following the method of Ishida *et al.* (1981).

Each group having both the sexes of *Clarias magur* were fed @ 10-5% body weight in two split doses daily once in the morning and next in the evening. For first 90 days they were fed with 10% body weight and after that fishes were fed with 5% of their body weight till harvesting. The sampling was done at the day of initiation of the experiment and later every 15 days to record the fish weight, length as well as to check the health status and then the feed amount was adjusted accordingly. For digestibility studies method given by Cho *et al.* (1985) and Talbot (1985) were followed. The digestibilities of the experimental diets were determined by indirect method using chromic oxide as an external marker. The experimental feeds were prepared by incorporating 0.5% chromic oxide (Cr_2O_3). The feed containing Cr_2O_3 were fed to the experimental fishes for 20 days of the experimental period. For this digestibility study experiment, 20 no's of aquariums were stocked with 02 fishes each of almost uniform size (170-180 g).

RESULTS AND DISCUSSION

The proximal compositions of the feed ingredients are summarized in Table 1. The graphical representation of nutritional comparison between Fish meal, Vermi meal and Chicken viscera meal (Fig. 1) reflects that a highest value of protein retain in Fish meal (59%); lipid in Chicken viscera meal (14.2%); Fibre (3.2%), Ash (19.1 %), NEF (20.2 %) and Moisture (11.8 %) in Vermi meal and Dry matter in Chicken viscera (90.3). Similar trend of findings were seen in a study conducted by Mmanda *et al.* (2020); Bhaskar *et al.* (2014) and Pyne *et al.* (2015) in their studies on evaluation of chemical composition of different feed ingredients for fishes. They observed the crude protein and lipid percentage in poultry viscera was 60.67% and 12.05-12.5 % and fish meal it was 55.19% and 7.77 % respectively. In the studies of proximal composition of earthworm by Beg *et al.* (2016); Fadaee (2012) and Mamauag (2016) the experimental results were as high in protein (50-60%), lipids (7-9%) and sugars (11-14%).

The amino acid profiles of experimental diets are illustrated in Table 2. In the graphical representation of amino acid profile (Fig. 2) for all four diets it can be concluded that diet-4 contain comparatively high value of L-Aspartic acid, L-Glutamic acid, L-Serine, L-Histidine, Glycine, L-Threonine, L-Arginine, L-Alanine, L-Tyrosine, L-Cystine, L-Methionine, L-Phenylalanine, L-Leucine and L-Lysine. The high content of essential amino acids (mainly lysine and methionine) as comparable to fish or meat meal reported in earthworm meal (Bou *et al.*, 2013; Sakthika *et al.*, 2014; Vodounnou *et al.*, 2016). In the present study also noticeably good amount of lysine (2.58g %) and methionine (0.59 %) was recorded in Vermi meal based diet as compared to Chicken viscera and Fish meal. Very few studies have been conducted in India to study the amino acid composition in fresh water food fishes. In a holistic approach Mohanty *et al.* (2014) has documented amino acid composition and protein content of 27 food fishes from India. They concluded that the carps and catfishes are rich in glutamic acid and glycine. In the present study all diets contains considerable amount of glycine and glutamic acid which may lead to better growth and utilization of the diets by the experimental fishes. Nevertheless, the results of the present study are in complete agreement with several other studies using different fish species and poultry by product meals. Shapawi *et al.* (2007) reported that poultry by product meals could be successfully used as the sole animal protein source in diets for the humpback grouper, *Cromileptes altivelis*.

Table 1: Important biochemical parameters of feed ingredients and experimental diets

Ingredients	Protein (%)	Lipid (%)	Fiber (%)	Ash (%)	NFE (%)	Dry matter (%)	Moisture (%)
Fish meal	59.0±0.2	9.0±0.2	1.0±0.10	15.1±0.1	15.8±0.2	90.0±0.6	10.0±0.6
Vermi meal	51.0±0.2	6.5±0.1	3.2±0.1	19.1±0.1	20.2±0.4	88.2±0.1	11.8±1.0
Chicken viscera meal	58.0±1.0	14.2±0.9	1.9±0.1	8.1 ±0.2	17.7±1.4	90.3±0.8	9.7±0.8
Rice polish	9.1±0.2	12.0±0.1	19.0±0.2	10.2±1.0	49.7±1.2	89.7±0.6	10.3±0.6
Wheat flour	12±0.5	1.7±0.1	0.9±0.01	0.6 ±0.2	84.8±0.7	89.2±1.0	10.8±1.0
Diet-1	35±0.1	6.8±0.2	6.2±0.2	13.7±0.2	38.3±1.1	90.0±1.3	10.0±1.3
Diet-2	35±0.1	11.4±0.7	7±0.2	7.4±0.2	39±1.0	90.3±1.2	9.7±1.2
Diet-3	35±0.1	9.2±0.3	6.5±0.2	10.4±0.2	38.8±1.0	90.2±1.1	9.8±1.1
Diet-4	35±0.1	7.9±0.2	5.4±0.2	10.1±0.2	41.5±1.2	90.2±1.3	9.8±1.3

(Values are mean ± SD, n = 3)

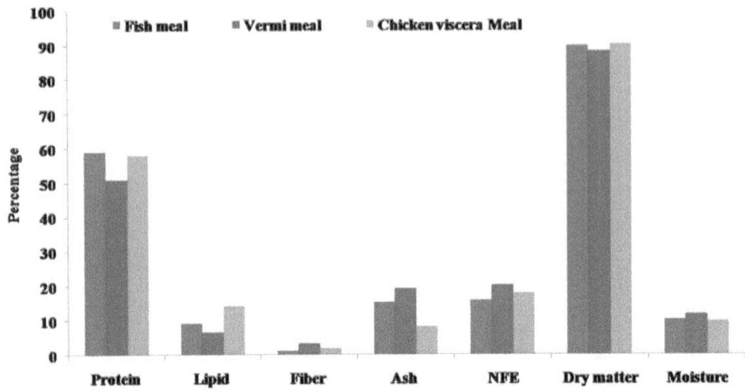

Fig. 1: Nutritional comparison between Fish meal, Vermi meal and Chicken viscera meal

Among the prepared diets D_2 was found to be having highest lipid percentage i.e. 11.4 % followed by D_3 (9.2%), D_4 (7.9%) and D_1 (6.8%) (Table 1). The high lipid content in diet D_2 may be attributed to the reason that diet 2 is formulated by 100% replacement of Fish meal with Chicken Viscera meal. Chicken Viscera meal was found to have high lipid content (14.2 ± 0.9%) among the feed ingredients.

The water quality parameters are the key factors that influence the growth of fishes. Hence, studies of the water quality parameters are necessary for interpret of results of growth performance. Variation of water quality parameters influences the growth of fishes (Sahoo *et al.* 2021). The data of water temperature of experimental tanks recorded during the experimental period ranged between 22.1 °C to 32.5 °C. During the experimental period seasonal variation of water temperature (Fig. 3) of the tanks were observed but there was no significant difference ($P < 0.05$) in water temperature between the tanks for each month of observation. The other water quality parameters like, DO, depth, pH, $CO_{2,}$ total alkalinity, total hardness and ammonia did not record any marked trend in the treatments during the culture period for both the experimental location, this may be attributed to good water quality management practices.

Although the initial average weights of the fishes are not significantly different for all the treatments, fishes fed with fishmeal based diet have showed significantly highest (Fig. 4) final weight 204. 93 g. The results reflects that growth performance of the group fishes fed with diet 4 were the best followed by diet 2, diet 3 and diet 1. Similar trend was also recorded for % weight gain, SGR, FCR and PER (Table 3). The better FCR and highest PER was recorded 1.50 and 1.90 respectively in the group of fishes fed with diet 1 as compared to those with 100% replacement of Fish meal with Vermi meal, 100% replacement of Fish meal with Chicken Viscera meal and 100% replacement of Fish meal with mixture of Vermi meal and Chicken Viscera meal diet. The results obtained in the present study for body weight gain, SGR, FCR and PER are similar to those obtained by Cayen *et al.* (2016) in their study on replacement of fish meal with broiler chicken viscera on growth of catfish *Clarias gariepinus*. Samad *et al.* (2014) also reported a FCR (2.02) where the *Clarias batrachus* was fed with formulated diet of 30% protein containing Poultry viscera, Mustard oil cake and Rice polish. Dedeke *et al.* (2013) reported that fish meal can be substituted with earthworm meal up to 25% in the diet of *Clarias gariepinus* fry without adverse effects on growth and nutrient utilization. Beg *et al.* (2016) opined that 50% replacement of fish meal with earthworm meal in the diet may be used as an ideal alternate protein source for better growth performance of Indian major carps. Significant higher rate of growth was recorded for *Claris batrachus* fed with earthworms as compared to molluscan meal and goat liver meal (Verma *et al.* 2011). Oke *et al.* (2016) recommended that up to 30 % of Chicken viscera meal could be incorporated in the diets of *C. gariepinus* without negative effects on growth and whole body composition. Giri *et al.* (2010) reported that dried chicken viscera could be used as a complete and superior substitute of marine by catch fishmeal without adversely affecting the performances of *C. batrachus* fingerlings, when incorporated in a 500 g kg^{-1} diet.

Significantly highest apparent digestibility coefficient (32.64 %) was recorded for diet-4 and lowest in diet-1 (Table 3). The apparent digestibility coefficient for diet-2 and diet-4 was found to be on per. It indicated that both the diets are equally suitable. The result obtained in the digestibility revealed that digestibility is also better in the group where FCR is observed better.

Table 2: Amino Acid Composition in experimental diets (g %)

Sl. No.	Amino Acid	D-1	D-2	D-3	D-4
1.	Tryptophan	0.40	0.30	0.47	0.35
2.	L-Aspartic acid	3.52	2.80	2.96	3.88
3.	L-Glutamic acid	5.95	5.41	5.56	7.44
4.	L-Serine	1.58	1.59	1.48	1.93
5.	L-Histidine	0.96	0.95	0.87	1.31
6.	Glycine	1.71	2.31	1.88	2.92
7.	L-Threonine	1.72	1.49	1.53	1.90
8.	L-Arginine	1.93	2.31	2.02	2.84
9.	L-Alanine	2.31	1.77	1.90	2.46
10.	L-Tyrosine	1.17	0.96	1.05	1.32

11.	L-Cystine	0.18	0.17	0.16	0.20
12.	L-Valine	2.13	1.92	1.92	2.12
13.	L-Methionine	0.59	0.56	0.57	1.08
14.	L-Phenylalanine	1.60	1.35	1.34	1.82
15.	L-Isoleucine	1.86	1.49	1.56	1.83
16.	L-Leucine	3.07	2.56	2.62	3.28
17.	L-Lysine	2.58	2.32	2.06	3.57
18.	L-Proline	1.03	1.39	1.17	1.35

Table 3: Digestibility Coefficient, Percentage weight gain, SGR, FCR and PER

Parameters	**Group fed with**			
	Diet-1	**Diet-2**	**Diet-3**	**Diet-4**
% Weight Gain	1986.3	2216.1	2195.8	2369.0
SGR	1.69	1.75	1.74	1.78
FCR	1.71	1.54	1.62	1.50
PER	1.67	1.86	1.76	1.90
Apparent digestibility coefficient (%)	24.53	30.48	28.36	32.64
Apparent digestibility coefficient (%) (Protein)	73.72	76.21	75.89 ± 2.05	78.89
Apparent digestibility coefficient (%) (Lipid)	77.63	80.30	77.80 ± 2.98	83.62

Values are mean of replications, n = 5 (Digestibility study), n=10 (for other parameters)

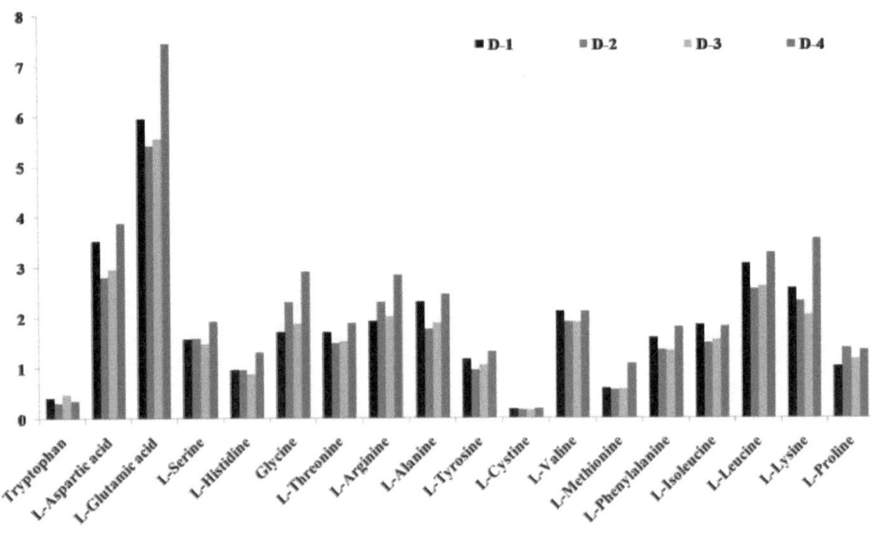

Fig. 2: Amino acid (g/100g) in different experimental diets

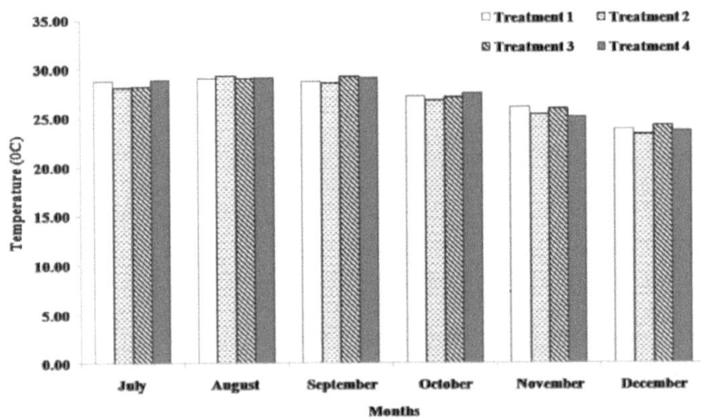

Fig. 3: Water temperature of the tanks during culture period

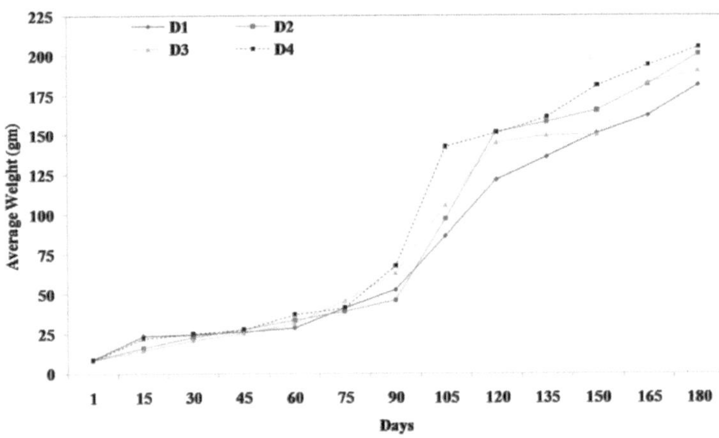

Fig. 4: Average weight of C. magur reared using different experimental diet

CONCLUSIONS

The results of the present study indicate that protein from Vermi meal, Chicken viscera meal, Rice polish and Wheat flour is well digested by *Clarias magur*. The digestibility of protein and lipid in diet 4 and diet 2 was found to be on per ($P < 0.05$). The results also suggested that it may be possible to use Chicken viscera meal, which is locally available, to replace imported expensive fish meals in diets for magur.

REFERENCES

Ahmad, R., Pandey, R.B., Arif, S.H., Nabi, N., Jabeen, N. and Hasnain, A. 2012. Polymorphic β and γ lens crystallins demonstrate latitudinal distribution of threatened walking catfish Clarias batrachus (Linn.) populations in North-western India. Journal of Biological Sciences, 12: 98–104.

AOAC.2005. Association of official analytical chemist. In Official methods of analysis, 18th ed. (ed. By S. Williams), AOAC, Arlington, VA,USA.

APHA.2017. Standard Methods for the Examination of Water and Wastewater.21st Edition. American Public Health Association/American WaterWorks Association/Water Environment Federation, Washington, DC.

Beg, M. M., Mandal, B. and Moulick, S.2016. Potential of earthworm meal as a replacement of fish meal for Indian major carps. International Journal of Fisheries and Aquatic Studies, 4(3): 357-61.

Bhaskar, P., Pyne, S. K. and Ray, A. K. 2014. Evaluation of poultry viscera as potential fish feed ingredient, compared to fishmeal. International Journal of Current Research, 6 (02): 5241-43.

Binoy, V. 2010. Catfish Clarias is vanishing from the waters of Kerala. Current Science, 99: 714.

Boaru, A., Georgescu, B., Struti, D. and Ladosi,D. (2017). The potential of Eisenia foetida usage in aquaculture. ABAH Bioflux, 9 (2): 55-58.

Bou, M. E., Loupiac, C., Loison, A., Rollin, B., Cayot, P., Cayot, N., Marquez, E. and Medina A. L.2013.Impact of preparation process on the protein structure and on the volatile compounds in Eisenia foetida protein powders. Food and Nutrition Sciences, 4:1175-83.

Cayen, A. S., Oke, V. and Abou, Y. 2016. Effect of replacement of fish meal with broiler chicken viscera on growth, feed utilization and production of African catfish Clarias gariepinus (Burchell, 1822). International Journal of Fisheries and Aquatic Studies, 4(6): 182-86.

Chakrabarty, D., Das, S. K., Das, K. M., Biswas, P. K. and Karmegam, N. (2009). Application of vermitechnology in aquaculture. Dynamic Soil, Dynamic Plant, 3(2): 41-44.

Cho, C. Y., Cowey, C. B. and Watanabe, J. 1985. Finfish Nutrition in Asia. Methodological Approaches to Research and Development. Canada, IDRC. pp.154.

Dedeke, G. A., Owa, S. O., Olurin, K. B., Akinfe, A. O. and Awotedu, O. O. 2013. Partial replacement of fish meal by earthworm meal (Libyodrilus violaceus) in diets for African catfish, Clarias gariepinus. International Journal of Fisheries and Aquaculture, 5(9): 229-33.

Dong, X. H., Guo, Y. X., Ye, J. D., Song, W. D., Huang, X.H. and Wang, H. 2010. Apparent digestibility of selected feed ingredients in diets for juvenile hybrid tilapia, Oreochromis niloticus× Oreochromis aureus. Aquaculture Research, 41(9): 1356-64.

Fadaee, R. 2012. A review on earthworm Esienia fetida and its applications. Annals of Biological Research, 3(5): 2500-6.

Ghazala, R., Tabinda, A. B. and Yasar, A. 2011. Growth response of juvenile grass carp (Ctenopharyngodon idella) fed with isocaloric diets with variable protein levels. Journal of Animal Plant Science, 21: 850-56.

Gimenez, A. V. F., Diaz, A. C., Velurtas, S. M. and Fenucci, J. L. 2009. Partial substitution of fishmeal by meat and bone meal, soybean meal, and squid concentrate in feeds for the prawn, Artemesia longinaris: Effect on digestive proteinases. The Israeli Journal of Aquaculture, 61(1): 48-56.

Giri, S. S., Sahoo, S. K., and Mohanty, S. N. 2010. Replacement of by-catch fishmeal with dried chicken viscera meal in extruded feeds: effect on growth, nutrient utilisation and carcass composition of catfish Clarias batrachus (Linn.) fingerlings. Aquaculture International, 18(4): 539-44.

Giri, S. S., Sahoo, S. K., Sahu, A. K. and Mukhopadhyay, P. K. 2000a. Growth, feed utilization and carcass composition of catfish Clarias batrachus (Linn.) fingerlings fed on dried fish and chicken viscera incorporated diets. Aquaculture Research, 31(10): 767-71.

Giri, S. S., Sahoo, S. K., Sahu, A. K. and Meher, P. K. 2003. Effect of dietary protein level on growth, survival, feed utilisation and body composition of hybrid Clarias catfish (Clarias batrachus × Clarias gariepinus). Animal Feed Science and Technology, 104: 169–78.

Hardy, R. W. and Tacon, A. G. 2002. Fish meal: historical uses, production trends and future outlook for sustainable supplies. Responsible Marine Aquaculture, 311-25.

Hussain, S. M., Singh, I. J., Ram, R. N., Pathak,M. and Yadav, H.S. (2016). Reproductive Cyclicity of Clarias batrachus (magur) under Tarai conditions of Uttaranchal. Journal of Experimental Zoology, 19 (1):1457-64.

Ishida, Y., Fugita, T. and Asai, K. 1981.New detection and separation method for amino acid by high performance liquid chromatography. Journal of Chromatography, 204:143-48.

Kaushik, S. J. 1998. Whole body amino acid composition of European seabass (Dicentrarchus labrax), gilthead seabream (Sparus aurata) and turbot (Psetta maxima) with an estimation of their IAA requirement profiles. Aquatic Living Resources, 11(5): 355-58.

Khedkar, G. D., Tiknaik, A. D., Shinde, R. N., Kalyankar, A. D., Ron, T. B. and Haymer, D. 2014. High rates of substitution of the native catfish Clarias batrachus by Clarias gariepinus in India. Mitochondrial DNA, 27(1): 569-74.

Krishnankutty, N. 2005. Plant proteins in fish feed: An additional analysis. Current Science, 89(6): 934-36.

Mamauag, R. E. P.2016. Supporting ASEAN good aquaculture practices: utilization of alternative protein sources for aquafeed to minimize pressure on fishery resources. Fish for the people, 14(2): 83-89.

Mishra, S. and Mukhopadhyay, P. K. 1996. Effect of some formulated diets on growth, feed utilization and essential amino acid deposition in Clarias batrachus fry. Indian Journal of Fisheries, 43(4): 333-39.

Mmanda, F. P., Lindberg, J. E., Halden, A. N., Mtolera, M. S. P., Kitula,R. and Lundh T. 2020. Digestibility of Local Feed Ingredients in Tilapia Oreochromis niloticus Juveniles, Determined on Faeces Collected by Siphoning or Stripping. Fishes, 5:32.

Mohanta, K. N., Sankaran, S. and Veeratayya, S. K. (2016). Potential of earthworm (Eisenia fetida) as dietary protein source for rohu (Labeo rohita) advanced fry. Cognitive Food and Agriculture, 2:(1): 1385-94.

Mohanty, B., Mahanty, A., Ganguly, S., Sankar, T. V., Chakraborty, K., Rangasamy, A., Paul, B., Sarma, D., Mathew, S., Kurukkan, K. A., Behera, B., Aftabuddin, Debnath, D., Vijayagopal, P., Sridhar, N., Akhtar, M. S., Sahi, N., Mitra, T., Banerjee, S., Paria, P., Das, D., Das, P., Vijayan, K. K., Laxmanan, P. T. and Sharma, A. P. 2014. Amino Acid Compositions of 27 Food Fishes and Their Importance in Clinical Nutrition. Journal of Amino Acids, 1–7.

Nahar, Z., Azad, S. A. K. M., Bhandari, R. K., Ali, M. H. and Dewan, S. 2000. Effect of different feeds on growth, survival and production of African catfish (Clarias gariepinus Burchell). Bangladesh Journal of Fisheries Research, 4(2):121-26.

NRC (National Research Council) .1993. Nutrient requirements of fish. National Academy Press. Washington, DC, USA.

Nyina, W. L., Wathelet, B., Richir, J., Rollin, X. and Kestemont, P. 2010. Partial or total replacement of fish meal by local agricultural by-products in diets of juvenile African catfish (Clarias gariepinus): growth performance, feed efficiency and digestibility. Aquaculture Nutrition, 16(3): 237-47.

Oke, V., Odountan, H. O. and Abou, Y. 2016. Chicken Viscera Meal as a Main Component in Diet for African Catfish Clarias gariepinus (Burchell 1822) Reared in Earthen Ponds. Journal of Food and Nutrition Research, 4(12): 799-05.

Pandian, T. J. 2001. Sustainable Indian Fisheries. National Academy of Agricultural Sciences, Indian Council of Agricultural Research, Ministry of Agriculture.

Paul, B. N., Chanda, S., Sridhar, N., Saha, G. and Giri, S. 2015. Proximate and Mineral Composition of Magur (Clarias batrachus) and Singhi (Heteropneustes fossilis), Indian Journal of Animal Nutrition, 32(4): 453-56.

Pyne, S. K., Bhaskar, P. and Ray, A. K. 2015. Poultry viscera as an alternative of fishmeal in compound Diet confirms better growth performances of magur fish, Clarias batrachus (Linn). International Journal of Bio-resource, Environment and Agricultural Sciences, 1(1): 92-97.

Sahoo, P. R., Das, P. C., Nanda, S., Mohanta, K. N. and Sahu, B. 2021. Influence of Photoperiod and Rearing Density of Fry in Production of Stunted Fingerlings of Catla catla (Hamilton). Indian Journal of Animal Research, 55(4): 479-85.

Sakthika, T., Ronald, J., Kumar, S. V. and Felicitta, J. 2014. Growth of Mystus montanus fed with two different earthworm meals. International Journal of Environmental Science, 4(4): 551–57.

Samad, A., Imteazzaman, A. M., Hossain, I. and Reza, M. S. 2014. Effects of three different low cost feeds on growth performance of walking catfish (Clarias batrachus L.) in earthen ponds. Rajshahi University Journal of Life & Earth and Agricultural Sciences, 42:1-10.

Shapawi, R., Ng, W. K., and Mustafa, S. 2007. Replacement of fish meal with poultry byproduct meal in diets formulated for the humpback grouper, Cromileptes altivelis. Aquaculture, 273(1):118-26.

Sobana, K. and Jegadeesan, M. 2016. Effect of vermicompost-bred earthworms on fish growth. Advances in Applied Science Research, 7(1): 184-88.

Tabinda, A.B. and Butt, A. 2012. Replacement of fish meal with poultry by-product meal (chicken intestine) as a protein source in grass carp fry diet. Pakistan Journal of Zoology, 44(5):1373-81.

Talbot, C. 1985. Laboratory methods in fish feeding and nutritional studies. In: Fish energetic, Springer, Dordrecht, 125-54.

Tucker, A., Yigit, M., Ergun, S., Karaali, B. and Erteken, A. 2005. Potential of poultry by-product as a substitute for fish meal in diets for black sea turbot, Scopthalmus maeoticus: growth and nutrient utilization in winter. Israeli Journal of Aquaculture-Bamidgeh, 57(1): 39-61.

Van Weerd, J. H.1995. Nutrition and growth in Clarias species-a review. Aquatic Living Resources, 8(4): 395-01.

Verma, S. K., Gupta, D. K. and Sinha, M.P. 2011. Growth performance of the air breathing cat fish, Clarias batrachus (Linn.) fed with verm meal as fish food. The Ecoscan, 209-14.

Vishwanath,W.2010. Clarias magur. The IUCN Red List of Threatened Species 2010: e.T168255A6470089.http://dx.doi.org/10.2305/IUCN.UK.2010-4. RLTS. T 168255 A 6470089. en.

Vodounnou, D. S., Juste, V., Kpogue, D. N. S., Apollinaire, M. G. and Didier, F. E. 2016. Culture of earthworm (Eisenia fetida), production, nutritive value and utilization of its meal in diet for Parachanna obscura fingerlings reared in captivity. International Journal of Fisheries and Aquatic Studies, 4(5): 01–05.

Yigit, M., Erdem, M., Koshio, S., Ergün, S., Turker, A. and Karaali, B. (2006). Substituting fish meal with poultry by product meal in diets for black Sea turbot Psetta maeotica. Aquaculture Nutrition, 12 (5): 340-47.

10

Effect of Dietary Carotenoids on Body Pigmentation of Blue Gourami *Trichogaster trichopterus*

Anurag Protim Das and Shyama Prasad Biswas

ABSTRACT

A feeding trial of 45 days was conducted to observe the effects of dietary natural source of Carotenoids. This study examines the effects of broccoli (Brassica oleracea var. italica) on body pigmentation, body indices, and survival of Blue Gourami, Trichogaster trichopterus. Using readily accessible feed ingredients locally, five isonitrogenous experimental meals were formulated, with the addition of carotenoids meal at four different concentrations (1%, 2%, 3%, and 4%). Analyzing the growth data, it was found that 3% diet had a considerably (P<0.05) greater weight increase % than other treatments. The growth rate specific (SGR) was showed a substantial (P<0.05) decrease in 1% group and a significant increase in 3% and 2% groups. The survival percentage was 100% in all the treatment groups. At the end of the feeding trial, fish fed with the meal containing 3% broccoli diet had a considerably higher (p < 0.05) total carotenoid content in their muscle (4.24±0.27 μgg-1) than the female (3.12±0.13 μgg-1). Overall, the study showed that 3% of natural sources of carotenoid, produced good growth and coloration. However, in this particular study, the fish group fed with 3% diet was more successful in boosting the growth and coloration of Blue Gourami, Trichogaster trichopterus.

Keywords: *Broccoli, carotenoids, Trichogaster trichopterus, Skin pigmentation*

INTRODUCTION

An expanding aquaculture industry that offers high incomes and job opportunities to people worldwide is ornamental fish farming. With over 2 billion live ornamental fish traded globally, the ornamental fish trade has become a very lucrative sector (Das & Biswas, 2016; Katia, 2001) with global net worth of more than 20 billion USD. According to Saxena (1994), the most significant factor influencing an ornamental fish's market value is its vibrant body colouration. Fish are incapable to synthesize carotenoids De novo (Goodwin 1951),and rely on dietary carotenoids for its body pigmentation (Hata and Hata, 1973). So, there is hence a direct correlation between pigmentation and dietary carotenoids(Halten *et al.,* 1995). Aquatic animals' skin, meat, shell, and exoskeleton are coloured red, orange, and yellow due to a class of naturally occurring lipid-soluble organic pigments called carotenoids (Pailan*et al.*, 2012). Fish skin and muscle colour regulation is significantly influenced by dietary carotenoids (Ahilan *et al.*, 2008). The freshwater ornamental fish industry has encountered the issue of fadingfish body colouration, particularly in cases where the fish are reared in captivity for extended periods of time and subjected to rigorous cultivation conditions (Das and Biswas, 2016). According to Das and Biswas (2016), an ornamental fish's food should generally be nutritionally balanced, appetizing, resistant to crumbling, water stable, buoyant, and enhance or retain the fish coloration.

The phenomena of coloration in fish are quite significant and have a direct impact on its aesthetic value. If pigment-enriched feed can be prepared and used to increase coloration, the fish will undoubtedly have better quality and a cheaper price tag. Detailed research on colour enrichment in native ornamental fish using formulated feed is lacking (Das and Biswas, 2016). For the purpose of colour enrichment in fish, native plant sources have been used. Choubert (1979), Boonyaratpalin and Phromkunthong (1986), and Alagappan *et al.* (2004), utilized spirulina as a source of carotenoid colours for fancy carp and rainbow trout. Enhancement of pigmentation of *Cyprinus carpio* and *Carassius auratus* with microalgal biomass supplementation has demonstrated that *Chlorella vulgaris* is just as effective as synthetic colours (Gouveia *et al.*, 2003). Alagappan *et al.* (2004) observed increased body pigmentation in *Trichogaster trichopterus* using *Spirulina sp*. algae as a source of carotenoid pigment. The body pigmentation of goldfish (*Carassius auratus* L.), red swordtail (*Xiphophorus helleri*), and tiger barb (*Puntius tetrazona*) was enhanced by feeding with marigold petal diet (Boonyarapatin and Lovell, 1977; Ezhil *et al.*, 2008; Alma *et al.,* 2013). Using natural sources of carotenoid pigments to improve the colour of ornamental fish is a cost-effective and ecofriendly solution. Due to the aforementioned information, the current study was carried out to assess how dietary natural carotenoids from source such broccoli (*Brassica oleracea* var. italica) effects the development and coloration of Blue Gourami, *Trichogaster trichopterus*, one of the important ornamental fishes with high market value.

MATERIALS AND METHODS

Design of experimental feeding

For the experiment, Blue Gourami, *Trichogaster trichopterus* of a homogeneous size group was purchased from a local breeder of Dibrugarh district. Following collection, the fish were bathed in 5 ppm KMnO4 to disinfect them until they began to exhibit signs of stress. The fish treated in the bath were moved to 50-liter aquariums with constant aeration after dead and weak fish were discarded. Before the experiment, the fish were acclimated for ten days in the aquariums. The feeding trial was conducted in 12 aquariums of 50-liter capacity over the course of forty-five days. These twelve aquariums were divided into six sets, with two duplicates in each set. Five female fish and ten male fish were placed in each trough, totaling fifteen fish per tank. Fish weight and length were recorded on a weekly basis. The fish were fed at a rate of 3% of their body weight twice a day, in the morning and the evening. Every ten days, the meal was changed based on the fish body weight. Daily syphoning off of feed particles and metabolic wastes from the bottom allowed the aquariums to be adequately cleaned. Once a day, there was a water exchange to ensure the fish had an adequate quantity of oxygen.

Feed formulation

Dry pelleted feed was formulated for the study using Hardy's (1980) trial-and-error and Pearson Square method. Basic feed ingredients such as fish meal (25%), soy bean meal (22%), groundnut oil cake (15%), rice bran (20%), and wheat flour (12%) were used to formulate the control diet. The chosen components were ground and well combined before the necessary amount of water was added to make the dough. This dough was steam-cooked for 40 minutes at 70°C in a water bath. Following a period of cooling, the dough was combined well with additional ingredients, including 2% soy oil, 3% starch powder, and 1% vitamin and mineral premixes. In order to prepare the treatment diets (T1 to T4), broccoli was dried in a lypholizer, coarsely powdered, and then added in the appropriate concentrations of 1, 2, 3, and 4g/100g of the baseline diet by substituting the same amount of rice bran. In order to shield these micronutrients from heat, this was done. Using a hand pelletizer, the dough was formed into 2.0 mm pellets. Before being used the pelleted feed was air dried and stored in airtight containers.

Proximate composition of the diet

In accordance with AOAC, (2005), the trial diets proximate components were examined for moisture, crude protein, ether extract, and total ash. As per Das and Biswas (2019), a quantitative assessment of crude fat was carried out.

Assessment of carotenoid concentration

Olson (1979) recommended that the total carotenoid concentration (TCC) in the muscle tissue of the fish, both male and female, be measured at 15-day intervals and after the experiment was over. Due to the heightened sensitivity of carotenoids to light, temperature, and oxygen, all processes were carried out under low light.

Growth study

Fish were weighed at the start and termination of the experiment on the 60th day. The growth performances of Blue Gourami, *Trichogaster trichopterus* were evaluated in terms of weight gain (%), specific growth rate (SGR), feed conversion ratio (FCR) by using following equation:

Weight gain (%) = (final weight − initial weight) / initial weight x 100;

SGR = 100 x (log average final weight − log average initial weight)/number of culture days;

FCR = Total feed given (dry weight) (g) / weight gain (wet weight) (g)

The survival rate was evaluated as, total survival rate (S) = the number alive at the end of time t /the number alive at the start of time t.

Water quality analysis

Water quality metrics, including temperature, pH, dissolved oxygen, total hardness, and alkalinity, were assessed biweekly for all aquariums using the APHA (1998) standard techniques. The Microsoft Lumia programmed Lux Meter was utilized to monitor light intensity of the aquaria.

RESULTS

Table 2 presents the FCR, growth performance and survival ratesof the Blue Gourami at the conclusion of the experiment. The control group's feed conversion ratio (FCR) was 1.80±0.05, whereas the experimental groups' equivalent values varied from 1.72to 1.78, with T3 having the highest value and T4 having the lowest. Two metrics that are used to measure feed utilization are the feed conversion ratio and the feed efficiency ratio. The current study's findings unequivocally shown that a higher dose of carotenoids supplementation in diets significantly decreased fish FCR. The growth rate of Blue Gourami was not significantly affected by the addition of broccoli diet. On the other hand, the T3 treatment with 3% broccoli diet showed comparatively higher net weight growth values (0.73±0.01g) when compared to the other treatments and control group. The Blue Gourami's survival rate with treatment meals and control diets ranging from 90% to 100% was not significantly affected by the dietary addition of marigold oleoresin (Table 3).

Total carotenoid concentration (TCC)

The concentration of total carotenoids in Blue Gourami's muscle and skin was 1.21µg/g wet weight at the beginning of the trial. Following a 45-day experimental feeding trial, the total carotenoids concentration in the muscle and skin of Blue Gourami clearly demonstrated an increase with the addition of broccoli meal to the diet. The higher levels of carotenoids concentration in skin and muscle were observed at 3% level (4.24±0.27µg/g wet weights) in the case of the male and 3.12±0.13µg/g wet weights) in the case of the female. A further significant increase in carotenoids content was not found in either gender after that (Table 3). There is a positive correlation was found in both male (0.91) and female (0.89) fishes between the amount of carotenoids in their diet and the concentration of carotenoids in their skin and muscles.

Physico-chemical parameters

Table 1 displays the water quality metrics for each experimental group. The range of water temperatures in each group is 23°C to 28°C. The pH was discovered in the between 7.5 and 8.5. There was a slight variation in the dissolved oxygen content, which ranged from 5.4 to 7.6 mg/L. The various experimental groups' total ammonia contents were measured within the range of 0.15 to 0.27 mg L-1. Hardness ranged from 249 to 264 mgL-1 in the various treatment groups.

Table 1: Water quality parameters recorded in experimental tanks

Parameter	C	T1	T2	T3	T4
Temperature (°C)	24.5-25.6	26.0-27.5	26.4-28.0	24.0-26.5	23.5-25.6
pH	7.6-8.1	7.5-8.2	7.4-8.2	7.5-8.2	7.6-8.2
Dissolved oxygen (mgL -1)	5.3±7.9	5.8-7.6	5.6-7.4	5.8-7.9	5.6-7.5
Ammonia (mgL -1)	0.15-0.22	0.22-0.27	0.24-0.26	0.16-0.27	0.23-0.26
Light intensity (Lux)	320.6-331.5	320.6-331.5	320.6-331.5	320.6-331.5	320.6-331.5

Table 2: Biochemical composition of experimental diets

Diet	Moisture	Crude fat	Crude protein	Crude Fibre	NFE	Total Ash
Control	7.51±0.06	8.2±0.01	30.63±0.03	6.26±0.03	35.73±0.02	11.43±0.01
T1	7.56±0.02	8.1±0.03	30.65±0.07	6.29±0.01	36.11±0.01	11.42±0.01
T2	7.60±0.07	8.1±0.01	30.61±0.04	6.25±0.03	35.63±0.03	11.44±0.02
T3	7.55±0.05	8.3±0.02	30.54±0.07	6.29±0.02	35.81±0.02	11.42±0.02
T4	7.58±0.06	8.2±0.01	30.63±0.03	6.26±0.01	35.71±0.03	11.43±0.01

T- Carotenoids feed with numbers referring to percentage of carotenoids incorporated per 100gm of basal ingredients, FCR- Feed conversion ratio, PER- Protein energy ratio

Nitrogen free extract = 100 - (% moisture + % crude protein + % crude lipid + % crude fibre + % ash).

Table 3: Mean survival, weight, length, and FCR of Blue Gouramiafter 45 days of trial

Diet	Increase in weight (gm)	Increase in length (cm)	FCR	Survival %	Carotenoids% (µg/gm wet weight)	
					Male	Female
Control	0.45±0.01	0.29±0.02	1.80±0.05	90	2.61±0.06	2.11±0.09
T1	0.66±0.01	0.42±0.01	1.75±0.06	100	3.48±0.11	2.63±0.06
T2	0.70±0.01	0.45±0.02	1.77±0.03	100	3.93±0.12	2.71±0.03
T3	0.73±0.01	0.52±0.02	1.79±0.04	100	4.24±0.27	3.12±0.13
T4	0.67±0.01	0.46±0.01	1.72±0.04	100	3.89±0.04	2.81±0.05

(The means with different superscript in each row indicate a significant difference ($p < 0.05$))

DISCUSSION

The intricate behaviour of feeding is strongly linked to food consumption. High protein diet is often not needed for the upkeep and growing of ornamental fish. In young Atlantic salmon given a diet supplemented with astaxanthin, Christiansen and Torrissen (1996) achieved the optimal FCR. Likewise, it was shown that feeding prawn on a diet containing spirulina (*Arthrospira platensis*) improved FCR in *Penaeus monodon*. The body indices of Blue Gourami were not significantly affected by broccoli supplementation. The current study's results are similar to those of recent investigations by Ramamoorthy *et al.* (2010) and Alma *et al.* (2013), which found no evidence to support the idea that adding dietary carotenoid such as marigold meal to a fish's diet enhances growth. The findings demonstrated that the amount of broccoli supplemented at varied levels in the treatment meals had no impact on the survival percentage of fish. When fed diets without astaxanthin supplementation, Atlantic salmon (*Salmo salar*) fry (Christiansen *et al.*, 1995) and juveniles (Christiansen and Torrissen, 1996) were shown to have low survival rates than the groups that were fed diets containing astaxanthin. Blue Gouramies are a fairly hardy species, the current investigation's results may be explained by lower densities and/or hardiness of the species.

During the feeding trail, the diet that included carotenogenic biomass showed stabilization and improvement of body pigmentation. Fish species-specific carotenoid sources and their impact on pigment deposition in the body are distinct (Ha *et al.*, 1993). Numerous fish species also often exhibit sexual dichromatism. According to Gogoi *et al.* (2013), males tend to have more colourful than that of female. Owing to an enhanced total carotenoid content in their bodies, even the female fish in the current feeding study that were fed carotenoid diets also displayed vibrant colouration. Different fish species may have different pathways for the metabolism of carotenoids (Matsuno, 2001).

According to Alma *et al.* (2013), 200 mg of carotenoids from marigold meal is the ideal amount to enhance the skin pigmentation of goldfish, and didn't observe further accumulation of carotenoids beyond that amount. Comparably, fish fed a food containing 3% broccoli in the current study had higher levels of carotenoids in their skin and muscular tissue than fish fed in comparison to other treatment diets. The fish's colour did not change despite the high amounts of dietary carotenoids. The increment in overall skin carotenoid content was found in similar work on *Etroplus maculatus* fed a meal with marigold oleoresin (Jagadeesh *et al.*, 2014). The effects of feeding *Pagruspagrus* with varying carotenoid sources according to their dietary amounts on development and skin colour were reported by Kalinowski *et al.* (2005). When *Calendula officinalis* was added to *Xiphophorus helleri* diet, Ezhil *et al.* (2008) observed an increase in pigmentation in the fish, indicating that lutein from this plant is utilized as a source of pigmentation.

A variety of natural plant sources can be used to improve or preserve the colour of ornamental fish kept in captivity, while synthetic carotenoids have a negative impact on aquatic environments. It will help open doors for the growth of the feed sector for color-enhanced fish as well as the development of jobs (Das & Biswas, 2016). The experiment suggests that, in order to achieve the appropriate body coloration in Blue Gouramies, broccoli meal, a natural source of carotenoid, may be added to the diet at a level of 3% without having a negative impact on growth, feed conversion efficiency, or survival.

ACKNOWLEDGEMENT

Authors are thankful Department of Life Sciences, Dibrugarh University for providing the laboratory facilities.

REFERENCES

Ahilan, B., Jegan, K., Felix, N. and Ravaneswaran, K. 2008. Influence of botanical additives on the growth and colouration of adult goldfish, Carassius auratus (Linnaeus). Tamil Nadu Journal of Veterinary and Animal Sciences, 4(4): 129-134.

Alagappan, M., K. Vijula and Sinha, A. 2004. Utilization of spirulina algae as a source of carotenoid pigment for blue gouramis (Trichogaster trichopterus Pallas). Journal of Fisheries and Aquatic Science, 10: 1-11.

Alma, A. del Villar-Martinez., Juan, Orbe-Rogel., E.C, Vanegas-Espinoza., P, Quintero-Gutierrez, A. G. and Lara-Flores, M. 2013. The effect of marigold (Tagetes erecta) as natural carotenoid source for the pigmentation of goldfish (Carassius auratus L.). Research Journal of Fisheries and Hydrobiology, 8(2): 31-37.

AOAC, 2005. Official Methods of Analysis of AOAC, 18th edition. (Horwitz, W. Ed), AOAC, Washington. D.C. 1094.

APHA, 1998. Standard Methods for the Examination of Water and Waste Water. Association of Official Waste Water. 20th edition. American Public Health Association, Washington, DC, USA.

Boonyarapatin, M. and Lovell, R. T. 1977. Diet preparation for aquarium fish. Aquaculture, 12: 53-62.

Boonyaratpalin, M. and Phromkunthong, W. 1986. Effects of carotenoid pigments from different sources on colour changes of fancy carp, Cyprinus carpio Linn. Journal of Food Science and Technology, 8: 11-20.

Choubert, G. 1979. Tentative utilization of spirulina algae as a source of carotenoid pigments for rainbow trout. Aquaculture, 18:135-143.

Christiansen, R., Lie, O. and Torrissen, O. J. 1995. Growth and survival of Atlantic salmon, Salmo salar L., fed different dietary levels of astaxanthin. First-feeding fry. Aquaculture Nutrition, 1: 189-198.

Christiansen, R. and Torrissen, O. J. 1996. Growth and survival of Atlantic salmon, Salmo salar L. fed different dietary level of astaxanthin. Juveniles. Aquaculture Nutrition, 2, 55-62.

Das, A. P. and Biswas, S.P. 2016. Carotenoids and Pigmentation in Ornamental Fish. Journal of Aquaculture & Marine Biology,4(4): 00093.

Das, A. P. and Biswas, S.P. 2019. Improvised protocol for quantitative determination of crude fat in fish feeds. Journal of Aquaculture & Marine Biology, 8(4): 144–146.

Ezhil, J., Jeyanthi, C. and Narayanan, M. 2008. Marigold as a carotenoid source on pigmentation and growth of red swordtail, Xiphophorus helleri. Turkish Journal of Fisheries and Aquatic Sciences, 8: 99-102.

Gogoi, R., Behera, S., Chetia Borah, B. and Bhuyan, S. 2013. Sexual dimorphism and gonadal development of a rare murrel species Channa bleheri (Bleher) in Assam. The Bioscan, 8(4): 1265-1269.

Goodwin, T. W. 1951. Carotenoids in fish. In: The biochemistry of fish. Biochemical Society Symposia.

Gouveia, L., Rema, P., Pereira, O. and Empis, J. 2003. Colouring ornamental fish (Cyprinus carpio and Carassius auratus) with micro algal. Aquaculture Nutrition, 19: 123-129.

Ha, B. S., Kang, D. S., Kim, J. H., Choi, O. S. and Ryu, H. Y. 1993. Metabolism of dietary carotenoids and effects to improve the body colour of cultured flounder and red sea bream. Bulletin of the Korean Fisheries Society, 26: 91-101.

Halten, B., Aas, G. H., Jorgensen, E. H., Storebakken, T. and Goswami, U.C. 1995. Pigmentation of 1, 2 and 3 years old Arctic charr (Salvelinus alpinus) fed two different dietary astaxanthin concentrations. Aquaculture, 138: 303-312.

Hardy, R.1980. Fish feed formulation, In: Fish feed Technology, Agriculture development coordination programme. FAO/ADCP/Rep. 80(1-1): 233-239.

Hata, M. and M. Hata .1973. Studies on astaxanthin formation in some freshwater fishes. Tohoku. Journal of Agricultural Research, 24(4): 192-196.

Jagadeesh, T. D., Shivananda Murthy, H., Swain, H., Chetan, N., Manjunatha, A. R. and Baglodi, V. 2014. Effect of marigold oleoresin on growth, survival and pigmentation in orange chromide, Etroplus maculatus (Bloch, 1795). Fishery Technology, 51: 25-30.

Kalinowski, C. T., Robaina, L. E., Fernadez-Palacios, H., Schuchardt, D. & Izquierdo, M. S. 2005. Effect of different carotenoid sources and their dietary levels on red porgy (Pagruspagrus) growth and skin colour. Aquaculture, 244: 223-231.

Katia, O. 2001. Ornamental fish trade. Infofish International, 3: 14-17.

Matsuno, T.2001. Aquatic animal carotenoids. Fisheries Sciences, 67: 771-783.

Olson, J.A. 1979. A simple dual assay for Vitamin A and carotenoids in human liver. Nutrition Reports International. 19: 808-813.

Pailan, G. H., Sinha, A. and Kumar, M. 2012. Rose Petals Meal as Natural Carotenoid Source in Pigmentation and Growth of Rosy Barb (Puntius Conchonius). Indian Journal of Animal Nutrition, 29(3): 291-296.

Ramamoorthy, K., Bhuvaneswari, S., Sankar, G. and Sakkaravarthi, K.2010. Proximate composition and carotenoid content of natural carotenoid sources and its colour enhancement on marine ornamental fish Amphiprionocellaris (Cuveir 1880). World Journal of Fish and Marine Sciences, 2(6): 545-550.

Saxena, A.1994. Health; colouration of fish. Page No. 94 in Proceedings of International Symposium on Aquatic Animal Health. University of California, School of Veterinary Medicine, Davis, CA, U.S.A.

11

Fish Diversity in Maguri Motapung Beel, Tinsukia District of Assam A Brief Investigation

Sikha Sonowal and Monjeet Sonowal

ABSTRACT

The study conducted over a span of 7 months from October 2023 to April 2024, recorded a of total of 38 fish species belonging to 27 genera, under 16 families from 10 orders in Motapung Maguri Beel (wetland), Tinsukia District of Assam, India. The different families recorded are Belonidae, Notopteridae, Cyprinidae, Cobitidae, Bagaridae, Nandidae, Channidae, Tetraodontidae, Mastacembelidae, Heteropneustidae, Siluridae, Ambassidae, Clarridae, Synbranchidae, Anabantidae, and Osphronemidae. The calculated Shannon diversity index H is equal to 2.79 indicates normal range offish diversity. Among the recordedfish species 1 species is vulnerable, 32 species are lower risk least concern, 1 species is data deficient, 1 species is endangered, 1 species is lower risk near threatened and other 2 species are not evaluated. The Threat analysis using Questioner method participated by 160 respondents recorded high anthropogenic threats in overfishing, Agricultural contamination and Fish catching during breeding seasons and low threats to conservational measure and oil spillage from fringe oil wells. The study recorded no natural threats in the study area.

Keywords: *Ichthyofauna, Anthropogenic threats, Diversity Index.*

INTRODUCTION

Fish live in a variety of watery habitats and comprise around 50% of all vertebrate species worldwide. There are approximately 21,723 known living fish species out of a total of 39,900 vertebrate species, with 8,411 found in freshwater habitats and 11,650 in marine environments (Jayaram, 1999). Fish are used as markers of contamination and contribute significantly to the preservation of human history and welfare, either directly or indirectly. According to Mittermeier *et al.* (1997), India ranks eighth in the world for freshwater biodiversity, making it one of the nations with significant biodiversity. There are over 2,500 fish species in India; roughly 930 of them are found in freshwater environments, and roughly 1570 are found in marine ones (Kar *et al.*, 2006).

Fish diversity is particularly rich in the northeastern part of the nation; studies have shown that there are 267 freshwater fish species from 114 genera distributed over 38 families and 10 orders (Mahanta *et al.*, 2003). Fish plays a crucial role in Assam's cuisine, where a wide range of fish species may be found in the region's floodplain wetlands because of the subtropical climate, favorable ecological and topographical circumstances, and inherent capacity for stocking. Assam's waters, including its wetlands, are home to around 217 fish species, representing 104 genera, 37 families, and 10 orders, according to Goswami & Singha (2023). Assam's lakes and marshes are vital for supplying the state's domestic fish needs, fostering economic growth, and giving the general public access to fairly priced food. In Maguri Beel, several fish are in danger of going extinct because of pollution, habitat damage, and the introduction of non-native species, overexploitation and human activity (Moyle & Moyle, 1995; Malakar & Boruah, 2017). The 2020 oil spill incident might have had a major impact on the variety of the ichthyofauna's composition. To address the same, a comparison of the same with earlier research is desperately needed. As a result, the current study intends to examine the fish variety of Maguri Motapung Beel, close to Motapung Village in the Tinsukia region of Assam and Dibru-Saikhowa National Park.

MATERIALS AND METHODS

Study area: The current research was conducted over a span of 7 months , starting from october 2023 to april 2024, at Motapung- Maguri Beel, a wetland located 11 kilometers from the Tinsukia District Headquarter and on the southeast side of Dibru- Saikhowa National park. The Beel is located close and linked to Dibru Saikhowa National park and provides a source of livelihood to the local community (Kalita *et al.*, 2016). Thousands of migratory birds come to the wetland every year making it one of the most loved destinations of birdwatchers from throughout the world. It has grown in importance because of its diverse, aquatic life, plants species as well as various migratory birds. These wetlands have also been a crucial source of nourishment and sustenance for the local inhabitants (Bailung & Biswas, 2014).

Survey sites: Throught the study period, 2 sampling sites were selected for sample collection, based on the areas abundant ecological diversity (Kalita *et al.*, 2016). Information on the occurrence of various species was collected from local fisherman through questionnaire interviews. Fish caught for personal consumption was also documented. The nearby fish markets were also explored to gather additional information about the fish of Maguri Beel

(Kalita *et al.*, 2016). Fishes were photographed using mobile camera to aid in identification. Additionally, some collected fish were preserved in a 10% formaldehyde solution for identification purposes. The specimen was identified by consulting relevant literature (Talwar and Jhingran, 1991) and research papers (Kalita *et al.*, 2016). The conservation status was determined using the conservation Assessment and management plan (CAMP, 1998). Analysis of constraints was conducted through surveys and interactions with local residents in the area (Nath and Deka, 2012).

Questionnaires for the field study

1. About the study area.
2. How many fish landing sites are there?
3. Types of fishes found (Local Name).
4. Do they do overfishing?
5. Do they understand fish breeding season and do they fish during that period?
6. Impacts of Baghjan oil spillage incidence in 2020.
7. Any occurrence of draught in the beel during winter.
8. Are there any agricultural and tea crop fields in the fringe area?
9. Any other human activity that is not good for the beel.
10. Is there any conservational measure for the beels?

Fig 1a: Study area map (Maguri Motapung Beel) (See colour version on page 343)

Fig 1b: Site 1 (Motapung Beel Ghat- Maguri Beel) (See colour version on page 344)

Fig 1c: Site 2 (Horu Beel Ghat- Maguri Beel)(See colour version on page 344)

Data analysis: - After the process of data collection, the collected data was set for data analysis. Here the method of the Shannon diversity index will be used which is given by the formula –

Shanon diversity index (H) = $- \sum [(pi) \times \log (pi)]$

Where H = Shannon diversity index

- pi = proportion of individuals of i- th species in a whole community.
- $\sum$ = sum symbol, and

log = natural logarithm

pi = n/N

n = individuals of a given type/ species and

N = total no of individuals in community.

The observed data were analyzed for threats using Microsoft Excel based on the frequency of occurrence and the severity of the impact on study area (Arslan *et al.*, 2021).The observed Physico-chemical parameters during the study are analyzed for validation using statistical analysis to observed their correlation and how they influence the diversity of the study area.

RESULTS AND DISCUSSION

A total of 38 fish species have been recorded from the study site belonging to 27 genera, 16 families and 10 orders with IUCN status (Table 1 & Plate I). Among the recorded species from two different landing sites of Maguri Motapung Beel, it has been noted that Cyprinidae family was the most prevalent, which includes 10 species, followed by Bagridae and Channidae with 4 species each; Ambassidae with 3 species and so on. By using the shannon diversity index H was equal to 2.79 which indicates normal range of fish diversity (Table 2).

Table 1: Fishes of Maguri – Motapung Beel and their biodiversity

Sl. No.	Order	Family	Scientific Name	Local Name	IUCN Status
1	Beloniformes	Belonidae	*Xenentodon cancila*	Kokila	LRlc
2	Beloniformes	Belonidae	*Mastacembelus armatus*	Bami	LRlc
3	Cypriniformes	Cyprinidae	*Rasbora daniconius*	Darikana	LRlc
4	Cypriniformes	Cyprinidae	*Puntius conchonius*	Puthi	LRlc
5	Cypriniformes	Cyprinidae	*Puntius sophore*	Puthi	LRlc
6	Cypriniformes	Cyprinidae	*Puntius sarana*	Seni puthi	LRlc
7	Cypriniformes	Cyprinidae	*Labeo rohita*	Rou	LRlc
8	Cypriniformes	Cyprinidae	*Labeo gonius*	Kui	LRlc
9	Cypriniformes	Cyprinidae	*Cirrhinus reba*	Lachim	LRlc
10	Cypriniformes	Cyprinidae	*Labeo bata*	Bhangan	LRlc
11	Cypriniformes	Cyprinidae	*Ctenopharyngodon idella*	Grasscarp	NE
12	Cypriniformes	Cyprinidae	*Cyprinus carpio*	Commoncarp	NE
13	Cypriniformes	Cobitidae	*Lepidocephalicthys guntea*	Batia	LRlc
14	Cypriniformes	Cobitidae	*Canthophyrs gongota*	Kkur botia	LRlc
15	Siluriformes	Bagridae	*Mystus cavasius*	Singora	LRlc
16	Siluriformes	Bagridae	*Mystus vittatus*	Singora	LRlc
17	Siluriformes	Bagridae	*Mystus tengara*	Singora	LRlc
18	Siluriformes	Bagridae	*Mystus nigriceps*	Singora	LRlc

19	Siluriformes	Heteropneustidae	*Heteropneustes fosilis*	Singhi	LRlc
21	Siluriformes	Siluridae	*Wallago attu*	Borali	VU
22	Siluriformes	Clarridae	*Clarias magur*	Magur	EN
20	Characiformes	Heteropneustidae	*Charax gibbosus*	Glass headstander	LRlc
23	Anabantiformes	Nandidae	*Badis badis*	Randhoni	LRlc
24	Anabantiformes	Osphronemidae	*Trichogaster lailus*	Kholihona	LRlc
25	Anabantiformes	Osphronemidae	*Trichogaster fasciata*	Kholihona	LRlc
26	Anabantiformes	Channidae	*Channa gachua*	Chengeli	LRlc
27	Anabantiformes	Channidae	*Channa punctata*	Goroi	LRlc
28	Perciformes	Channidae	*Channa striata*	Sol	LRlc
29	Perciformes	Channidae	*Channa marulius*	Sal	LRlc
30	Perciformes	Anabantidae	*Anabas testudineus*	Kawoi	DD
31	Osteoglossiformes	Notopteridae	*Notopterus notopterus*	Kandhuli	LRlc
32	Osteoglossiformes	Notopteridae	*Chitala chitala*	Chital	LRnt
33	Tetradontiformes	Tetradontidae	*Tetradon cutcutia*	Gangatop	LRlc
34	Synbranchiformes	Mastacembelidae	*Macrognathus pancalus*	Tura	LRlc
35	Synbranchiformes	Symbranchidae	*Monopterus cuchia*	Kuchia	LRlc
36	Perciformes	Ambassidae	*Pseudambassis lala*	Aina puthi	LRlc
37	Perciformes	Ambassidae	*Chanda nama*	Chanda	LRlc
38	Perciformes	Ambassidae	*Pseudambassis beculis*	Chanda	LRlc

Plate-I: Some glimpse of edible & ornamental fish species from the Maguri – Motapung Beel

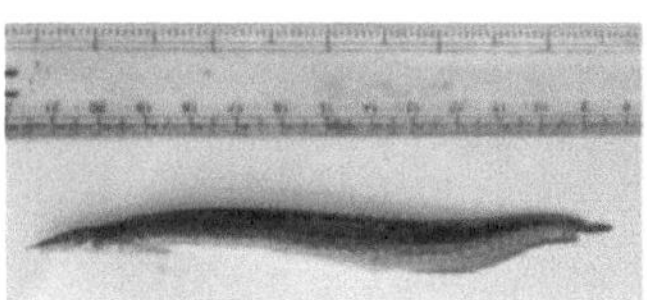

Macrognathus pancalus

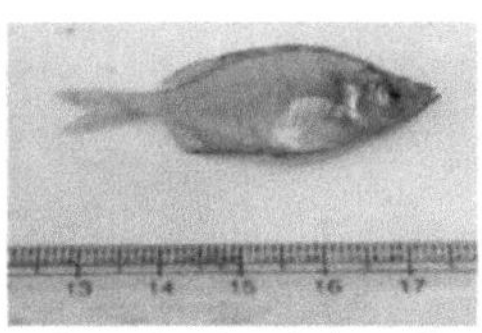

Pseudambassis beculis

Puntius sophore

Wallago attu

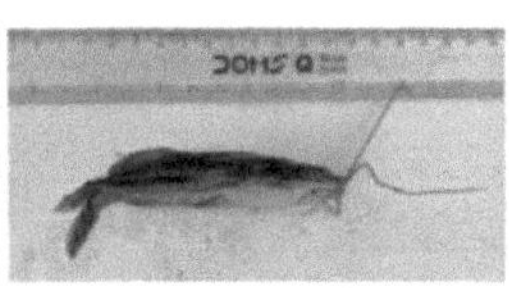

Mystus cavasius

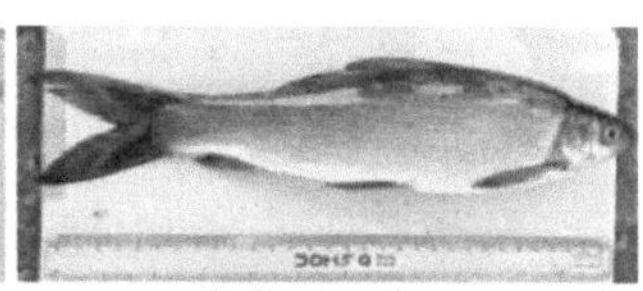

Labeo gonius

(See colour version on page 344)

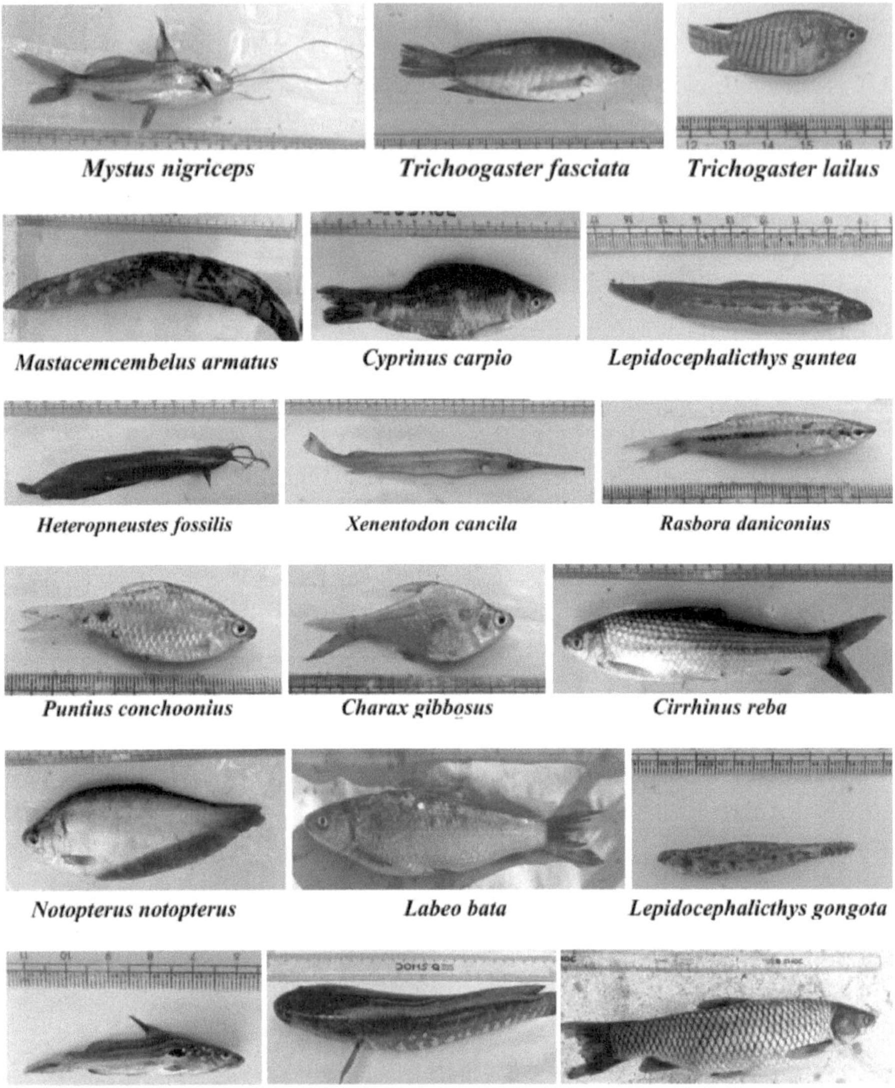

Mystus nigriceps | *Trichoogaster fasciata* | *Trichogaster lailus*
Mastacemcembelus armatus | *Cyprinus carpio* | *Lepidocephalicthys guntea*
Heteropneustes fossilis | *Xenentodon cancila* | *Rasbora daniconius*
Puntius conchoonius | *Charax gibbosus* | *Cirrhinus reba*
Notopterus notopterus | *Labeo bata* | *Lepidocephalicthys gongota*
Mystus vittatus | *Channa punctata* | *Ctenopharyngodon idella*

(See colour version on page 344)

Table 2: Sampling area shannon diversity index

Sl.No.	Scientific name of species	Number of individuals	p_i	$ln(p_i)$	$p_i * ln(p_i)$
1	*Xenentodon cancila*	2	0.0057	-5.16	-0.029
2	*Mastacembelus armatus*	2	0.0057	-5.16	-0.029
3	*Rasbora daniconius*	39	0.11	-2.2	-0.242
4	*Puntius conchonius*	89	0.25	-1.38	-0.345

5	*Puntius sophore*	2	0.0057	-5.16	-0.029
6	*Labeo rohita*	2	0.0057	-5.16	-0.029
7	*Labeo gonius*	5	0.014	-4.26	-0.059
8	*Cirrhinus reba*	5	0.014	-4.26	-0.059
9	*Labeo bata*	2	0.0057	-5.16	-0.029
10	*Ctenopharyngodon idella*	3	0.0086	-4.75	-0.040
11	*Cyprinus carpio*	4	0.011	-4.5	-0.049
12	*Lepidocephalicthys guntea*	13	0.037	-3.29	-0.121
13	*Mystus cavasius*	7	0.02	-3.91	-0.078
14	*Mystus vittatus*	19	0.055	-2.9	-0.159
15	*Mystus tengara*	15	0.043	-3.14	-0.135
16	*Mystus nigriceps*	5	0.014	-4.26	-0.059
17	*Heteropneustes fosilis*	7	0.02	-3.91	-0.078
18	*Wallago attu*	8	0.023	-3.77	-0.086
19	*Clarias magur*	5	0.014	-4.26	-0.059
20	*Badis badis*	5	0.014	-4.26	-0.059
21	*Trichogaster lailus*	9	0.026	-3.64	-0.094
22	*Trichogaster fasciata*	11	0.031	-3.47	-0.107
23	*Channa gachua*	22	0.063	-2.76	-0.173
24	*Channa punctata*	32	0.092	-2.38	-0.218
25	*Notopterus notopterus*	5	0.014	-4.26	-0.059
26	*Tetradon cutcutia*	3	0.0086	-4.75	-0.040
27	*Micrognathus pancalus*	5	0.014	-4.26	-0.059
28	*Monopterus cuchia*	1	0.002	-6.21	-0.012
29	*Pseudambassis lala*	4	0.011	-4.5	-0.049
30	*Chanda nama*	3	0.0086	-4.75	-0.040
31	*Pseudambassis beculis*	2	0.0057	-5.16	-0.029
32	*Channa striata*	1	0.0028	-5.87	-0.016
33	*Charax gibbosus*	1	0.0028	-5.87	-0.016
34	*Chitala chitala*	1	0.0028	-5.87	-0.016
35	*Channa marulius*	1	0.0028	-5.87	-0.016
36	*Anabas testudineus*	1	0.0028	-5.87	-0.016
37	*Systomus sarana*	2	0.0057	-5.16	-0.029
38	*Canthophyrs gongota*	2	0.0057	-5.16	-0.029
	Total	345			H=2.79

The order Cypriniformes represented 12 species, Siluriformes represented 7 species, Perciformes represented 6 species, Anabantiformes 5 species, Beloniformes, Osteoglossiformes, and Synbranchiformes represented 2 species each, and Characiformes and Tetradontiformes are recorded with

1 species each (Fig. 2). The IUCN status of the recorded fishes have been classified into 6 category, LRnt= Lower risk near threatened, LRlc= Lower risk least concern, VU= Vulnerable, DD= Data deficient, NE= Not evaluated, EN= Endangered; 32 species is lower risk least concern, 1 species is vulnerable, 1 species is data deficient, 1 species is endangered, 1 species is lower risk near threatened and other 2 species are not evaluated (Fig. 3). The family Cyprinidae represented 10 species; Bagridae and Channidae represented 4 species each, Ambassidae represented 3 species, Belonidae, Cobitidae, Heteropneustidae, Osphronemidae, Notopteridae represented 2 species each; Symbranchidae, Mastacembelidae, Tetradontidae, Anabantidae, Siluridae, Clarridae and Nandidae represented 1 species each Fig. 4).

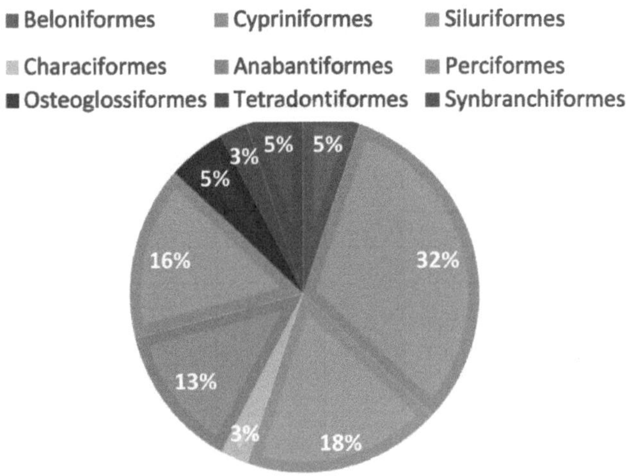

Fig. 2: Order wise distribution of fish species

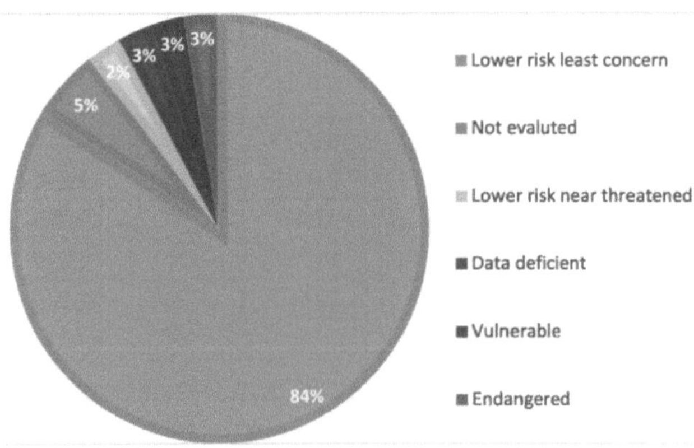

Fig. 3: Percentage distribution of conservation status of recorded fish species

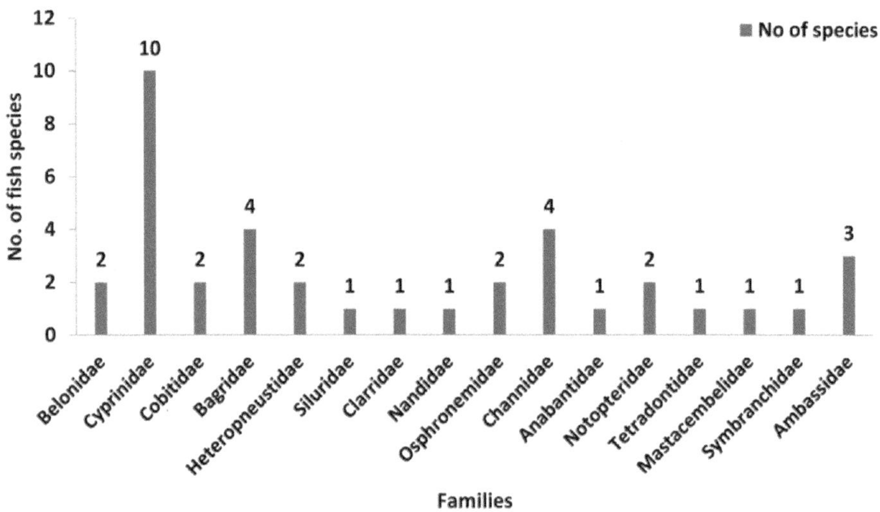

Fig. 4: Family wise abundance of fish in Maguri Motapung Beels

The data on threats to the aquatic study area is tabulated with respect to their positive and neagtive responseby respondants (Table 3). Both the local people and fishermen some participated in the survey and only anthropogenic threats has been reported. Such as overfishing, fishing during breeding season, nearby agricultural and tea crop fields, lack of conservational measure and impact of baghjan oil spillage incident. A total of 162 respondant participated and most of the respondents have reported that people are overfishing more now a days. Majorly due to agricultural and tea crop fields the beel is more affecting. Due to lack of knowledge about the breeding season the fish population decreases. According to respondents there is no any conservation measure from the government. And all the respondents are reported as that there has never been a drought till date.

The analysis shows Higher positive response to threats in overfishing, Agricultural contamination and Fish catching during breeding seasons and low positive response to conservational measure and oil spillage from fringe oil wells (Fig. 5). This reflects higher threats influence to the study area which all come from anthrpogenic activites. There is no any natural threat observed for the study area as the ecosystem has not reported any draught so far.

Table 3: Observed threats at the study area

SL. No.	Threats	Positive Response	Negative Response	Total
1.	Do they do overfishing	18	15	33
2.	Are there any agricultural and tea crop fields in the fringe area	31	0	31
3.	Do they do fishing during breeding period	27	12	39
4.	Is there any conservational measure for the beel	5	15	20
5.	Is there any impact of baghjan oil spillage 2020 in the beel	5	19	24
6.	Is there any occurrence of drought in the beel during winter	0	15	15
	Total	86	76	162

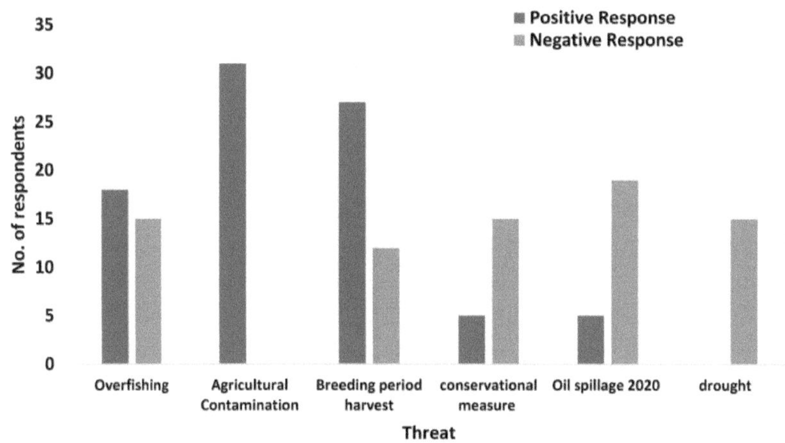

Fig. 5: Observed Threats analysis at the study area

During the study period the physicochemical parameters are checked such as pH, air temperature, water temperature and dissolved oxygen. Correlation between parameters and collected fish and within the parameters during sampling was observed (Table 5). The result observed moderatepositive correlation between the water and air temperature and moderate negative correlation between the water temperature and dissolved oxygen. All the other parameters shows weak correlation between the parameters and collected specimen.

Table 5: Correlation between observed physicochemical parameters and specimen frequency during sampling

	Fish specimen collected	**pH**	**Air temperature(°C)**	**Water temperature(°C)**	**Dissolved oxygen**
Fish specimen collected					
pH	-0.205503632				
Air temperature (°C)	0.119561214	0.381097961			
Water temperature (°C)	0.124373781	0.291527877	0.598986339		
Dissolved oxygen mg/l	-0.024058538	0.164918563	-0.717119231	-0.579410032	

Fish play a crucial role in various aspects of our lives and the environment and also a critical part of aquatic ecosystems. They help maintain balance by controlling populations of other species. The study recorded 345 individuals of 38 fish species, affiliation to 17 families, 10 orders were identified. Family Cyprinidae was the most predominant followed by the other 15 families. By using the shannon diversity index H was equal to 2.79 revealing normal range of icthyofaunal diversity in the Maguri Motapung Beel. However, the several fish species are in danger due to habitate destruction because of different threats particularly anthropogenic in nature as observed in the study area (Kalita *et al.*, 2016). Some of the important anthropogenic activities were discussed below-

i) **Agricultural activities:** Near the wetland, making agricultural practices the most harmful activity for both fish populations and the entire ecosystem community. The use of artificial fertilizers, insecticides, and pesticides in agriculture has resulted in water pollution, eutrophication, increased water turbidity, and the formation of algal blooms impacting numerous species.

ii) **Over fishing during breeding season:** Fishing during breeding season can lead to a reduction in the number of individuals reproducing, which can result in decreased recruitment of young fish into the population.

iii) **Over exploitation:** Due to over exploitation the most immediate impact is a decline in the no of fish. Continuous over fishing often results in a shift in the age and size structure of the fish population. Fishermen tend to target bigger, elder fish, which can disrupt the natural balance of the population.

iv) **Accidental oil spillage:** The oil spillage incidence in 2020 may have influenced significant change leading to its composition in icthayofaunal diversity. Because oil can coat fish and their habitats, leading to suffocation, reduced oxygen levels, and disruption of feeding and reproduction. Also it can contaminate the food chain, affecting fish health and reproductive success. Long term effects depend on the size of the spill, the type of oil, and the effectiveness of cleanup efforts (Kalita *et al*, 2016).

During the study period a total of 162 respondant participated including both local people and fishermen. Most of the respondents have reported that people are overfishing now a days and also saying that there has never been a drought till date. According to local people and fishermen there is no any protection system from the government in the Maguri – Motapung beel so far. Majorly due to agricultural and tea crop fields the beel is more affecting. Due to lack of knowledge about the breeding season the fish population decreases.

During the study period the physicochemical parameters are checked such as pH, air temperature, water temperature and dissolved oxygen. Correlation between parameters and collected fish and within the parameters during sampling was observed. The correlation determine between the collected fish specimen and ph, the value observed was -0.205503632, it means the relationship was weak. The correlation determine between the collected fish specimen and air temperature, the value observed was 0.119561214, it means the relationship was weak, The correlation determine between the collected fish specimen and water temperature, the value observed was 0.1243737819561, it means the relationship was weak, The correlation determine between the collected fish specimen and dissolved oxygen the value observed was -0.024058538 it means the relationship was weak (Suganthi *et al.* 2018). Also correlate within the parameter. From Table 5, the result observed was that the relation between the water and air temperature and between the water temperature and dissolved oxygen was moderate. Rest all the relation within the parameters was weak.

Conservation status of the fish species

The IUCN status of the recorded fish species have been classified into 6 category, Lower risk least concern 32 species, Lower risk near threatened 1 species, vulnerable 1 species, Data deficient 1 species, Endangered 1 species and Not evaluated 2 species. Conservation of fish species is crucial for maintaining biodiversity and healthy ecosystems. Measures include controlling pollution, awareness program among the local people, restriction of fishing during breeding season, implement of certain laws in the oil rigs and pesticides controlmethods, promoting sustainable fishing practices and need more research orientation works in that area (Nath and Deka, 2012).

CONCLUSIONS

The present study reveals that the Maguri Motapung Beel, is elevated in fish diversity by the presence of various species from different families and orders. The record of 38 species affiliation to 27 genera, 16 families and 10 orders. Present study noted that Cyprinidae family was the most predominant, which includes 10 species. Accounting the percentage distributin of recorded fishes are Cypriniformes represented 32%, Siluriformes represented 18%, Perciformes represented 16%, Anabantiformes represented 13%, Beloniformes, Osteoglossiformes and Synbranchiformes represented 5% each, and Characiformes and Tetradontiformes are recorded 3% both orders. The study also report fish diversity in the study area in danger due to habitate destruction because of different threats particularly anthropogenic in nature as observed. The analysis shows high frequency occurrence of overfishing, agricultural contamination and fish catching during breeding seasons and lack of conservational measure as major threat to the area.

ACKNOWLEDGEMENTS

We acknowledge the Department of Zoology, Arunachal University of Studies, Namsai, Arunachal University of Studies for the immense support in carrying out the study. We are also thankful to the local people and the local fishermen for their co-operation and on field assistance during this research work.

DECLARATION

The authors declare no conflict of interest as this is an original research work and has not been published anywhere else.

REFERENCES

Arslan, D., Cicek, K., Donduren, O. and Ernoul, L. 2021. Threat Ranking to Improve Conservation Planning: An Example from the Gediz Delta, Turkey, Land, 10 (1382), Retrieved from: https://doi.org/10.3390/land10121381

Bailung, B. and Biswas, S. P. 2014. Checklist, status and role of fish diversity as a measure of alternative livelihood in dihing river basin, Assam. International journal of applied biology and pharmacutical technology, 5(3): 1-4.

Conservation Assessment and Management Plan (CAMP) Workshop Reports .1998. Reshwater Fishes of India, Lucknow, 156 pp.

Goswami, P. and Singha, S. 2023. Fish diversity of the Beki river, Assam, India: present status and conservation needs . Scientific journal of the European ecocycles society , 9 (12): 1-6.

Jayaram, K. C. 1999. The freshwater fishes of the Indian region. Narendra Publishing House, Delhi, India, 551pp.

Kalita, P., Pathak, S. and Deka, P. 2016. A preliminary study on ichthyofaunal resources of motapung- maguri beel of Tinsukia dist of Assam, India. International Journal of Fauna and Biological Studies, 3(4): 97-102.

Kar, D., Nagarathna, A. V., Ramachandra, T. V. and Dey, S. C. 2006. Fish diversity and conservation aspects in an aquatic ecosystems in northeastern India. Zoos print journal, 2(7): 2308-2315.

Mahanta, P. C., Tyagi, L. K., Kapoor, D. and Ponniah, A. G.2003.Integration of fish Biodiversity conservation and development of fisheries in North Eastern Region: Issues and Approach, In: Participatory Approach for fish Biodiversity conservation in North East India. NBFGR, Lucknow, India.

Malakar, M. and Boruah, S.2017. Diversity and present status of fish species in three floodplain wetlands of central Assam, India. IOSR Journal of Environmental Science, Toxicology and Food Technology ,11(1): 54-59.

Mittermeier, R. A., Robles-Gil, P. and Mittermeier, C. G. 1997.Megadiversity: Earth's Biologically Wealthiest Nations. Cemex, Mexico.

Moyle, P. B. and Moyle, P. R.1995. Endangered fishes and economics: intergenerational obligations. Environmental Biology of Fishes, 43(1): 29-37.

Nath, B. and Deka, C. 2012. A study on fish diversity, conservation status and anthropogenic stress of chandubi tectonic lake, Assam, India. Journal of Bio Innovation, 1(6): 148-155. Retrieve from www.jbino.com

Talwar, P. K. and Jhingram, A. G. 1991. Inland fishes of India and adjacent countries. Oxford and IBH publishing co. Pvt. Ltd., New delhi, 2, 1158.

12

A Study on Ichthyofaunal Diversity and Influence of Physico-chemical Parameters on Fish Species of the Maguri-Motapung Beel, Tinsukia District, Assam

Januka Sharma and Tribeni Hazarika

ABSTRACT

This study was conducted in the Maguri Motapung Wetland, located in the Tinsukia district of Assam, India, to explore fish diversity and analyze specific physiochemical parameters of the wetland. A total of 36 fish species belonging to 25 genera under 17 families from 8 orders has been recorded from the present study area. Most fish species collected from the wetland fall under the order Cypriniformes. The different families recorded are Belonidae, Belontidae, Cyprinidae, Cobitidae, Ambassidae, Anabantidae, Channidae, Osphronemidae, Bagridae, Clariidae, Heteropneustidae, Schilbeidae, Siluridae, Mastacembelidae, Notopteridae, Nandidae and Tetradontidae. The taxonomic study shows that Cyprinidae is the most dominant family with 11 numbers of species contributing about 37% followed by Bagridae 15% which is followed by Mastacembelidae comprising 9%, Channidae comprising 7%. As per IUCN status, 29 species are categorized under Least Concern, 5 species under Not evaluated, 1 species each from Near Threatened and Endangered. The study reveals that the Maguri-Motapung beel is high in fish diversity. Ranges of the physic-chemical parameter of water within the permissible limit of a healthy ecosystem. The study found varying degrees of correlation between different environmental factors and fish species. Water temperature and fish species showed a strong, positive and significant correlation (r=0.95, p=0.185). Water depth had a moderate, positive and significant correlation (r=0.49, p=0.67) with fish species, pH levels exhibited a weak, positive and significant correlation (r=0.18, p=0.87), total dissolved solid had a negative, strong and significant correlation with the fish species (r= -0.98, p= 0.09) and water transparency showed a strong, positive and significant correlation (r=0.77, p=0.43) with the fish species.

Keywords: *Maguri beel; Physicochemical parameters, Fish species, Ecosystem*

INTRODUCTION

The aquatic ecosystem is highly dependent on water quality and biodiversity. The water quality parameters play an important role in the biology and physiology of fish (Dhawan and Kaur, 2002). Fish communities are an important aspect of the aquatic ecosystem. Moreover, they also serve as an indicator of the quality of the aquatic ecosystem (Sharma *et al.*, 2014). Ramsar Convention on wetlands in 1971, characterized wetlands as "Areas of Marsh, fen, peatland, whether natural or artificial, permanent or temporary with water that is static or flowing, brackish or salt, including areas of marine water the depth of which at low tide should not exceed". At the surface, freshwater serves as a habitat for numerous species and comprises a substantial portion of Earth's biological diversity. Wetlands cover approximately 6.4% of the Earth's total surface area. Being a large biodiverse region, India accounts for 11.72% of the world's fish diversity and 40% of fish live in freshwater ecosystems (Kar *et al.*, 2006, Balasubramanium, 2017). Fish are key species that determine the distribution and abundance of other organisms and have a significant impact on the ecosystem and are considered good indicators of water quality in the ecosystem (Moyle and Leidy, 1992). Assam, located in the northeastern region, possesses extensive and diverse freshwater resources, earning it recognition as one of the richest biodiversity sites in India. There are 2,500 species of fish in India; 930 of them live in fresh water and 1570 in the sea (Kar, 2003). The entire northeastern region, including Assam, has large and diverse freshwater resources and is considered one of the richest biodiversity hotspots of India. The NE region has unique topographic conditions and is blessed with large and varied water resources in the form of rivers (19,150 km); swimming pools (23,792ha); lakes and marshes (143,740 ha); ponds and mini-dams (40,808 ha) and shallow spawning systems paddy with fish farming systems (2780 ha) (Mahanta *et al.*, 2003). This region of the country is very diverse in terms of fish. So far, 267 freshwater fish species belonging to 114 genera 38 families and 10 orders have been reported from the region (Mahanta *et al.*, 2003). A total of 266 species belonging to 114 genera under 38 genera and 10 orders have been reported from the Northeast including Assam (Sen, 2010). A survey of the fish fauna of the Brahmaputra in Assam revealed 41 commercially important species (Talwar and Jhingran, 1991). Assam has 1,392 beels in over 100,000 hectares, which constitutes 61 percent of the state's lentic water supply. Water is one of the most important compounds in the ecosystem. Better water quality is characterized by its physical, chemical and biological characteristics. Polluted water is the biggest source of disease. With industrialization, urbanization, and technological development, water sources in all sectors are becoming more and more seriously polluted. Several physiological or biological factors can act as stressors and inhibit fish growth and reproduction which ensures that a regular monitoring of physiological parameters is necessary to determine the condition of fish in the aquatic ecosystem. Changes in physico-chemical parameters can lead to changes in living organisms, especially their abundance, diversity and distribution in an ecosystem (Mishra *et al.*, 2018). Concerning this, the present study was carried out to evaluate the monthly variations in water physico-chemical parameters and fish diversity in the Maguri-Motapung beel of Tinsukia District, Assam.

MATERIAL AND METHODS

***Description of study sites*:** The Maguri-motapung beel, covering an area of 6.56 sq.km. and located 27.3432^{0}N and 95.2343^{0}E (Fig. 1) is one of the major wetlands of the upper Brahmaputra basin in the Tinsukia district of Assam. It is a part of the Dibru Saikhowa, an important birding area lying just at the edge of Dibru Saikhowa National Park and Biosphere Reserve (DSNPBR). The DSNPBR with a record of about 385 bird species also harbors a population of 108 fish species as major aquatic bioresources. It is located 3.8 km away from the Guijan Ghat, the gate way of Dibru Saikhowa National Park. Thousands of migratory birds come to the wetland every year making it one of the most loved destinations for bird watchers worldwide. The six villages, Baghjan Gaon, Purani Motapung, Rangagara Te Natungaon, Gotong Gaon, Na motapung Gaon, and Dhelakhatgaon, are located on the east, west, and north sides of the wetland, with the Dibru-Saikhowa National Park to the south.

Fig 1: Map of the Study Area(See colour version on page 346)

Fish Sampling and Identification: To find out the fish diversity of the Maguri beel area two sampling sites were selected (site 1- Mohar nola ghat, and site 2- Bebejia ghat) on the basis of rich ecological diversity (Fig. 1). During the field survey, fish samples were collected once a month from local fishermen through interviews conducted with the aid of questionnaires (Suleiman *et al.,* 2018).

For identification, fishes were collected in live condition and photographs were taken by digital camera placing them in a clean paper and keeping a scale along its length. Some collected fish were brought to the lab and preserved in a 5% formaldehyde solution for identification. Identification of fish species was done by following various literature of Datta and Srivastava, (1988); Talwar and Jhingran, (1991); Jayaram (1999) and Biswas and Das (2008).

***Sample collection of Water Quality Parameters*:** Surface water samples were collected and analyzed in the field and Laboratory from two sites (Site 1 and Site 2) every month from September 2023 to April 2024. During the study period, a total of five parameters were analysed including pH, temperature, water depth, total dissolved solids (TDS), and water transparency (WTS) by following the standard procedures of APHA (1995).

Data Analysis: The Shannon diversity index formula was used to evaluate the fish diversity (Shannon & Weaver, 1949). The formula of the Shannon Diversity Index is (H)= - $\sum_{i=1}^{s} pi \ \ln pi$.

The environmental parameter data were analyzed using annual mean seasonal bases i.e., post-monsoon (September, October, and November), winter (December, January and February) and pre-monsoon (March and April). Correlations of fish diversity and physiochemical parameters were analyzed by using SPSS (Kamboj *et. al.*, 2022).

RESULT AND DISCUSSION

Fish Diversity: The study recorded a total of 432 individual fish belonging to 36 species, 25 genera, 18 families, and 8 orders from the two sampling sites (Table 1 & Fig. 2). The fish diversity at Maguri Motapung Wetland is represented by H = 3.27 indicating high species diversity. As the fish species were collected from local fishermen, site-specific fish distribution could not be determined. It has been observed that among the 17 families, the Cyprinidae family was the most dominant, which includes 11 species, accounting for a percentage of 37% followed by Bagridae family with 3 species holding 15% share, Mastacembelidae comprising 9%, Channidae comprising 7%, Notopteridae 6% and Claridae comprising 4%. Whereas families Heteropneustedae, Ambassidae, Cobtidae comprise each with 3%. Families Belontiidae, Osphronemidae, Siluridae, and Ananbantidae comprising each with 2% and Nandidae, Schilbeidae and Tetradontidae were observed to be least dominating with 1% each (Fig. 3). The recorded fishes were *Channa striata, Channa marulius, Polyacanthus labiosus, Trichogasterlalius, Xenentodon cancila, Puntius sophore, Esomusdanricus, Labeo calbasu, Cirrhinus reba, Telestesmuticellus, Labeo bata, Labeo gonius, Puntius javanicus, Lepidocephalichthys guntea, Labeo rohita, Rasbora daniconius, Ctenopharyngodon idells, Notopteus notopterus, Channa punctatus, Trichogaster fasciata, Badis badis, Pseudambassis lala, Anabas testudineus, Channa gachua, Chanda nama, Heteropneustes fossilis, Wallagu attu, Mystus cavasius, Pseudeutropius atherinoides, Mystus carcio, Mystus tengara, Clarias magur, Macrognathus aculeatus, Mastacembelus armats, Macrognathus pancalus* and *Tetradon cutcutia.* The dominant fish species of Maguri- Motapung Beel were *Punctiousjavanicus, Mystus tengara, Notopterus notopterus* and *Rasbora daniconius.*

Table 1: List of fish species from Maguri-motapung beel

SL. NO.	Order	Family	Species	Local/Common Name	IUCN Status
1	Anabantiformes	Channidae	*Channa striata*	Sal	Least concerned
2		Channidae	*Channa marulius*	Sol	Least concerned
3		Belontidae	*Polyacanthus labiosus*	Gourami	Least concerned
4		Osphronemidae	*Trichogaster lalius*	Kholihona	Least concerned
5	Beloniformes	Belonidae	*Xenentodon cancila*	Kokila	Least concerned
6	Cypriniformes	Cyprinidae	*Puntius sophore*	Puthi	Least concerned
7		Cyprinidae	*Esomus danricus*	Darikona	Least concerned
8		Cyprinidae	*Labeo calbasu*	Kaliya jura	Least concerned
9		Cyprinidae	*Cirrhinus reba*	Lasim	Least concerned
10		Cyprinidae	*Telestes muticellus*	Vairone	Least concerned
11		Cyprinidae	*Labeo bata*	Bhangone	Least concerned
12		Cyprinidae	*Labeo gonius*	Kuri	Least concerned
13		Cyprinidae	*Puntius javanicus*	Java puthi	Not evaluated
14		Cobitidae	*Lepidocephalichthys guntea*	Botia	Least concerned
15		Cyprinidae	*Labeo rohita*	Rou	Least concerned
16		Cyprinidae	*Rasbora daniconius*	Darikona	Least concerned
17		Cyprinidae	*Ctenopharyngodon idells*	Grass carp	Not evaluated
18	Osteoglossiformes	Notopteridae	*Notopteus notopterus*	Kandhuli	Least concerned

19	Perciformes	Channidae	*Channa punctatus*	Goroi	Least concerned
20		Belontiidae	*Trichogaster fasciata*	Banded gourami	Not evaluated
21		Nandidae	*Badis badis*	Randhoni	Least concerned
22		Ambassidae	*Pseudambassis lala*	Lal chanda	Not evaluated
23		Anabantidae	*Anabas testudineus*	Kawoi	Least concerned
24		Channidae	*Channa gachua*	Cheng	Least concerned
25		Ambassidae	*Chanda nama*	Chanda	Least concerned
26	Siluriformes	Heteropneustedae	*Heteropneustes fossilis*	Singhi	Least concerned
27		Siluridae	*Wallagu attu*	Borali	Near threatened
28		Bagridae	*Mystus cavasius*	Hingora	Least concerned
29		Schilbeidae	*Pseudeutropius atherinoides*	Bordowa	Least concerned
30		Bagridae	*Mystus carcio*	Singora	Least concerned
31		Bagridae	*Mystus tengara*	Hingora	Least concerned
32		Clariidae	*Clarias magur*	Magur	Endangered
33	Synbranchiformes	Mastacembelidae	*Macrognathus aculeatus*	Tura	Not evaluated
34		Mastacembelidae	*Mastacembelus armats*	Zig zag eel	Least concerned
35		Mastacembelidae	*Macrognathus panculus*	Tuwa	Least concerned
36	Tetradontiformes	Tetradontidae	*Tetradon cutcutia*	Gangatope	Least concerned

Fig. 2: Some glimpse of edible & ornamental fish species from the Maguri – Motapung Beel (See colour version on page 346)

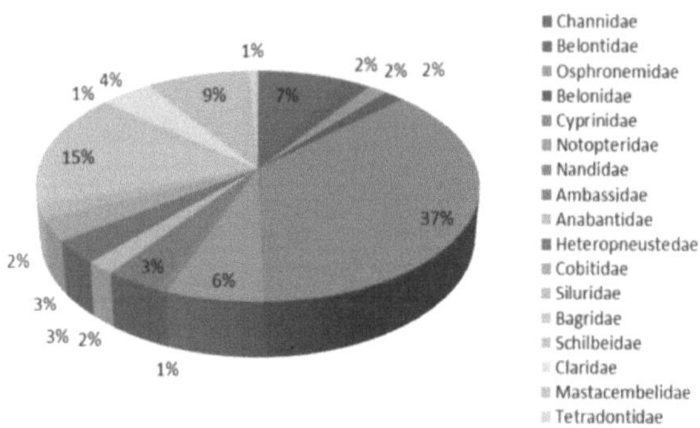

Fig. 3: Family wise distribution of fish species

The rich ecosystem of Maguri-Motapung Beel also supports various fish habitats for ornamental fishes. According to Kalita *et al.* (2016), this wetland produces more than 200 metric tons of fish per year and thus plays an important role in the livelihood of local people. The construction of a certain oil pipeline by the oil company beneath the Maguri-motapung beel from Bebejia to Baghjanhas shown a major environmental safety concern (Kalita *et al.*, 2016). Any leaks or spillage in the tubing under the beelcan have far-reaching effects on this beel.

The fish diversity of the beels of lower Assam like Chanddubi (57 species), Dora (62), Deepor (41) and Tamranga beel (63) was reported by previous workers (Dey, 1981; Lahon, 1983; Agarwala, 1996) which more or less corroborate with the findings of the present study. The fish diversity of the Maguri Motapung wetland reported by previous workers was 48 species (Kalita *et al.*, 2016) which is more than the present study findings.

During the study period, it was also revealed that the distribution of fishes was not found to be equal. The highest number of fish was recorded during the winter seasoni.e., 219 individual fish followed by 126 individuals during pre-monsoon and the lowest number of fish was recorded during post-monsoon i.e., 89 individuals (Fig. 4). According to the IUCN status, 29 species were classified as Least Concern, 5 species were Not evaluated,1 species was Near Threatened and 1 species as Endangered. The highest number of fish species was recorded during the post-monsoon season and pre-monsoon season with 35 species and lowest was recorded during the winter season with 30 species (Fig. 5)

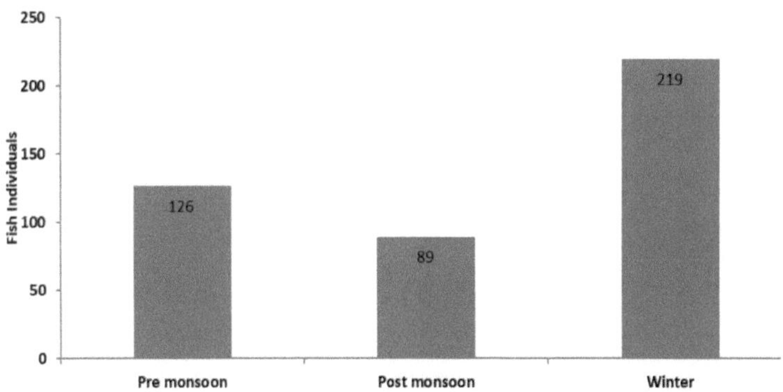

Fig. 4: Seasonal Distribution of fish in Maguri Motapung Beel

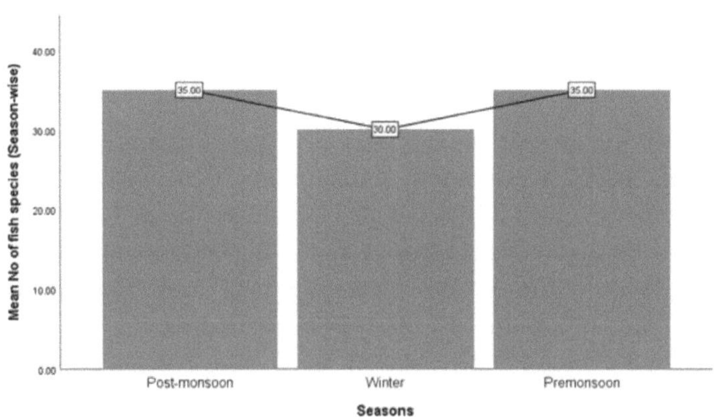

Fig. 5: Histogram showing season-wise distribution of fish species

Study of Water Parameters

Water Temperature: The surface water temperature in Maguri-Motapung Beel ranges from 21.9^0±1.59 to 30.4±1.30^{0}C. The average value of temperature by combining both the sites (site 1 and site 2) was recorded as maximum during post-monsoon i.e. (30.4±1.30^{0}C) and the minimum temperature was (21.9±1.59^{0}C) during the winter season. The water temperature values were found in similar trends and slightly high during post-monoon season. It is important for fish metabolism, reproduction, and immune function. Dhakal (2015) has made a similar observation in the Reu River, Nepal.

Water Depth: Water depth is important for oxygen distribution and temperature regulation. The depth of the Maguri motapung Beel was comparatively the same in both sites. The average depth was found 60±1.63 cm by combining both sites during winter and 58±1.46 cm in pre-monsoon but increased up to 235±4.08 cm during post-monsoon.

pH: The pH of Maguri motapung Beel was almost neutral to slightly alkaline. The average pH from both sites was recorded a maximum of 8.0±0.08 during September/October and a minimum 7.7±0.1 during the winter season. pH is considered an indicator of overall productivity that causes habitat diversity (Raj & Jayasekhar, 2007). The pH of most natural water is varied between 6.0 and 8.5. In the present study, the seasonal mean pH was close to neutral. According to Goldman & Horne (1983), low pH < 5.0 can severely reduce aquatic species diversity. However, the range of water pH was reasonably good for the fish population.

Total Dissolved Solids: The average value of total dissolved solids (TDS) of Maguri Motapung beel was found 47 mg/l in pre-monsoon, 51 mg/l in post-monsoon and 71 mg/l in the winter season. The highest TDS was measured as 71±1.63 mg/l during December and January and lowest was 47±1 mg/l during March and April. Total dissolved solids (TDS) refer to the sum of all components dissolved in water. Water containing too much dissolved matter is not suitable.

Water Transparency: The transparency of water is an important factor that reveals the level of sunlight penetration into the water. The water transparency in Maguri motapung Beel ranges from 35±3.55 to 63±0.81cm. The minimum transparency of both siteswas recorded during December/January and the maximum during September and October.

Correlation of Water Quality Parameters and Fish Species at 0.05% level of significance: There is a strong, positive and significant relationship between water temperature and fish species and no significance between them i. e., r=0.95, p=0.185. There is a moderate, positive, and significant relationship between water depth and fish species i.e., (r=0.49, p=0.67). pH levels exhibited a weak positive and significant correlation (r=0.18, p= 0.87) with the fish species, total dissolved solid showed a negative, strong and significant correlation with the fish species (r= -0.98, p= 0.09) whereas water transparency and fish species showed a strong positive and significant correlation between them (r=0.77, p= 0.43) (shown in Table 2).

Table 2: Showing season-wise correlation between the water quality parameters and number of fish species

	Water temperature	**Water Depth**	**pH**	**Total Dissolved Solids**	**Water Transparency**	**No of fish species (Season-wise)**
Water temperature	1					
Water Depth	0.720	1				
pH	0.462	0.948	1			
Total Dissolved Solids	-0.902	-0.350	-0.034	1		
Water Transparency	0.922	0.932	0.769	-0.665	1	
No of fish species (Season-wise)	0.958	0.491	0.189	-0.988	0.773	1

Conclusion and Recommendations

Maguri Motapung wetland is one of the largest floodplain wetlands in Tinsukia district of Assam covering an area of about 1000 ha. Conservation involves the study of biodiversity to safeguard both species and their habitats. Due to the overexploitation of natural resources, and inappropriate management practices to restore them,species are becoming extinct as per the studies. When a species becomes extinct, this combination of genome is lost forever, humans cannot restore it for future use. Therefore, the conservation of fish species is very important. The present work evaluated some physicochemical parameters like Temperature, pH, water depth, TDS and transparency of the Maguri Motapung wetland whose ranges are within the permissible limit of healthy ecosystem holding a high diversity of fish species. The incidence of oil spills in the Maguri Motapung beel in 2020 can potentially led to a substantial alteration in its ichthyofauna diversity. Additionally, factors such as rising population, intensified agricultural fertilizer usage, and excessive fishing may induce fluctuations in water quality parameters and the depletion of aquatic life (Manjare *et al.*, 2010). For successful management and conservation of existing fish species farmers should be made aware of the losses in fish diversity to the population due to excessive use of pesticides. Awareness programs on fish growth, population and conservation be propagated through media. The government should effectively implement legislation, policies and strategies towards fish growth, population and conservation (Bhagwan, 2010).

ACKNOWLEDGEMENT

We acknowledge the Department of Zoology, Arunachal University of Studies, Namsai, Arunachal University of Studies for the immense support in carrying out the study. We are also thankful to the local people and the local fishermen for their co-operation and on field assistance during this research work.

DECLARATION

The authors declare no conflict of interest as this is a original research work and has not been published anywhere else.

REFERENCES

Agarwala, N. K. 1996. Limnology and fish productivity of Tamranga wetland in productivity indicators. Ph.D. Thesis, Gauhati University.

APHA.1985. Standard Methods for Examination of water and waste water,20th Edition, American public health Association, Washington D.C

Balasubramanium,A.2017.Biological Profile of India, Technical Report. DOI:10.13140/RG.2.2.10664.57601

Bhagwan, 2010. Studies in ichthyofaunal diversity in Marathwada region, Dr. Babasaheb Ambedkar Marathwada University.

Biswas, S. P. and Das, J. N. (2008). A Handbook of Ornamental Fishes of the Brahmaputra basin, EBH Publishers (India), ISBN: 978-81-903834-8-6.

Datta, M. J. S. and Srivastava, M. P. 1988. Natural History of fishes and systematics of freshwater fishes in India, Narendra Publishing Co. Delhi, India.

Dey, S. C. 1981. Studies on the hydrobiological conditions of some commercial lakes (Beels) of Kamrup District of Assam, their bearing on fish production. Final Technical Report, North Eastern Council, pp: 177.

Dhakal, D. P. 2015. Species diversity and distribution of fish community of Reu River, Kirtipur, Kathmandu.

Dhawan, A. and Kaur, S. 2003. Effect of pig dung on water quality and polyculture of carp species during winter and summer. Aquaculture International, 10: 297-307.

Goldman C. R and Horne A. J. 1983. Diversity and its evenness, a unifying notation and its consequences. Ecology, 54: 427 432

Jayaram, K. C. 1999. The Freshwater Fishes of the Indian Region, Narendra Publishing House, Delhi, pp: 551.

Kar DA, Kumar, C Bohra, LK Sigh, (Eds). Fishes of Barak drainage, mizoram and Tripura; In: Environment, pollution and management, APH publishing corporation, New Delhi, 2003:604: 203-211.

Kar, D., Nagarathna, A. V., Ramachandra, T. V. and Dey, S. C. 2006. Fish diversity and conservation aspects in an aquatic ecosystems in northeastern India. Zoos Print Journal, 2(7): 2308-2315.

Kalita, P., Pathak, S. and Deka, P. 2016. A preliminary study on ichthyofaunal resources of Motapungmaguribeel of Tinsukia district of Assam, India, International Journal of Fauna and Biological Studies, 3(4): 97-102

Kamboj, V., Kamboj, N. Sharma A. and Bisht, A.2022.Fish diversity associated with environmental parameters in impacted area of Ganga river, India.

Lahon, B. 1983. Limnology and fisheries of some commercial beels of Assam, India. Ph.D. Thesis, Gauhati University, Assam. 349pp.

Mahanta, P. C., Tyagi, L. K., Kapoor, D. and Ponniah, A. G. 2003. Integration of fish Biodiversity conservation and development of fisheries in North Eastern Region: Issues and Approach, In: Participatory Approach for fish Biodiversity conservation in North East India. NBFGR, Lucknow, India.

Mishra, A., Raveendar, B., Sharma, A. P., Grurjar, U. and Gugulothu, R. 2018. Assessment the present status of fish diversity in relation to physicochemical characteristics of Navaksagar reservoir of Uttrakhand, Journal of Entomology and Zoology Studies, 6 (2): 477-484.

Moyle, P. B. and Leidy, R. A. 1992. Loss of biodiversity in aquatic ecosystems: evidence from fish faunas. Conservation Biology, 127-169.

Manjare, S. A., Vhanalakar, S. A. and Muley, D. V. 2010. Analysis of water quality using physiochemical parameters Tamdalge tank in Kolhapur district, Maharashtra, International Journal of Advanced Biotechnology and Research, 1(2):115-119.

Raj, A. D. S. and Jayasekher, T. 2007. Hydrogeochemistry of the river basins of Kanyakumari district. Indian Journal of Environmental Protection, 27(2): 145-152.

Sen, N. 2000. Occurrence, Distribution and Status of Diversified fish fauna of Northeast India: In A.G Poniah and U.K. Sarkar (Eds). Fish Biodiversity of Northeast India. Proc. National workshop, Abstracts NATP Publication I, NBFGR, Lucknow, pp. 31- 48.

Shannon, C. E. and Weaver, W. 1949. A Mathematical Theory of Communications. University of Illinois Press, Urbana, USA, pp: 117.

Sharma, J, Parashar, A. and Bagare, P. 2014. Icthyofaunal divergence of Narmada river in Hoshangabad region and their conservation measures, International Journal Applied Bioresearch, 22: 5-11.

Suleiman, M., Ahmad, J. N. and Ezra, A. G. 2018. Impact of physico-chemical parameters on Fish Diversity in Gubi Reservoir, Bauchi State, Nigeria, International Journal of Innovative Research and Development, 7 (11): 128-133.

Talwar, P. K. and Jhingram, A. G. 1991. Inland Fishes of India and Adjacent Countries. Oxford and IBH publishing Co. Pvt. Ltd.

13

Pallisentis bishnupurensis n. sp. from the Intestine of a Fresh Water Fish *Channa striatus* (Bloch) from Awangsoi Lake, Bishnupur, Manipur

H. Puinyabati, M. Shomorendra and Devashish Ka

ABSTRACT

Pallisentis bishnupurensis n. sp. from the intestine of a fresh water fish Channa striatus (Bloch) has been described and compared with the other known species of the genus Pallisentis Van Cleave, 1928.

***Keywords**: Pallisentis bishnupurensis n. sp., Channa striatus, Awangsoi Lake, Manipur.*

INTRODUCTION

The genus *Pallisentis* was erected by Van Cleave, 1928 with *P. umbellatus* as its type species. The other species reported under this genus are *P. gaoes* (Mac Callum, 1918) Van Cleave, 1928; *P. cleatus* Van Cleave, 1928; *P. ophiocephali* (Thapar, 1930) Baylis, 1933; *P. nagpurensis* (Bhalerao, 1931) Baylis, 1933; *P. nandai* Sarkar, 1953; *P. colisai* Sarkar, 1956; *P. basiri* Farooqi, 1958; *P. allahabadi* Agarwal, 1958; *P. buckleyi* Tadross, 1966; *P. pandei* Rai, 1967; *P. guntei* Sahay *et al.*, 1967; *P. garuai* n. comb. *Davendrosentis garuai* Sahay *et al.*, 1971; *P. magnum* (Saeed & Bilqees, 1971); *P. tetradontae* Troncy, 1978; *P. clupei* Gupta and Gupta, 1979; *P. cavasi, P. fasciati, P. gomtii* Gupta and Verma, 1980; P. *indica, P. croftoni* Mithal and Lal, 1981; *P. kalrai* Khan and Bilqees, 1985; *P. guptai, P. mehrai* Gupta and Fatma, 1985; *P. fotedari* Gupta and Sinha, 1991 and *P. jagani* Koul *et al.*, 1991.

MATERIALS AND METHODS

The fish hosts were collected from the study area and brought to the laboratory for parasitic examination. The external body surface as well as the internal body organs were thoroughly examined for the parasites. The collected acanthocephalan parasites were washed and relaxed in normal saline. Then the specimen was fixed and preserved in AFA (alcohol-formalin-acetic acid) solution. To facilitate identification of the worms the parasites were cleared in lactophenol and mounted in glycerine jelly. Diagrams were made with the help of camera lucida and measurements were taken with ocular micrometer. All the measurements are in millimetres. The collected parasites were identified following Yamaguti (1963) and Bhattacharya (2007).

RESULTS AND DISCUSSION

During the present investigation a single specimen of the present form has been collected from the intestine of *Channa striatus* (Bloch).

Description: Body elongated, cylindrical, spinose, tapering to posterior end. Proboscis globular armed with 4 circles of 10 recurved hooks. The hooks of the first circle largest, of the second circle larger than that of the third and the fourth circles. Size of the hooks of third and fourth circles are equal. Neck short. Proboscis sheath cylindrical, single layered. Lemnisci unequal. Body spination consists of collar spines and trunk spines. Collar spines are followed by a spineless area which is again followed by trunk spine.

Male: Body 3.22 mm long, 0.27 mm wide. Proboscis 0.16 mm long, 0.16 mm wide. Neck 0.15 mm long, 0.16 mm wide. Proboscis hooks of first circle 0.06 mm long, of second circle 0.05 mm long, of third circle 0.03 mm long, of fourth circle 0.03 mm long. Proboscis sheath 0.32 mm long, 0.15 mm wide. Lemnisci unequal. Shorter lemniscus 0.50 mm long, 0.04 mm wide, longer lemniscus 0.62 mm long, 0,04 mm wide extending far above to anterior testis. Collar spine 16 rows, each with 16 spines, 0.015 mm long. Trunk spines 28 rows, extending upto anterior half of the body. Testes elongated, cylindrical, sub equal, overlapping, post equatorial. Anterior testis 0.25 mm long,0.11 mm wide. Posterior testis 0.21 mm long, 0.10 mm wide. Cement gland 0.35 mm long, 0.07 mm wide, with 28 nuclei Cement reservoir 0.10 mm long, 0.07 mm wide. Seminal vesicle 0.40 mm long, 0.06 mm wide.

The present form is referred to the genus *Pallisentis* Van Cleave, 1928 with *P. umbellatus* as its type species. The present form differs from all the known species of this genus except *P. umbellatus, P. cleatus, P. ophiocephali, P. nagpurensis, P. colisai, P. nandai, P. guntei* and *P. garuai* in the absence of Saefftigen's pouch.

The present form differs from *P. garuai* in the number of proboscis hooks. Proboscis armed with 4 circles of 6 hooks each in *P. garuai* while 4 circles of 10 hooks each in the present form.

The present form differs from *P. umbellatus* in having 14-16 collar spine rows instead of 9 rows and in having 28 cement gland nuclei instead of 16.

The present form also differs from *P. cleatus* in the number of collar spine rows and cement gland nuclei. 5-6 collar spine rows and 16 cement gland nuclei are found in *P. cleatus*.

Collar spine rows are 11 in *P. ophiocephali*, 12-14 in *P. nagpurensis*, 13-15 in *P. garuai*, 16-17 in *P. guntei*, while 14-16 in the present form. In the present form the number of cement gland nuclei are 28 while nuclei are 4-5 in *P. ophiocephali*, 20-30 in *P. nagpurensis* and 23-25 in *P. nandai*. The present form further differs from *P. nandai* in having unequal lemnisci and unequal testes.

The present form has a close resemblance with *P. colisai* in having unequal lemnisci and unequal testes but differs in the number of cement gland nuclei. The present form has 28 nuclei instead of 8-15 nuclei in the later.

Accordingly, it is regarded as new with the specific name *P. bishnupurensis* n. sp. after the name of its locality.

Host: *Channa striatus* (Bloch)

Location: Intestine

Locality: Awangsoi Lake, Bishnupur, Manipur

Table 1: A comparative chart of the species of genus *Pallisentis* Van Cleave, 1928 which have no Saefftigen's pouch (All measurements in mm)

Structures	Present form	*P. ophiocephali* (Thapar, 1930) Baylis, 1933	*P. nagpurensis* (Bhalerao, 1931) Baylis, 1933	*P. colisai* Sarkar, 1956	*P. nandai* Sarkar
Size of body					
Male	3.22 x 0.27	5.5-6.00 x 0.32	2.4-19.00 x 0.43-0.9	4.15 x 0.385	5.6-9.9 x 0.37-0.63
Female	-	-	-	5.4-12.9 x 0.61-0.62	6.3-10.4 x 0.35-0.56
Size of proboscis	0.16 x 0.16	0.18	0.35 x 0.3	0.132 x 0.154	0.17-0.48 x 0.19-0.32
Length of proboscis hooks					
1st row					
2nd row	0.06	1.10	0.076	0.076	0.09
3rd row	0.05	0.05	-	0.068	0.08-0.09
4th row	0.03	0.04	-	0.0306	0.06
Size of neck	0.03	0.03	0.03	0.0255	0.03
Lemnisci	0.15 x 0.16	-	-	0.264 x 0.165	-
I					
II	0.50 x 0.04	-	-	2.2 x 0.05	x 0.04
Size of proboscis sheath	0.62 x 0.04	-	-	-	0.76 x 0.04
Size of anterior testis	0.32 x 0.15	-	-	0.605 x 0.11-0.165	0.44-0.84 x 0.12-0.25
Size of posterior testis	0.25 x 0.11	0.65	0.63-1.82 x 0.16-0.37	0.385 x 0.165	0.57 x 0.14
Size of cement gland	0.21 x 0.10	0.65	0.49-1.28 x 0.17-0.36	0.352 x 0.165	0.53 x 0.14
Size of cement reservoir	0.35 x 0.07	-	1.5-1.18	0.44 x 0.187	0.77 x 0.14
Size of seminal vesicle	0.10 x 0.07	-	-	0.20-0.42 x 0.12-0.13	0.32 x 0.13
Egg	0.40 x 0.06	-	-	0.27-0.38 x 0.11-0.13	0.30 x 0.13
	-	-	0.116-0.214x0.083-0.115	-	-

(Abbreviations used: AT – Anterior testis; B - bursa; CG – Cement Gland; CLS - Collar spine; CR – Cement reservoir; LEM - Lemnisci; N – Nucleus; NK - Neck; P - Proboscis; PH - Proboscis hook; PS - Proboscis sheath; PT – Posterior testis; SV – Seminal vesicle; TS - Trunk spine).

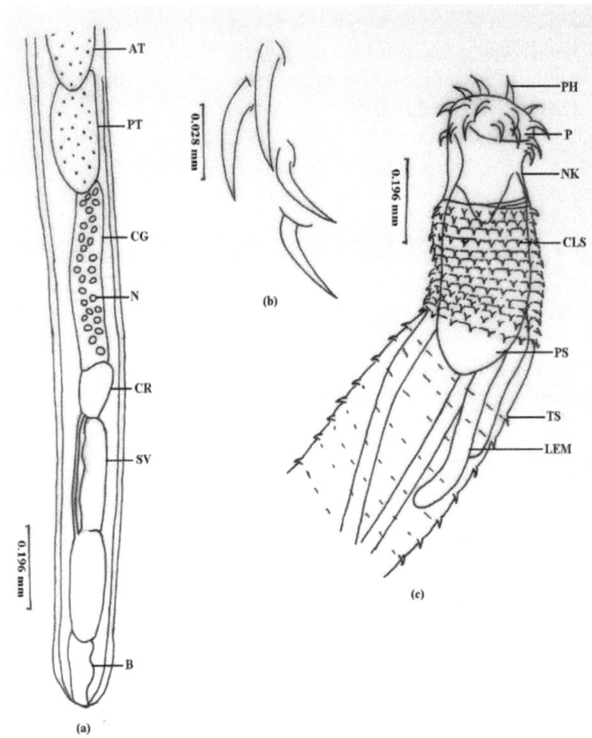

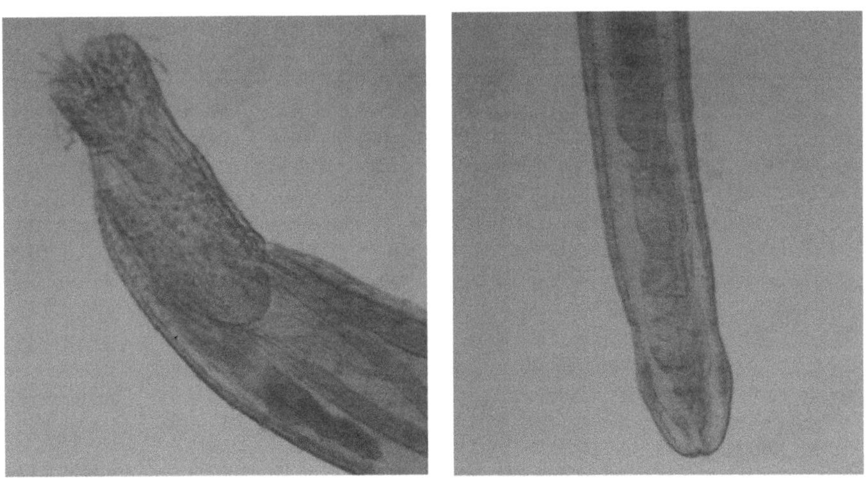

(a) (b)

REFERENCES

Agarwal, S. C. 1958. A new species of the genus Pallisentis (Acanthocephala). Current Science, 27(3): 107.

Amin, O.M. 1985. Biology of the Acanthocephala: Classification. Cambridge University Press. 27-72.

Baylis, H. A.1933. On some parasitic worms from Java with Remarks on the Acanthocephalan genus Pallisentis. Annals and Magazine of Natural History, 10: 559-573.

Bhalerao, G. D. 1931. On a new species of Acanthocephala from Ophiocephalus striatus. Annals and Magazine of Natural History, 10 (7): 569-573.

Bhattacharya, S. B. (2007). Handbook of Indian Acanthocepha. Zoological survey of India. pp. 225.

Farooqi, H. U. 1958. A new species of the genus Pallisentis from a fresh water eel. Journal of Parasitology, 18: 457-464.

Gupta, S. P. and Fatma, S. 1985. On some Acanthocephalan parasites (Quadrigyridae) from fishes of Uttar Pradesh and Tamil Nadu. Indian Journal of Helminthology, 37(2): 149-180.

Gupta, S. P. and Gupta, R. C. 1979. On six new Acanthocephalan parasites from marine fishes of Arabian sea at Quilon, Kerala. Indian Journal of Helminthology, 31: 135-156.

Gupta, S. P. and Sinha, 1991. On a new Acanthocephalan parasite Pallisentis fotedari sp. nov. from the intestine of a marine fish Clupea longiceps Gunther from Puri coast. Indian Journal of Helminthology, 43(1): 19-26.

Gupta, S. P. and Verma, S. L. 1980. On three new Acanthocephalan parasites of the genus Pallisentis Van Cleave, 1928 from fresh water fishes of Lucknow. Helminthologia, 17(4): 269-282.

Khan, A. and Bilqees, F. M. 1985. Two new Acanthocephala species from fresh water fishes of Kalari lake. Pakistan Journal of Zoology, 19: 263-271.

Koul, P. L. 1991. Pallisentis jagani sp. nov. from Channa channa in Jammu. Indian Journal of Helminthology, 43(2): 124-128.

Mac Callum, G. A. 1918. Notes on the genus Camallanus and other nematodes from various hosts. Zoopathologica, 1: 123-134.

Mithal, R.P. and Lal, S.S. 1981. Two new Acanthocephalan worms Pallisentis croftoni n. sp. and P. indica n. sp. from fresh water fishes of the genus Ophiocephalus. Indian Journal of Zoology, 17: 165-175.

Rai, P.1967. On four Acanthocephalan genera parasitic in fresh water fishes with description of three new species. Indian Journal of Helminthology, 19(1): 27-44.

Saeed, R. and Bilqees, F. M.1971.Pallisentis magnum new species (Acanthocephala: Quadrigyridae) from the fish Wallago attu of Kalri lake, West Pakistan. Pakistan Journal of Zoology, 3: 221-223

Sahay, U., Nath, A. and Ghos, A. K. 1971. On Devendrasentis garuai gen.et.sp.nov. (Neoechinorhynchidae Southwell and Macfie, 1925, Quadrigyridae Van Cleave, 1920, Acanthocephala from a fresh water fish Clupisoma garua (Ham). Annals of Parasitology (Paris), t. 46(1): 69-80.

Sahay, U., Nath, S. and Sinha, A. 1967. On Acanthocephala from a hill stream fish Lepidocephalichthys guntea (Ham) from Patna. Zoologischer Anzeiger,178: 348-353.

Sarkar, H. L. 1953. On a new Acanthocephala, Pallisentis nandai, from the fish Nandus nandus (Hamilton), with notes on the other species of the genus. Proceeding of Zoological Society Bengal, 6(2): 139-147.

Sarkar, H. L. 1956. On a new Acanthocephala, Pallisentis colisai, from the fish Colisa fasciatus (Bloch. And Schn.), with a note on Acanthogyrus acanthogyrus Thapar, from the fish Labeo rohita (Hemilton). Record of Indian Musuem, 52(2-4): 349-362.

Tadross, G.1966. On three new Acanthocephala of the genera Pallisentis Van Cleave, Saccosentis gen. nov. and Acanthocephalus Koelreuther, from fish. Journal of Helminthology, 40(1/2): 155-180.

Thapar, G. S. 1930. On Farzandia, A new genus of acanthocephalid worms, from the intestine of Ophiocephalus marulius. Annals and Magazine of Natural History, (10)6: 76-81.

Troncy, P. M. 1978. New observation on the parasites of the fish of the Chad basin Nouvelles observation Sur Les parasites des poisons du basin tchadien. Bulletin de l'instutut Fandamental d'Afrique Noire, 40: 528-552.

Van Cleave, H. J. 1928. Two new genera and species of Acanthocephala from fishes of India. Record of Indian Musuem, 30(2): 147-149.

Van Cleave, H. J.1928. Acanthocephala from China. 1. New species and new genera from China fishes. Parasitology, 20: 1-9.

Yamaguti, S. 1963. Systema Helminthum Vol. 5: Acanthocephala. Interscience Publish. New York, London. 1-423.

14

Limnological Studies of Lakes of Nagpur City, Maharashtra, India

Shahnoor Khan and Sharda Dhadse

ABSTRACT

Lakes play a crucial role in sustaining ecological balance and supporting diverse forms of life, making them indispensable components of terrestrial landscapes. The lakes in Nagpur hold paramount importance due to their multifaceted contributions to the region's environment, economy, and social fabric. These lakes serve as crucial water reservoirs, supporting the city's water supply for domestic, agricultural, and industrial purposes. Additionally, they play a pivotal role in maintaining local biodiversity and providing habitats for various flora and fauna. The recreational value of Nagpur lakes is significant, offering spaces for leisure, tourism, and cultural activities that contribute to the well-being of the community. Nagpur's lakes face a myriad of challenges that threaten their ecological health and utility. Pollution, resulting from industrial discharges, agricultural runoff, and improper waste disposal, poses a severe threat to water quality. Eutrophication, caused by excessive nutrient influx, leads to algal blooms and oxygen depletion, further deteriorating the lakes' ecosystems. Plankton serves as a vital tool for studying the health of lakes, offering valuable insights into their ecological condition. As microscopic organisms suspended in water, plankton are highly responsive to changes in environmental factors such as nutrient levels, temperature, and pollution. Therefore, the present study is dedicated to examining the current status of plankton diversity in the lakes of Nagpur. According to Shannon Weiner Diversity Index Lakes Naik, Binaki, and Sakkardara were eutrophic, while the other lakes (Ambazari, Pandharabodi, Futala, Sonegaon, and Godewada) were mestrophic. However, it is noteworthy that Ambazari Lake is also on the verge of pollution. The result concluded that there is an urgent need for the conservation and restoration of the lakes of Nagpur.

Keywords: *Lake, Naik, Binaki, Sakkardara, Pandharabodi, Futala, Sonegaon, Godewada phytoplankton, zooplankton, pollution, SWDI.*

INTRODUCTION

Lakes are of immense ecological, economic, and cultural significance, representing integral features within our planet's landscapes. Studies show that there are around 117 million lakes worldwide, constituting about 4% of the Earth's non-glaciated land surface (Verpoorter *et al.,* 2014). These lakes play a crucial role in providing essential services to nearby communities and serve as important centres for recreation and overall well-being. However, the increasing demand for natural resources has led to widespread degradation of these lakes (Jenny *et al.,* 2020). Despite efforts to restore lakes, there is a prevailing perception that such programs often encounter high failure rates (Søndergaard *et al.,* 2007; Gulati *et al.,* 2008; Poikane *et al.,* 2024). This concern is particularly pronounced in less economically developed countries, where populations are highly exposed to the consequences of lake degradation, and monitoring, assessment, and restoration programs may be in nascent stages. Additionally, lakes play a pivotal role in climate regulation, influencing local weather patterns and contributing to global processes such as water and carbon cycles. Recognizing and preserving the significance of lakes is imperative for sustaining ecosystems, ensuring water security, and nurturing the well-being of both the natural environment and human societies. 21° 8' 55" 79° 4'46"E.

Nagpur situated at coordinates 21° 8' 55" N and 79° 4'46"E, holds the distinction of being the second capital of Maharashtra. Renowned as the "orange city" and the "city of lakes," Nagpur relies significantly on its lakes as a valuable resource base. The current water supply for the city is primarily sourced from, Kanhan River, Gorewada Lake, and Pench Dam. Futala Lake serves an irrigation purpose, benefitting 84 acres of cultivated agricultural land. Ambazari Lake, on the other hand, caters to the water needs of local residents in the MIDC colony, Hingna area, Nagpur, and supports industrial purposes. Other notable water bodies in Nagpur include Sonegaon, Gandhisagar, Naik, Sakkardara, and Lendi. However, these lakes face severe pollution due to anthropogenic activities such as bathing, washing clothes, and improperly disposing of plastic, paper, oil, and worship materials. Moreover, the lakes Gandhisagar and Lendi are now dried.

The Nagpur Environmental Status Report for 2017-18 identified untreated domestic wastewater and agricultural run-offs as the primary contributors to eutrophication and nutrient increase in the lakes, leading to detrimental effects on aquatic biodiversity. Remote sensing data further highlighted a concerning deterioration of Nagpur's ecosystem. Reports from 2021 indicate poor water quality in Gandhisagar, Futala, Ambazari, Sonegaon, Sakkardara, and Naik lakes, as emphasized by an article in Times of India. Urgent attention and comprehensive management strategies are imperative to address pollution, preserving these vital water resources for the well-being of current and future generations.

In the ecosystems of lakes, plankton assumes a central role as the primary producer, and alterations in its community composition have far-reaching impacts on the structure, function, and resilience of the entire ecosystem (Padisak *et al.,* 2006). Plankton communities are characterized by factors such as species composition, biomass, abundance, and diversity. These parameters are generally affected by abiotic factors such as pH, temperature, and nutrient levels (Geng *et al.,* 2022; Rombouts *et al.,* 2019). The process of eutrophication, characterized by the excessive proliferation of phytoplankton, disrupts the balance of zooplankton predation, leading to ecological instability and a decline in biodiversity within lakes (Znachor *et al.*, 2020). Consequently, substantial research efforts have been directed towards understanding the diversity of species and the stability of plankton communities within lake environments (Edwards *et al.,* 2016; Geng *et al.,* 2022). Investigating the mechanisms that govern the establishment and continuity of plankton diversity is crucial because of its broad impact on the functionality of complete lake systems (Reiss *et al.*, 2009).

Changes in environmental gradients lead to responses in the stability of plankton groups, revealing complex patterns in community structure (Arab *et al.*, 2019). Importantly, the stability of community structure is notably affected by the diversity of plankton species; a reduction in species diversity amplifies the loss of species, leading to a decrease in the stability and functionality of the community within aquatic ecosystems (Ptacnik *et al.,* 2008). The stability of organisms within higher plant communities relies significantly on the diversity of species (Ipsen and Schomerus, 2016; Geng *et al.*, 2022). The interconnected relationship among species diversity, abundance, and mutual variation within the phytoplankton community emphasizes a strong connection between diversity and stability. Additionally, zooplankton dynamics can be stabilized by the diversity of algal prey resistant to herbivorous zooplankton, and through many different methods, the temporal stability of total competitor biomass can be improved by raising the species richness of competitors (Cui *et al.,* 2021).

Limnology, a branch of aquatic sciences, is the systematic examination of inland waters, encompassing lakes, rivers, streams, ponds, and wetlands. This field delves into comprehending the physical, chemical, geological, and biological attributes of freshwater ecosystems. In this context, the current study specifically addresses the biological limnology status, particularly focusing on plankton diversity in Nagpur's lakes. The exploration of biological parameters serves to identify various elements contaminating or polluting aquatic environments. Phytoplankton and zooplankton emerge as practical indicators of water quality, chosen for their ease of sampling, widespread distribution, and the wealth of available information about these groups. The monitoring of biological parameters is both rapid and cost-effective, relying solely on the biological examination of water samples. The survival status of flora and fauna directly reflects the impact of pollution, with biological data providing a comprehensive overview of the subsequent effects of environmental pollution.

Methodology

Study Area

The eight lakes of Nagpur were selected for study of plankton diversity. These are Sonegaon, Gorewada, Sakkardara, Binaki Mangalwari, Ambazari, Futala, Pandharabodi, and Naik Talao (Table 1). The multiple samples of eight lakes were taken for phytoplankton and zooplankton analysis. The study area map is given in Fig 1.

Table 1: Sampling Sites

Location	Lakes Name	Coordinates	
Site 1	Sonegaon	21.09845744^0 N	79.05030722^0 E
Site 2	Futala lake	21.15167^0 N.	79.04381667^0 E.
Site 3	Gorewada	21.19383167^0 N	79.04106667^0 E
Site 4	Ambazari	21.12978^0 N	79.04551^0 E
Site 5	Binaki Mangalwari	21.16922^0 N	79.12158^0 E
Site 6	Naik talao	21.16249^0 N	79.11327^0 E
Site 7	Sakkardara	21.12106^0 N	79.11502^0 E
Site 8	Pandharabodi	21.14259^0 N	79.05082^0 E

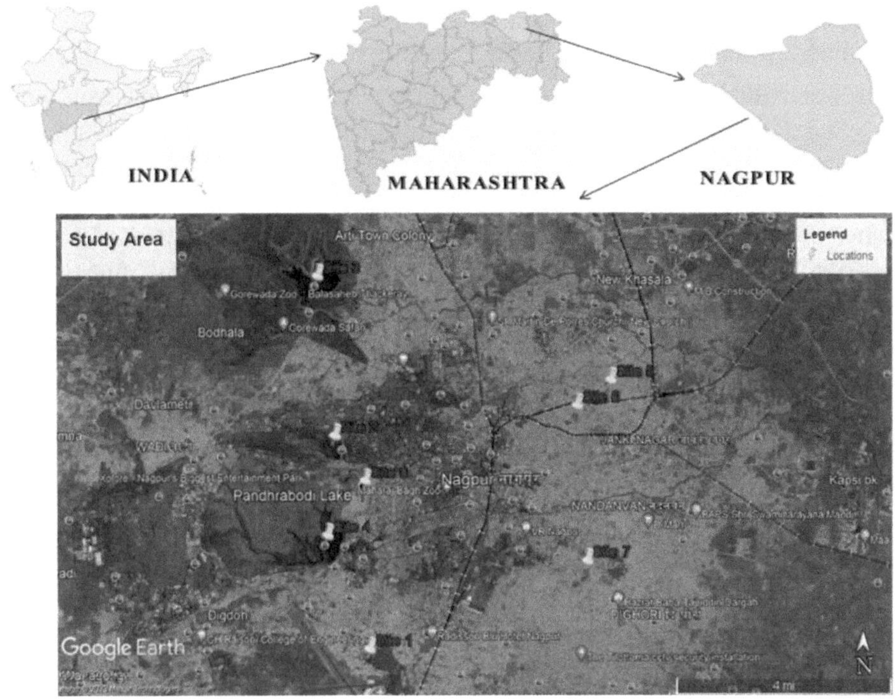

Fig. 1: Study Area Map (See colour version on page 347)

Sampling for Phytoplankton and Zooplankton

The procedure utilized for both the sampling and analysis of phytoplankton and zooplankton adhered to the established standard methods (APHA, 2012). The imaging of plankton specimens was conducted using a Medicams Microscope Camera.

Plankton diversity Analysis

To assess plankton diversity, (the diversity of species observed at each sampling site) was calculated. The Shannon Weiner diversity Index (SWDI) formula was used to determine which are as follows:

$$H = -\sum P_i(\ln P_i)$$

Where H is the Shannon Weiner diversity Index and P_i is the proportion of each species in the sample

Results and Discussion

Phytoplankton

Under the careful analysis of Phytoplankton in the water samples of Sonegaon, Gorewada, Sakkardara, Binaki Mangawari, Ambazari, Futala, Pandharabodi, and Naik talao, the sample consisted of different species of algal and diatom population which are given in Table 1 and the pictures are given in Fig 1.

The water quality varied among the lakes, with Godewada, Sonegaon, and Futala exhibiting relatively fresher and clearer water compared to the others. In contrast, Binaki Lake and Naik Lake were in notably poor conditions, resembling more of drains than lakes. Sakkardara Lake displayed green-colored water, and Naik, Binaki, and Sakkardara Lakes were covered with widespread weed growth. Algal blooms were particularly prominent in Naik and Binaki Lakes, with the water in Binaki emitting a foul odor. Additionally, some parts of Ambazari Lake showed observable weed growth. The assessment using the SWDI (Table 2) for phytoplanktons revealed that Naik (2.82), Binaki (2.75), and Sakkardhara (2.57) were categorized as having eutrophic water quality. In contrast, Godewada (1.60), Sonegaon (1.87), Futala (1.92), Pandharabodi (1.99), and Ambazari (2.03) were classified as having mesotrophic water quality.

Table 1: Phytoplanktons species observed from lakes of Nagpur

S.No	Location	Phylum	Genus
1	**Sonegaon**	Chlorophyceae	*Chlorella*
			Pandorina
		Cynophyceae	*Cylindospermopsis*
			Anthrospira
			Synecoccus
		Bacillariophyceae	*Nitzschia*
			Tabellia
2	**Futala**	Chlorophyceae	*Chlamydomonas*
			Pandorina
		Cynophyceae	*Oscillatoria*
			Anthrospira
			Anabaena
		Bacillariophyceae	*Navicula*
			Bacillaria
3	**Gorewada**	Chlorophyceae	*Ankistrodesmus*
			Sphaerocystis
		Cynophyceae	*Cylindrospermopsis*
			Synechococcus
		Bacillariophyceae	*Nitszchia*
4	**Ambazari**	Chlorophyceae	*Spirogyra*
			Scenedesmus

			Tetradesmus
		Cynophyceae	*Anabaena*
			Oscillatoria
			Anthospira
		Bacillariophyceae	*Navicula*
			Pinnularia
		Euglenophyceae	*Euglena*
5	**Binaki Mangalwari**	Chlorophyceae	*Chlorella*
			Spirogyra
			Chlamydomonas
			Closterium
			Pandorina
			Pediastrum duplex
			Ankistrodesmus
			Scenedesmus
		Cynophyceae	*Oscillatoria*
			Anthospira
			Merismopedia minima
			Microcystis
		Bacillariophyceae	*Bacillaria*
			Navicula
			Nitszchia
			Synedra
			Pennularia
		Euglenophyceae	*Euglena*
6	**Naik Talao**	Chlorophyceae	*Chlorella*
			Pandorina
			Chlamydomonas
			Scenedesmus
			Spirogyra
		Cynophyceae	*Oscillatoria*
			Cylindrospermopsis
			Microcystis
			Anthrospira
			Merismopedia
		Bacillariophyceae	*Bacillaria*
			Navicula
			Nitszchia
			Synedra

			Pennularia
			Tabellia
		Euglenophyceae	*Euglena*
7.	**Sakkardara**	Chlorophyceae	*Chlorella*
			Crucigenia
			Tetradesmus
			Teraedron
			Chlamydomonas
			Scenedesmus
		Cyanophyceae	Gomphosphaeria
			Microcystis
			Cylindrospermopsis
			Anaebaenopsis
			Merismopedia minima
		Bacillariophyceae	*Bacillaria*
			Navicula
			Nitszchia
			Synedra
8.	**Pandharabodi**	Chlorophyceae	*Spirogyra*
			Chlorella
			Chlamydomonas
		Cyanophyceae	*Cylindrospermopsis*
			Oscillatoria
			Merismopedia minima
		Bacillariophyceae	*Navicula*
			Synedra
		Euglenophyceae	Euglena

Table 2: Biological Parameter: Phytoplankton

Sr. No.	**Samples**	**Phyto-plankton No/Lit**	**Percentage Composition of Algal Group**				**SWDI**
			Chloro phyceae	**Bacillario-phyceae**	**Cyano phyceae**	**Eugleno phyceae**	
1.	Sonegaon	12857	28.5	43	28.5	-	1.87
2.	Futala lake	10000	28.5	28.5	43	-	1.92
3.	Gorewada	7143	40	20	40	-	1.60
4.	Ambazari	20000	36	27	28	9	2.03
5.	Binaki Mangalwari	44285	44	28	22	6	2.75
6.	Naik talao	47142	40	30	25	5	2.82
7.	Sakkardara	41428	40	27	33	-	2.57
8.	Pandharabodi	17143	33	22	34	11	1.99

Aquatic ecosystems greatly depend on the production of plankton, and the structure of plankton communities is influenced by various environmental factors (Engel *et al.,* 2019; Purina *et al.*, 2018; Li *et al.,* 2021). The majority of the phytoplankton group in this study consisted of Chlorophyta, Cyanophyta, and Bacillariophyta. The Euglenophyta was only found in Binaki, Naik, and Pandharbodi.

The proliferation of phytoplanktons is attributed to rising temperatures, changes in the water column (Recknagel *et al.,* 2014; Kosankar, 2020), and disturbances in sediments caused by both biotic and abiotic factors (Chang *et al.,* 2015; Kosankar, 2020). The concentration of Microcystin is dependent on the quantity of biotic material present in the lake (Pick, 2016; Kosankar, 2020), and the presence of Microcystis was observed in Naik, Binaki, and Sakkardara. Research by Nandan and Aher (2005) revealed that diatoms such as Nitzschia sp. and Navicula sp. are indicative of organic pollution and were found in almost all lakes in Nagpur during this study. The presence of the Ankistrodesmus and Scenedesmus genera is considered indicative of a eutrophicated environment (Jose and Kumar, 2011; Vidya *et al.,* 2014). Eutrophication promotes elevated algal and planktonic biomass, creating complexity in the nutrient composition of the lake and reducing its stability (Fu *et al.,* 2022). Seasonal variations exacerbate the impact of eutrophication by not only amplifying algal growth and planktonic biomass but also impeding the formation of stability in the turnover of a community (Cui *et al.*, 2021). The study on phytoplankton indicated poor water quality at the three sites (Naik, Binaki, and Sakkardara).

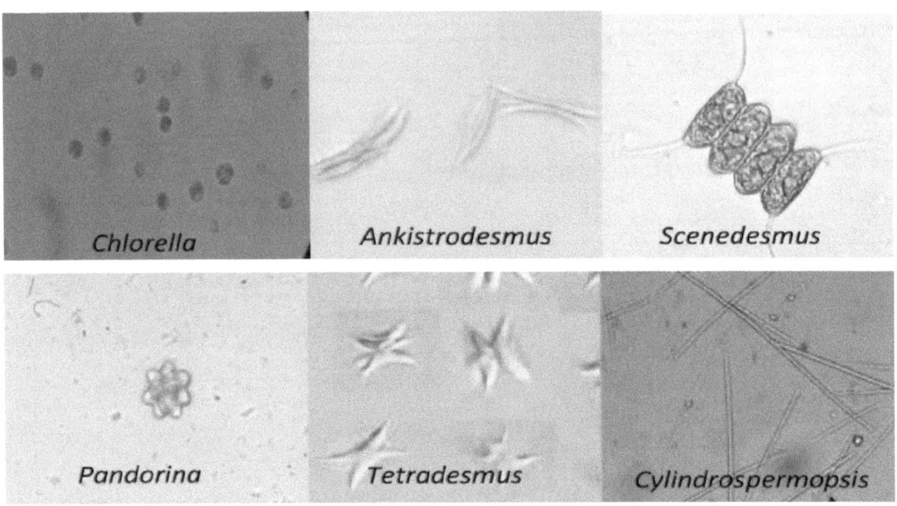

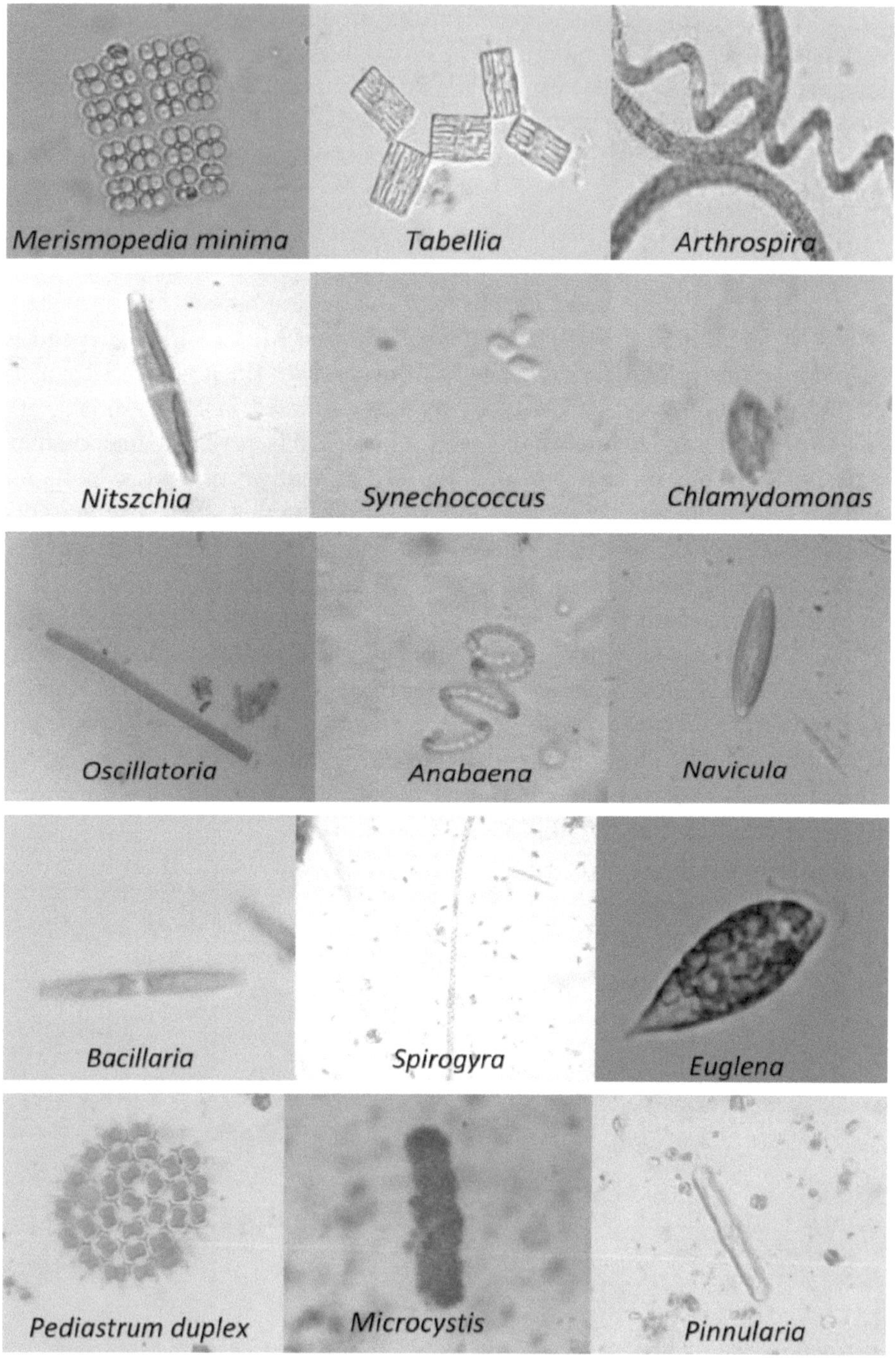
Merismopedia minima
Tabellia
Arthrospira
Nitszchia
Synechococcus
Chlamydomonas
Oscillatoria
Anabaena
Navicula
Bacillaria
Spirogyra
Euglena
Pediastrum duplex
Microcystis
Pinnularia

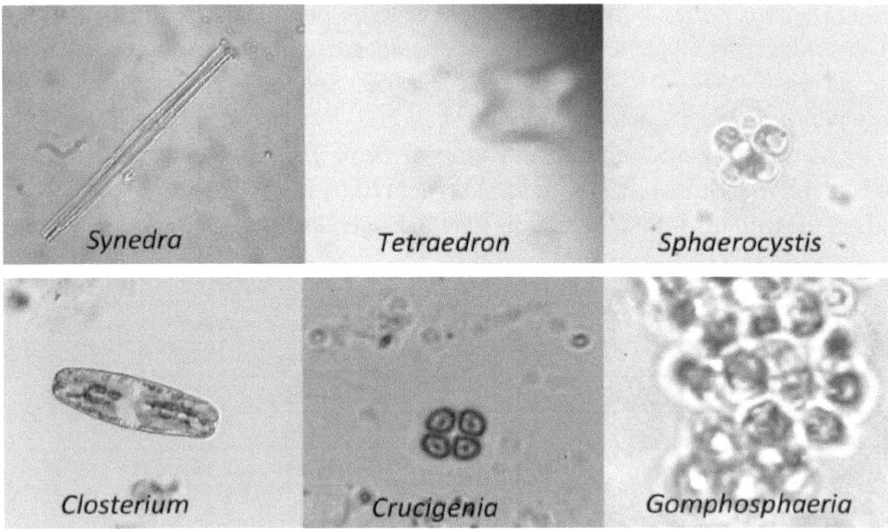

B

Fig. 1: Phytoplankton Observed in Lakes of Nagpur (A and B) (See colour version on page 349)

Zooplankton Analysis

The analysis of zooplankton samples from Sonegaon, Gorewada, Sakkardara, Binaki Mangawari, Ambazari, Futala, Pandharabodi, and Naik waters revealed the presence of various species, including rotifers, copepods, cladocerans, ostracods, and protozoans, as detailed in Table 3 and depicted in Fig 2. Protozoan members were exclusively observed in Futala and Naik. However, cladoceran members were not identified in Ambazari, Binaki, and Naik. Notably, the water flea (Daphnia), of cladoceran typically found in clean water, was absent in Binaki, Naik, and Sakkardara. The ostracod was found only in Naik, Binaki, and Sakkardara. However, the species *Cypria ophthalmica* is known to be particularly tolerant of high levels of organic pollution (Boomer and Attwood, 2007), and it was found in Naik and Sakkardara.

Zooplankton undergoes a transitional phase and plays a role in transforming energy from primary consumers that can withstand stress (Sulata and Devashish, 2013). Serving as comestible organic matter, zooplankton aids in purifying water bodies (Derevenskaya and Umyarova, 2016). Rotifers and copepods predominantly constituted the majority of the zooplankton community in the study area, and these phyla are mostly found in eutrophic lakes (Jia *et al.*, 2020). The zooplankton populations in Naik, Binaki, and Sakkardara lakes exhibited a more pronounced diversity compared to other lakes. The percentage diversity of Rotifera surpassed that of copepod and cladocera, with the rotifera phylum constituting 50% (Binaki), 57% (Sakkardara), 40% (Pandharabodi), and 34%

(Naik), among which polluted species were more prevalent. The measurement of *B. calyciflorus* species showed an elongated expansion correlated with the increase in food levels (Xi *et al.,* 2002). Species such as *Brachionus calyciflorus, Brachionus falcatus, Brachionus diversicornis,* and *Keratella tropica* are considered reliable indicators of eutrophication. The SWDI values of Naik (2.43), Binaki (2.21), and Sakkardara (2.0) were higher than the SWDI values of Ambazari, Pandharabodi, Futala, Sonegaon, and Godewada.

Table 3: Zooplanktons species observed from lakes of Nagpur

S.No	Location	Phylum/Class	Genus/Species
1	**Sonegaon**	Cladocera	*Daphnia*
			Moina
		Copepoda	*Cyclops*
		Rotifera	*Branchionus plicatilis*
			Branchionus quadridentus
2	**Futala**	Cladocera	*Daphnia*
		Copepoda	*Cyclops*
		Protozoa	*Paramecium*
		Rotifera	*Branchionus diversicornis*
			Branchionus quadridentus
3	**Gorewada**	Cladocera	*Daphnia*
			Leptodora kindtii
			Moina
		Copepoda	Cyclops
			Nauplius
			Mysis
		Rotifera	*Filinia longoseta*
			Keratella cochlearis
			Branchionus pllicatlis
			Branchionus quadridentus
4	**Ambazari**	Cladocera	*Daphnia*
		Copepoda	Cyclops
			Nauplius
		Rotifera	*Keratella cochlearis*
			Asplancha priodonata
			Branchionus quadridentus
			Brichionus calycifrous
5	**Binaki Mangalwari**	Copepoda	*Cyclops*
			Nauplius

			Pseudodiaptomus sps.
		Ostracoda	*Stenocypris*
		Rotifera	*Asplachna priodonata*
			Keratella tropica
			Testudinella patina
			Branchionus quadridentus
			Brchionus falcatus
			Brachionus forficula
6	**Naik Talao**	Copepoda	*Cyclops*
			Mystus
			Nauplius
		Protozoa	*Entamoeba sps.*
			Balantidium sps.
		Ostracoda	*Cypria ophthalmica*
		Rotifera	*Asplachna priodonata*
			Keratella tropica
			Branchionus quadridentus
			Brachionus calyciflorus
			Brachionus diversicornis
7	**Sakkardara**	Copepoda	*Cyclops*
			Nauplius
		Ostracoda	*Cypria ophthalmica*
		Rotifera	*Keratella cochlearis*
			Filinia opoliensis
			Testudinella patina
			Branchionus quadridentus
			Brachionus calyciflorus
			Branchionus pllicatlis
8	**Pandharabodi**	Cladocera	*Daphnia*
		Copepoda	*Cyclops*
			Nauplius
			Filinia longiseta
		Rotifera	*Branchionus plicatilis*
			Branchionus quadridentus
			Brachionus calyciflorus

Table 4: Biological Parameter: Zooplankton

Sr. No.	Samples	Zooplankton No/m^3	Percentage Composition of Zooplankton groups					SWDI
			Cladocera	Copepoda	Ostracoda	Protozoa	Rotifera	
1	Sonegaon	1786	50	25	-	-	25	1.38
2	Futala	2143	25	25	-	25	25	1.55
3	Gorewada	2857	33	34	-	-	33	1.20
4	Ambazari	9285	14	29	14	-	43	1.90
5	Binaki	16428	-	37.5	12.5	-	50	2.21
6	Naik	20000	-	33	11	22	34	2.43
7	Sakkardara	14643	-	29	14	-	57	2.01
8	Pandharabodi	10000	20	40	-	-	40	1.89

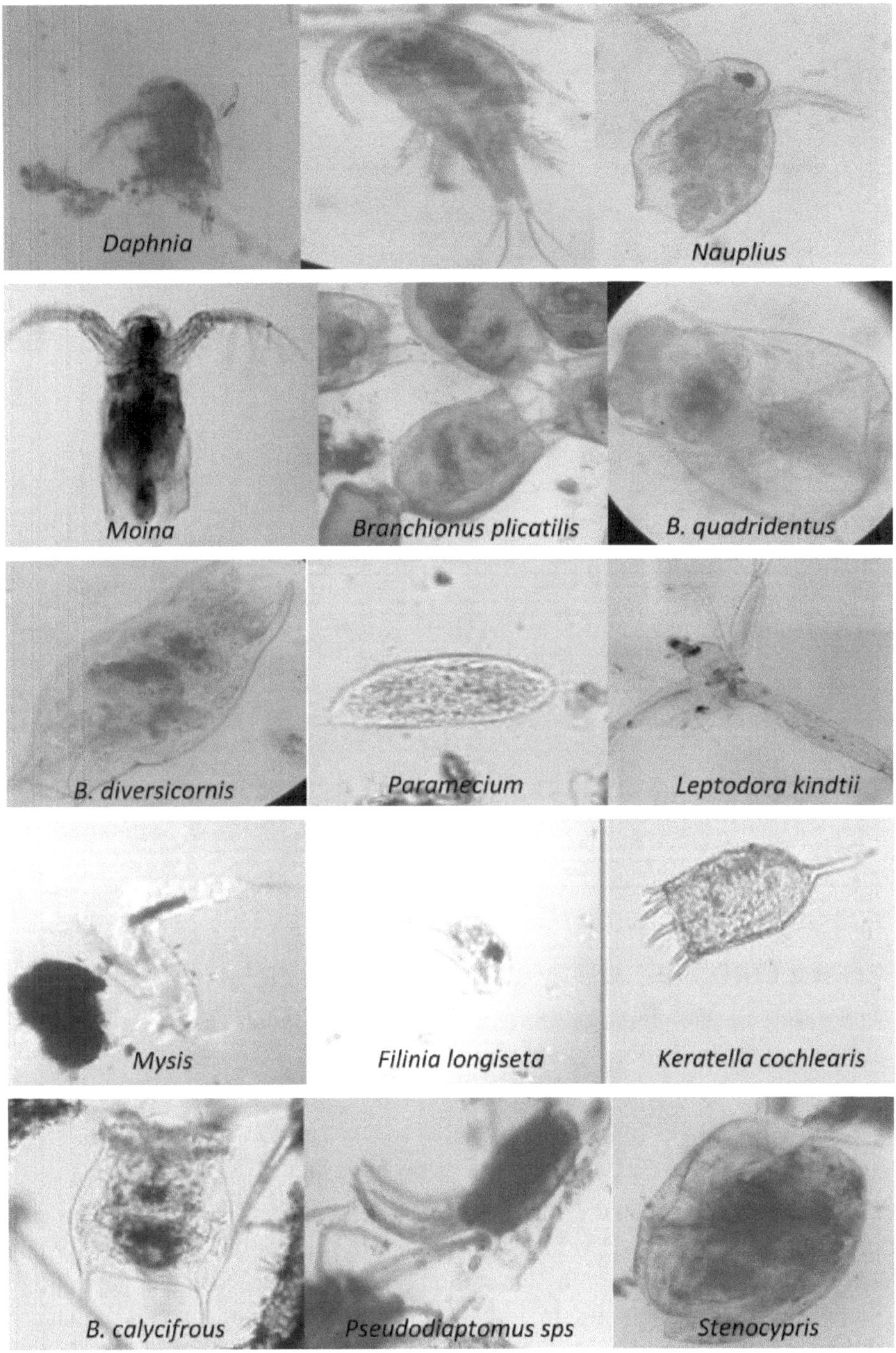
Daphnia
Nauplius
Moina
Branchionus plicatilis
B. quadridentus
B. diversicornis
Paramecium
Leptodora kindtii
Mysis
Filinia longiseta
Keratella cochlearis
B. calycifrous
Pseudodiaptomus sps
Stenocypris

A

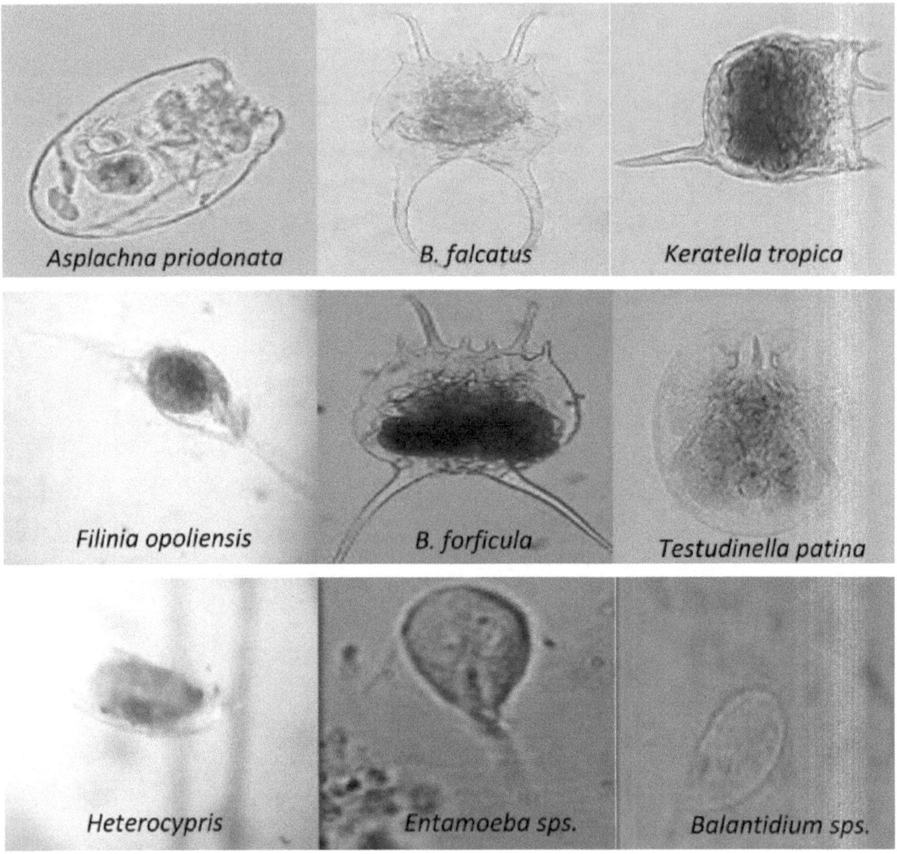

B

Fig. 2: Zooplankton Observed in Lakes of Nagpur (A and B) (See colour version on page 351)

CONCLUSION

According to the findings from the Shannon Wiener Diversity Index for phytoplankton and zooplankton, the trophic levels of the lakes were arranged in the following order:

Naik Talao > Binaki Talao > Sakkardara Lake > Ambazari Lake > Futala Lake > Pandharabodi >Sonegaon Lake > Godewada Lake.

The results derived from the Shannon Weiner Diversity Index for phytoplankton and zooplankton provide valuable insights into the trophic levels of Nagpur's lakes, establishing a hierarchy based on pollution levels. Notably, Naik Talao, Binaki Talao, and Sakkardara Lake exhibit higher trophic levels indicating elevated pollution levels. Ambazari Lake, a prominent water body in Nagpur, is also on the brink of pollution, underscoring the urgency for conservation measures. The responsibility for lake protection is a shared endeavor, extending beyond governmental agencies to the public. Often referred to as the "blue lungs" of the city, lakes play a crucial role in mitigating the urban heat island effect and maintaining groundwater levels. Collaborative efforts between the public and government entities are essential to safeguard these vital water bodies. Initiatives focusing on pollution reduction, nutrient control, and sustainable practices are imperative for the lakes' overall health. Preserving these natural resources is not only an environmental duty but also a crucial step toward ensuring a sustainable and healthy future for Nagpur.

REFERENCES

APHA, A. W., Greenberg, W. I. A., Clesceri, L. and Eaton, A. 2012. Standard methods for the examination of water and wastewater. Washington DC American Public Health Association.

Arab, S., Hamil, S., Rezzaz, M. A., Chaffai, A. and Arab, A. 2019. Seasonal variation of water quality and phytoplankton dynamics and diversity in the surface water of Boukourdane Lake, Algeria. Arabian Journal of Geosciences, 12: 1-11.

Boomer, I., and Attwood, F.2007. Ostracods as freshwater pollution indicators: a case study from the Ouseburn, a polluted urban catchment (Tyneside, NE England). Journal of Micropalaeontology, 26(2), 117-125.

Chang, C. H., Cai, L. Y., Lin, T. F., Chung, C. L., Van der Linden, L., and Burch, M. 2015. Assessment of the impacts of climate change on the water quality of a small deep reservoir in a humid-subtropical climatic region. Water, 7(4):1687-1711.

Cui, G., Wang, B., Xiao, J., Qiu, X. L., Liu, C. Q., and Li, X. D. 2021. Water column stability driving the succession of phytoplankton functional groups in karst hydroelectric reservoirs. Journal of Hydrology, 592:125607.

Deng, J., Qin, B., Sarvala, J., Salmaso, N., Zhu, G., Ventelä, A. M., ... and Vuorio, K. 2016. Phytoplankton assemblages respond differently to climate warming and eutrophication: a case study from Pyhäjärvi and Taihu. Journal of Great Lakes Research, 42(2): 386-396.

Derevenskaya, O. and Umyarova, R. 2016. Zooplankton as an indicator of river ecological condition.

Edwards, K. F., Thomas, M. K., Klausmeier, C. A. and Litchman, E. 2016. Phytoplankton growth and the interaction of light and temperature: A synthesis at the species and community level. Limnology and Oceanography, 61(4):1232-1244.

Engel, F., Drakare, S. and Weyhenmeyer, G. A. 2019. Environmental conditions for phytoplankton influenced carbon dynamics in boreal lakes. Aquatic Sciences, 81(2):35.

Fu, H., Chen, L., Ge, Y., Wu, A., Liu, H., Li, W., ... and Jeppesen, E. 2022. Linking human activities and global climatic oscillation to phytoplankton dynamics in a subtropical lake. Water research, 208, 117866.

Geng, Y., Li, M., Yu, R., Sun, H., Zhang, L., Sun, L., ... & Xu, J. 2022. Response of planktonic diversity and stability to environmental drivers in a shallow eutrophic lake. Ecological Indicators, 144, 109560.

Gulati, R. D., Pires, L. M. D. and Van Donk, E. 2008. Lake restoration studies: failures, bottlenecks and prospects of new ecotechnological measures. Limnologica, 38(3-4):233-247.

Ipsen, J. R., and Schomerus, H. 2016. Isotropic Brownian motions over complex fields as a solvable model for May–Wigner stability analysis. Journal of Physics A: Mathematical and Theoretical, 49(38): 385201.

Jenny, J. P., Anneville, O., Arnaud, F., Baulaz, Y., Bouffard, D., Domaizon, I., ... and Weyhenmeyer, G. A. 2020. Scientists' warning to humanity: rapid degradation of the world's large lakes. Journal of Great Lakes Research, 46(4):686-702.

Jose, L. and Kumar, C. 2011. Evaluation of pollution by Palmer's algal pollution index and physico-chemical analysis of water in four temple ponds of Mattancherry, Ernakulam, Kerala. Nature, Environment and Pollution Technology, 10(3): 471-472.

Kosankar, S. P. 2020. Biomonitoring of lakes of Nagpur city, Maharashtra, India. Journal of Indian Association for Environmental Management (JIAEM), 40(4):12-24.

Li, Z., Gao, Y., Wang, S., Lu, Y., Sun, K., Jia, J., and Wang, Y. 2021. Phytoplankton community response to nutrients along lake salinity and altitude gradients on the Qinghai-Tibet Plateau. Ecological Indicators, 128:107848.

Nagpur Environmental Status Report (2017-18). National Environmental Engineering Research Institute,

Nandan, S. N. and Aher, N. H. 2005. Algal community used for assessment of water quality of Haranbaree dam and Mosam river of Maharashtra. Journal of Environmental Biology, 26(2):223-227.

Padisák, J., Borics, G., Grigorszky, I. and Soróczki-Pintér, É. 2006. Use of phytoplankton assemblages for monitoring ecological status of lakes within the Water Framework Directive: the assemblage index. Hydrobiologia, 553:1-14.

Pick,F.R.2016. Blooming algae: a Canadian perspective on the rise of toxic cyanobacteria. Canadian Journal of Fisheries and Aquatic Sciences, 73(7):1149-1158.

Poikane, S., Kelly, M. G., Free, G., Carvalho, L., Hamilton, D. P., Katsanou, K., ... and Irvine, K.2024. A global assessment of lake restoration in practice: New insights and future perspectives. Ecological Indicators, 158:111330.

Ptacnik, R., Solimini, A. G., Andersen, T., Tamminen, T., Brettum, P., Lepistö, L., ... and Rekolainen, S.2008. Diversity predicts stability and resource use efficiency in natural phytoplankton communities. Proceedings of the National Academy of Sciences, 105(13):5134-5138.

Purina, I., Labucis, A., Barda, I., Jurgensone, I. and Aigars, J. 2018. Primary productivity in the Gulf of Riga (Baltic Sea) in relation to phytoplankton species and nutrient variability. Oceanologia, 60(4), 544-552.

Recknagel, F., Orr, P. T. and Cao, H.2014. Inductive reasoning and forecasting of population dynamics of Cylindrospermopsis raciborskii in three sub-tropical reservoirs by evolutionary computation. Harmful Algae, 31:26-34.

Reiss, J., Bridle, J. R., Montoya, J. M. and Woodward, G. 2009. Emerging horizons in biodiversity and ecosystem functioning research. Trends in Ecology & Evolution, 24(9): 505-514.

Rombouts, I., Simon, N., Aubert, A., Cariou, T., Feunteun, E., Guérin, L., ... and Artigas, L. F. 2019. Changes in marine phytoplankton diversity: Assessment under the Marine Strategy Framework Directive. Ecological Indicators, 102:265-277.

Søndergaard, M., Jeppesen, E., Lauridsen, T. L., Skov, C., Van Nes, E. H., Roijackers, R., ... & Portielje, R. O. B. 2007. Lake restoration: successes, failures and long-term effects. Journal of Applied ecology, 44(6): 1095-1105.

Sulata, K. and Devashish, K. 2013. Studies on Zooplankton Diversity of an Oxbow Lake of South Assam, India, International Journal of Current Research,5(12): 3652-3655.

Times of India- City's water bodies are drowning in pollution: ESR by Manka Behl .2021. https://timesofindia.indiatimes.com/city/nagpur/citys-water-bodies-are-drowning-in-pollution-esr/articleshow/81877948.cms

Verpoorter, C., Kutser, T., Seekell, D. A., and Tranvik, L. J. 2014. A global inventory of lakes based on high-resolution satellite imagery. Geophysical Research Letters, 41(18): 6396-6402.

Vidya, V., Sumathy, K., and Prasad, G. 2014. Phytoplankton as Pollution Indicators: A Case Study from Wetland Areas of Vembanad Lake Adjacent to Sea Food Processing Facilities. Journal of Aquatic Biology and Fisheries, 2:835-842.

Xi, Y. L., Liu, G. Y. and Jin, H. J. 2002. Population growth, body size, and egg size of two different strains of Brachionus calyciflorus Pallas (Rotifera) fed different algae. Journal of Freshwater Ecology, 17(2):185-190

Znachor, P., Nedoma, J., Hejzlar, J., Seďa, J., Komárková, J., Kolář, V., ... and Boukal, D. S.2020. Changing environmental conditions underpin long-term patterns of phytoplankton in a freshwater reservoir. Science of the Total Environment, 710:135626.

15

Assessing Plankton Diversity and Abundance in Anandwan Lakes Maharashtra: Implications for Water Quality

V.A. Meshram, Anurag Kr. Singh and S.R. Verma

ABSTRACT

The tiny aquatic microorganisms known as zooplankton are found in the water column and include aquatic mites, rotifers, larvae of open-water insects, and crustaceans. The fundamental food supply for fish and other aquatic animals in any aquatic habitat is plankton. When evaluating water quality, one of the most crucial ecological factors is plankton diversity. Because plankton communities are extremely sensitive to changes in the environment, research on them may be helpful in predicting long-term changes in lake ecosystems. Plankton communities are also sensitive to anthropogenic influences. Since they are most delicate floating population and is the primary target of water pollution, any unfavourable changes to the aquatic ecosystem have an impact on the community's biomass and variety. Quick generation period and rapid population renewal, phytoplankton are thought to be useful indicators of water quality and trophic conditions.

A study was carried out to examine the diversity and density of planktons in Anandwan, Warora Dist. Chandrapur (Maharashtra).The water sample were collected from three lakes in Anandwan region. The work was carried out for a period of three months January 2022 to March 2022.

The whole zooplankton species observed has been composed of Arthropoda, Copepods, Cladocera and Rotifera. About 14 genera of zooplankton were reported. The genera Copepod and Rotifers were more abundance during study period. Phytoplanktons, Chlorophyceae groups i.e. Chlorella Sp., Chara Sp. Chlymadomonas Sp., MicrosporaSp,. Ulothrix Sp., Oedogonium Sp.. Spirogyra Sp., and Spirogyra Sp. Shows dominance and flourishingly present in lake.

INTRODUCTION

Heterotrophic organisms, such as zooplankton, use phytoplankton as food, regenerating nutrients through metabolic processes and transferring energy to higher trophic levels. Zooplankton are involved in the conversion of waste matter into animal food that may be consumed, in addition to helping to shift food from the primary to the secondary level. These species act as a bridge in the food chain, carrying energy from the primary producers, planktonic algae, to the larger predatory invertebrates and fish that eat them (Kumari *et al.*, 2008). Macróplankton (200–2000 μm), microplankton (20–200 μm), nanoplankton (2–20 μm), picoplankton (0.2-2 μm), and femtoplankton (0.02-0.2 μm) are the different sizes of plankton (Kumar *et al.*, 2023).

The sensitivity of zooplankton to alterations in aquatic environments is significant. Differences in species composition, abundance, and distribution of body size can be used to identify the consequences of environmental disturbances(Joshi & Joshi, 2011). A number of variables, including temperature, sunlight, ocean currents, and nutrition availability, affect plankton populations. They are vital to ecosystems because they support carbon cycling, global biodiversity, and the general health of the oceans (Cottenie *et al.*, 2001). An evaluation of the pollution state heavily depends on the interaction between phytoplankton diversity and environmental conditions. The diversity of algae found in the Chlorophyceae family is a key sign of the quality of the water. The differences in water quality can be attributed to a combination of natural and human-caused factors. Owing to extensive human activity, anthropogenic inputs from various sources are typically the main variables influencing the water quality of the majority of rivers, lakes, estuaries, and seas, particularly those that are next to densely populated areas(Jeyaraj *et al.,* 2016). Living things, including plants, plankton, animals, and bacteria, are known as bioindicators, and they are used to monitor the state of the environment's natural ecosystem. They are known across the world as aquatic pollution indicator species. Every organic entity in a biological system gives information about the state of its surroundings. For example, plankton responds quickly to changes in the environment and is a valuable biomarker for determining the quality of water as well as a sign of pollution. Plankton is the best indicator of the health of aquatic flora and serves as an early warning system(Wadjikar *et al.*, 2017). An ecosystem that is in balance is one in which beneficial interactions occur between the environment and living organisms. Since water quality is essential to maintaining an environment in balance, it is evident that it plays a crucial role in this relationship(Jeyaraj *et al.*, 2016).

The varying characteristics of the water bodies, variances in the makeup of abiotic elements in the soil and water, and fluctuations in the productivity of various water bodies could all be contributing factors to the changes in the occurrence of peaks in zooplankton(Hasan and Tewari, 2023). From an ecological standpoint, zooplankton are among the most significant biotic elements that impact every aspect of an aquatic ecosystem's functioning, including food webs, food chains, energy exchange, and matter cycling(Ismail & Mohd Adnan, 2016). The pre-monsoon season had a strong phytoplankton dominance, but the monsoon season has the largest diversity compared to the other seasons, according to the examination of the diversity indices of the Margalef D index, Shannon-Weiner, and Simpson's index of diversity(Nath *et al.,* 2020). During the summer, zooplankton predominated, and the presence of indicator species at the sampling station, such as Moina micrura, suggests that the river is contaminated(Dahegaonkar, 2023). Summer was the season with the highest plankton population, followed by the monsoon and winter. It was also determined that the biological and physico-chemical parameters were appropriate for fish culture(Verma, 2020). The distribution of the zooplankton population is influenced by a variety of elements, including changes in the climate, physical and chemical characteristics, and amount of vegetation. The majority of planktonic organism species are found around the world(Perumal *et al.,* 2009). This study aims to find the diversity of plankton to analyze the environment of the aquatic ecosystem situated in Anandwan.

METHODOLOGY

Site Selection

Water sample were collected between 8:00 am to 10:00 am in January, February and March (Winter season) from three lake of Anandwan region. Thus, 3 sampling point for Zooplankton and 3 sampling point for Phytoplankton were selected for the convenience of sampling during the study period.

Physico-Chemical Characterization

Coolers were used to store the water samples as soon as they were collected to preserve their physical integrity. The samples were then brought to the lab for examination. A mercury-in-glass thermometer was used to measure the temperature on-site. A pH meter was also used on-site to measure pH. The remaining eight parameters were ascertained using the procedures outlined by APHA (1985): total suspended solids (TSS), total dissolved solids (TDS), chemical oxygen demand (COD), biological oxygen demand (BOD), conductivity, dissolved oxygen, alkalinity and Hardness (total).

Plankton sampling

Zooplankton samples were collected by passing water through plankton net of mesh size 50μm, 50 litres of water was passed through the plankton net and the final concentration of plankton sample is 20ml to this 10ml of 4% formalin is added which acts as fixative as well as preservative.

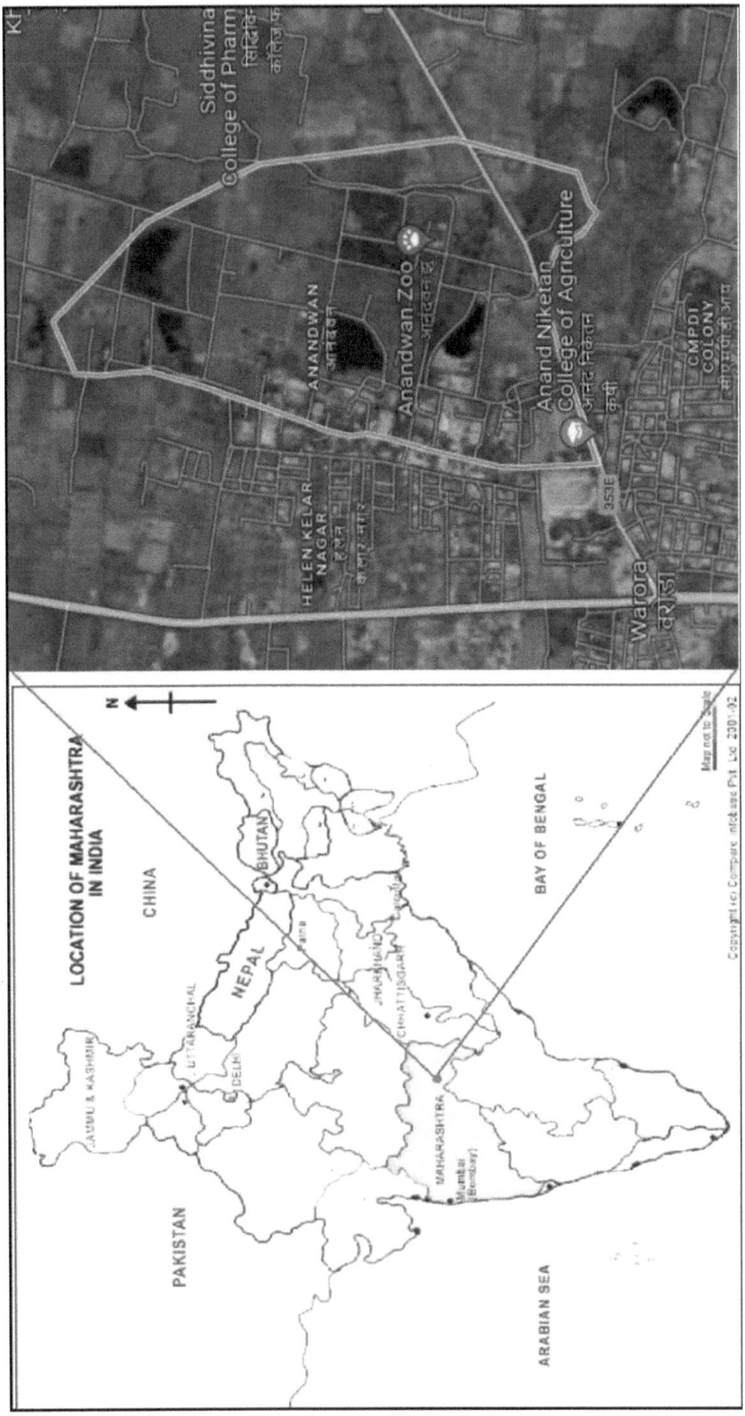

Fig. 1: Satellite View of Study area Anandwan

Likewise, Phytoplankton samples were collected by passing water through plankton net of mesh size 50 µm, 50 litres of water is passed through the plankton net and the final concentration of plankton sample is 20 ml to this 10ml of Lugol's solution is added which acts as fixative as well as preservative. Total 18 sample per month were collected. Out of 18, 9 samples for zooplankton and 9 samples for phytoplankton.

Identification and counting of plankton

i. **For Zooplankton:**Sedgwick rafter counting cell is used to count the plankton. Sedgwick rafter cell is approximately 50mm long, 200mm wide and 1mm deep. A trinocular compound microscope is used to count the zooplankton with different eyepieces such as 10x and 40x . The microscope is calibrated using micrometer. The following formula is used for quantities analysis of zooplanktons.

$$Number / ml = \frac{Cx1000}{AxDxF}$$

where,

C = no. of organism courted

A = area of field

D = depth of field(mm)

F = no. of fields counted

Density was expressed on individual/ liter. Finally to zooplankton were identified in laboratory with the help of keys provided by (Slotwinski *et al.*, n.d.; Witty, n.d.)

ii. **For phytoplankton**:For phytoplankton analysis, the drop count method was used. An ordinary 4ml dropper was taken and samples were sucked in the dropper and 2-3 drops were taken on the slide for observation. Organisms present in this drop were counted in high power number of deep forming 1ml for the dropper was taken into consideration the following formula was used. The phytoplankton density was expressed on units/liter.

Phytoplankton (density): N= abc/L

where,

N = no of phytoplankton/ lit.

a = average number of individuals in one drop

b = number of drops forming 1ml

c = volume of concentration

L = original water sample(liter)

Finally zooplanktons were identified in laboratory with the help of keys provided by(Bay Saldanha, 2001; Slotwinski *et al.*, 2014)

RESULTS

Throughout the study period, some alterations in various physico-chemical parameters were noted. The climatic, geochemical, geomorphologic, and pollution circumstances define the physical and chemical characteristics of freshwater bodies. The water's quality affects the quality of aquatic life. It is critical to research the physical and chemical elements that affect the water bodies' biological productivity. Physico-Chemical Characterization of samples were analysed as per the standard method (Baird B. Rodger *et al.,* 2017). The result of physico-chemical analysis of lake water are given in (Table 1).

Table 1: Physico-Chemical Characterization for Anandwan Lakes

Sr. No.	Test Parameter	Unit	Lake 1	Lake 2	Lake 3
1.	pH	-	7.69	7.26	8
2.	Temperature	°C	27	20	25
3.	Total suspended solids	mg/L	18	32	20
4.	Total dissolved solids	mg/L	688	614	690
5.	Chemical Oxygen Demand	mg/L	36	215	40
6.	Biochemical Oxygen Demand	mg/L	8	56	10
7.	Conductivity	µS/cm	847	747	847
8.	Dissolved Oxygen	mg/L	2.5	2.4	2.5
9.	Alkalinity	mg/L	337	321	337
10.	Hardness (total)	mg/L	272	239	272

In the present study, density and diversity of plankton are analyzed on monthly basis and presented in (Table 2) and (Table 1.3) In the study a total 11 genera belonging to 3 groups (Copepods, Cladocera and Rotifers) of zooplankton were identified from three sampling sites. Out of the 11 genera, 5 belongs to Copepods, 5 belongs to Cladocera and 1 belongs to Rotifers as shown in (Table 2).

Table 2: Species Composition of Zooplankton at Three Anandwan Lake During January 2022 to March 2022.

Species	January 2022			February 2022			March 2022		
	Lake 1	Lake 2	Lake 3	Lake 1	Lake 2	Lake 3	Lake 1	Lake 2	Lale 3
Copepods									
Diptomus	07	02	09	05	02	06	03	01	03
Aeanthocyclops sp.	04	02	04	04	02	05	03	00	02
Mesocyclops	05	05	07	04	02	04	03	01	03
Tropocyclops	04	02	08	03	03	04	04	02	07
Nauplius larva	25	15	21	12	14	16	08	08	08
Cladocera									
Ceriodaphnia dubia	09	04	05	04	02	05	03	02	02
Ceriodaphnia	03	06	08	03	03	02	01	01	01
Daphnia	04	02	05	04	05	03	01	03	04
Leptodioptomus sp.	02	04	03	01	01	02	02	02	01
Oithona	06	01	06	01	01	01	02	01	01
Rotifers									
Brachionus caudatus	20	10	25	14	10	25	15	8	14

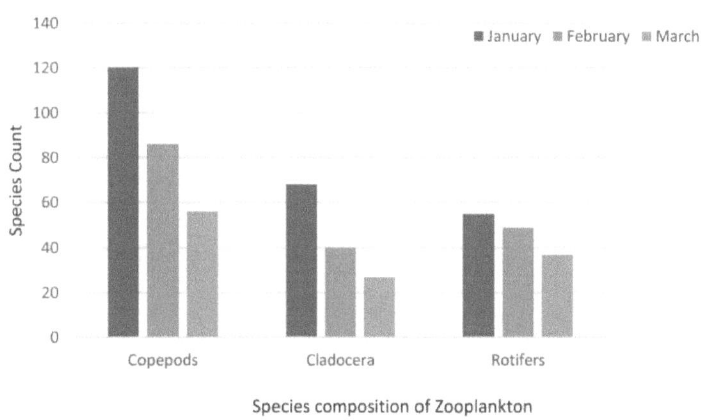

Fig. 2: Graphical representation of month wise composition of Zooplankton

In the study a total 19 genera belonging to 8 groups (Chlorophyceae, Cyanophyceae, Bacillariophyceae, Fragilariophyceae, Zygnematophyceae, Cryptophyceae, Euglenophyceae, Florideophyceae) of phytoplankton were identified from three sampling sites. Out of the 20 genera, 9 belongs to Chlorophyceae, 2 belongs to Cyanophyceae, 1 belongs to Fragilariophyceae, 2 belongs to Zygnematophyceae, 1 belongs to Cryptophyceae,1 belongs to Euglenophyceae and1 belongs to Florideophyceae as shown in (Table 1.3).

Table 3: Species Composition of Zooplankton at Three Anandwan Lake During January 2022 to March 2022

Class	Genus	January 2022			February 2022			March 2022		
		Lake 1	**Lake 2**	**Lake 3**	**Lake 1**	**Lake 2**	**Lake 3**	**Lake 1**	**Lake 2**	**Lake 3**
Chlorophyceae	*Chlamydomonus*	03	00	05	06	00	04	02	00	02
	Pediastrum duplex	04	01	04	03	00	03	02	00	04
	Ulothrix	05	00	05	02	00	03	02	00	05
	Cladophora	04	00	04	07	00	08	02	00	03
	Filamentous algae	04	01	04	08	01	06	02	00	04
	Non-filamentous algae	03	00	03	02	00	03	03	00	05
	Microspora	06	00	09	02	00	03	03	00	07
	Odogonium	05	01	05	06	02	06	09	00	07
	Spirogyra	06	02	07	09	02	09	05	01	08
Cyanophyceae	*Cylindrospermum*	25	03	30	30	03	25	21	02	17
	Anacystis	13	02	10	13	03	20	14	01	28
Bacillariophyceae	*Melosira*	14	00	16	19	04	27	16	00	07
	Diatom	18	05	14	16	03	15	16	01	29
Fragilariophyceae	*Fragilaria capucinax*	20	03	25	20	09	20	25	02	22
Zygnematophyceae	*Closterium*	15	07	19	21	01	19	16	00	26
	Desmidium	17	01	15	11	00	10	04	01	04
Cryptophyceae	*Chilomonas paramecium*	30	01	36	30	08	30	21	02	32
Euglenophyceae	*Phacus*	21	06	28	26	03	22	10	01	25
Florideophyceae	*Batrachospermum*	20	07	29	30	06	25	12	01	29

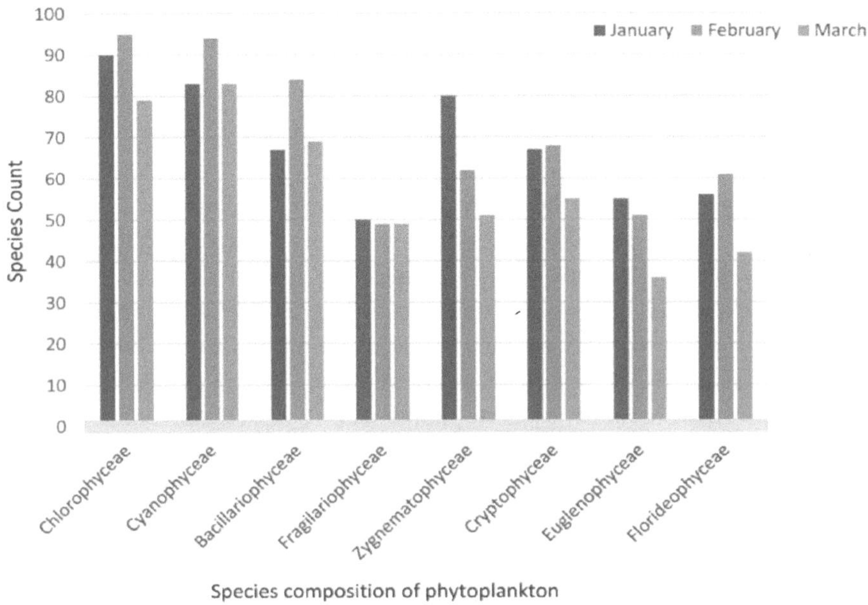

Fig. 3: Graphical Representation of Month Wise Composition of Phytoplankton

DISCUSSION

The average water temperature of the lakes was ranging between 25°C to 27°C during study period. The pH pf the lake was alkaline. D.O. is an important indicator of ability of a water body to support aquatic life(Reddy *et al.*, 2021). Dissolved oxygen levels in lakes vary according to their trophic levels, and depletion of DO in water probably is the most frequent result of water pollution(Jeyaraj *et al.*, 2016). Low D. O. Concentration i.e. <3 mg/lit in fresh water aquatic system indicates higher pollution in aquatic ecosystem.

In addition to being a common food source for many omnivorous and carnivorous fish, zooplanktons are significant aquatic animals that are found in large quantities in many kinds of aquatic habitats. The zooplankton community in the present study comprised of Copepods, Cladocera and Rotifers. Total 11 species of zooplankton were reported during the present study which belongs to four taxonomic groups that includes Copepods (5), Cladocera (5) and Rotifers (1). The Cladocera are the most dominant group in zooplankton. In clear waters, Cladocerans favoured to reside. The current investigation of Cladocerans revealed five species.Environmental conditions and phytoplankton distribution patterns were highly connected. In the present study, we found that, species from family Chlorophyceaewas dominating to others with 9 speciesEnvironmental factors had a strong correlation with the distribution patterns of phytoplankton. High percentage of dissolved oxygen is favourable for growth and development of Cyanophyceae and were recorded with two species. A few species, such as Phacus, were discovered in extremely contaminated waterways(Hemraj *et al.*, 2017).

CONCLUSION

The water quality of the lakes mentioned above is dangerous for human consumption; hence, significant attention is required. According to practically all of the water quality measures examined, the current study found that the lake's water quality is contaminated. According to WHO recommendations and drinking water regulations, the surface water of the lakes in our study area in the Anandwan region is unfit for residential use. We can gather fundamental information on the diversity of phytoplankton and zooplankton from the current study, which advances our understanding of the condition and biodiversity of the aquatic ecosystem. A total 11 species of Zooplankton and 19 species of Phytoplankton were recorded Numerous environmental circumstances can support the existence of zooplankton. They are also excellent bioindicators for determining the level of pollution in any freshwater body(Wadjikar *et al.*, 2017). The presence of copepodes, Cladocera and rotifers reveals that the Lakes are being organically polluted. During the investigation presence of plankton maximum in the January month and minimum in February.

Table 4: Zooplankton identified during study and its indication

Sr. No.	Names of Zooplankton	Indications	References
1.	*Diptomus*	Eutrophic	(V, 2023)
2.	*Aeanthocyclops sp.*	Polluted	(Sinha, 2003)
3.	*Mesocyclops*	Polysaprobic	(Kumari *et al.*, 2008)
4.	*Tropocyclops*	Polluted	(Sinha, 2003)
5.	*Ceriodaphnia*	Eutrophic	(Ismail & Mohd Adnan, 2016)
6.	*Daphnia*	Polluted	(V, 2023)
7.	*Leptodioptomus sp.*	Polysaprobic	(Sinha, 2003)
8.	*Oithona*	Polluted	(Srichandan *et al.*, 2018a)
9.	*Brachionuscaudatus*	eutrophic	(Srichandan *et al.*, 2018b)

Table 5: Phytoplanktons identified during Study and its Indication

Sr. No.	Name of Phytoplankton	Indications	References
1.	*Chlamydomonas*	Polluted	(V, 2023)
2.	*Pediastrum duplex*	Organic polution	(Perumal *et al.*, 2009)
3.	*Ulothrix*	Polluted	(Brasil *et al.*, 2020)
4.	*Cladophora*	Polysaprobic	(Kumari *et al.*, 2008)

5.	Filamentous algae	Organic polution	(Sharma *et al.*, 2013)
6.	Non-filamentous algae	Eutrophic	(Gao *et al.*, 2018)
7.	*Microspora*	Organic polution	(Malik & Bharti, 2012)
8.	*Odogonium*	Polluted	(Thakur *et al.*, 2013)
9.	*Spirogyra*	Polluted	(Brasil *et al.*, 2020)
10.	*Cylindrospermum*	Sewage pollution	(Sharma *et al.*, 2013)
11.	*Anacystis*	Eutrophic	(Hemraj *et al.*, 2017)
12.	*Melosira*	Eutrophic	(Thakur *et al.*, 2013)
13.	*Diatom*	Deteriorate	(Malik & Bharti, 2012)
14.	*Fragilaria capucina*	Polluted	(Parmar *et al.*, 2016)
15.	*Closterium*	Polluted	(Gao *et al.*, 2018)
16.	*Desmidium*	Polysaprobic	(Bay Saldanha, 2001)
17.	*Chilomonas paramecium*	Organic polution	(Hemraj *et al.*, 2017)
18.	*Phacus*	Polluted	(V, 2023)
19.	*Batrachospermum*	Polysaprobic	(Brasil *et al.*, 2020)

FUTURE PRESPECTIVES

Continuous monitoring of plankton diversity and abundance in aquatic ecosystems can provide valuable insights into seasonal and long-term variations. This data can be used to understand the impact of climate change, pollution, and other environmental factors on the lakes' ecosystem over time. The study can contribute to the development of effective water quality management strategies. Understanding the relationship between plankton and water quality can help in devising measures to maintain or improve water quality, which is crucial for both ecological health and human use. Investigating plankton diversity can provide insights into the overall biodiversity of the lakes. Future research could explore how changes in plankton populations might affect other species in the ecosystem, including fish, birds, and aquatic plants.

The data collected can be used to assess the risk of eutrophication in the lakes. Eutrophication is the process of water quality deterioration caused by excess nutrients, and understanding how plankton responds to nutrient inputs can help in preventing or mitigating eutrophication. The findings can be used for educational purposes to raise public awareness about the importance of preserving water bodies and their ecosystems. Outreach and education programs can still foster a sense of responsibility and engagement in local communities. Research findings can inform local and regional policies and regulations related to water quality and ecosystem conservation. Scientists can collaborate with policymakers to implement measures aimed at protecting these lakes.

ACKNOWLEDGEMENT

The authors are thankful to IHLR & SS, Anand Niketan College Anandwan Warora Dist: Chandrapur affiliated to Gondwana University, Gadchiroli for providing necessary facilities, support and encouragement.

REFERENCES

Baird B. Rodger, Eaton D. Andrew and Rice W. Eugene.2017. Standard methods for the examination of water and wastewater 23rd edition. American Public Health Association. https://doi.org/10.2105/SMWW.2882.216

Bay Saldanha. 2001. Phytoplankton: Identification Catalogue. Global Ballast Water Management Programme.

Brasil, J., Santos, J. B. O., Sousa, W., Menezes, R. F., Huszar, V. L. M. and Attayde, J. L. 2020. Rainfall leads to habitat homogenization and facilitates plankton dispersal in tropical semiarid lakes. Aquatic Ecology, 54(1):225–241. https://doi.org/10.1007/s10452-019-09738-9

Cottenie, K., Nuytten, N., Michels, E., and De Meester, L. 2001. Zooplankton community structure and environmental conditions in a set of interconnected ponds. In Hydrobiologia (Vol. 442).

Dahegaonkar, N. R. D. 2023. Studies on seasonal variation and diversity of zooplankton in river Zarpat in Chandrapur,Maharashtra,India. International Journal of Researches in Biosciences and Agriculture Technology. https: //doi.org /10. 29369/ ijrbat.2023.02.1.0028

Gao, H., Zhang, S., Zhao, R., and Zhu, L. 2018. Plankton community structure analysis and water quality bioassessment in Jiulong Lake. IOP Conference Series: Earth and Environmental Science, 199(2). https://doi.org/10.1088/1755-1315/199/2/022031

Hasan, Z. and Tewari, D. D. 2023. Plankton diversity of Kuwano River in Balrampur city,(U.P.) India.8(8).www.ijnrd.org.

Hemraj, D. A., Hossain, M. A., Ye, Q., Qin, J. G. and Leterme, S. C.2017. Plankton bioindicators of environmental conditions in coastal lagoons. Estuarine, Coastal and Shelf Science, 184, 102–114. https://doi.org/10.1016/j.ecss.2016.10.045

Ismail, A. H., & Mohd Adnan, A. A. 2016. Zooplankton composition and abundance as indicators of eutrophication in two small man-made lakes. Tropical Life Sciences Research, 27, 31–38. https://doi.org/10.21315/tlsr2016.27.3.5

Jeyaraj, M., Ramakrishan, K., Jai Anandhi, A., Arunachalam, S., &Magudeswaran, P. N. 2016. Investigation of physico-chemical and biological characteristics of various lake water in Coimbatore district, Tamilnadu, India. Oriental Journal of Chemistry, 32(4), 2087–2094. https://doi.org/10.13005/ojc/320436

Joshi, P. and Joshi, P. S.2011. Studies on zooplanktons of Rajura Lake of Buldhana district, Maharashtra India. http://www.jsrr.in

Kumar, J., Dwivedi, A. C. and Pal, P. 2023. Comparative study on plankton diversity of ponds (culture and non-culture) ecosystem in Prayagraj, Uttar Pradesh: a note. In Journal of the Kalash Science (Issue 11). https://www.researchgate.net/publication/372883578

Kumar Verma, A. 2020. Limnological studies of Muntjibpur pond of Prayagraj (U.P.) in relation to planktons. Www.Faunajournal.Com IJFBS, 7(4): 27–30. www.faunajournal.com

Kumari, P., Dhadse, S., Chaudhari, P., and Wate, S. 2008. A Biomonitoring of Plankton to Assess Quality of Water in the Lakes of Nagpur City.

Malik, D. S., and Bharti, U. (2012). Status of plankton diversity and biological productivity of Sahastradhara stream at Uttarakhand, India. Journal of Applied and Natural Science, 4(1), 96–103. www.ansfoundation.org

Nath, B., Borah, D. K., Deka, C., and Bhabesh, N. 2020. A study of plankton diversity in Kumri Beel, Goalpara, Assam, India. International Journal of Life Sciences International Peer Reviewed Open Access Refereed Journal Int. J. of Life Sciences, 8(1):145–148. www.ijlsci.in

Parmar, T. K., Rawtani, D., & Agrawal, Y. K. 2016. Bioindicators: the natural indicator of environmental pollution. Frontiers in Life Science, 9(2):110–118. https://doi.org/10.1080/21553769.2016.1162753

Perumal, N. V., Rajkumar, M., Perumal, P. and Thillai Rajasekar, K. 2009. Seasonal variations of plankton diversity in the Kaduviyar estuary, Nagapattinam, southeast coast of India. In Journal of Environmental Biology.

Reddy, R. M., Reddy, A. G. S. and Naveen, T. 2021. Impact Assessment of Artificial Recharge Structures on Kakatiya University Campus, Warangal Urban District, Telangana. Journal of the Geological Society of India, 97(4):375–384. https://doi.org/10.1007/s12594-021-1695-1

Sharma, C., Jindal, R., Bhan Singh, U., Singh Ahluwalia, A., Jindal, R., Ahluwalia, A., and Thakur, R. 2013. Population dynamics and species diversity of plankton in relation to Population dynamics and species diversity of plankton in relation to hydrobiological characteristics of river Sutlej, Punjab, India. In Copyright@ EM International,19(3). https://www.researchgate.net/publication/261983194

Sinha, B. 2003. Seasonal variation in zooplankton population of two lentic bodies at Assam State Zoo cum Botanical Garden, Guwahati, Assam. Ecology Environment and Conservation, 9(3):391–397.

Slotwinski, A., Coman, F., & Richardson, A. J. (n.d.). Introductory Guide to Zooplankton Identification. https://www.facebook.com/pages/Australian-PlanktonSurvey/325879294227720Websitehttp://imos.org.au/australiancontinuousplankton.html

Srichandan, S., Baliarsingh, S. K., Prakash, S., Panigrahy, R. C., & Sahu, K. C. (2018a). Zooplankton Research in Indian Seas: A Review. In Journal of Ocean University of China,17(5): 1149–1158). Science Press. https: //doi. org/ 10. 1007/s11802-018-3463-4.

Srichandan, S., Baliarsingh, S. K., Prakash, S., Panigrahy, R. C., and Sahu, K. C. (2018b). Zooplankton Research in Indian Seas: A Review. In Journal of Ocean University of China,17(5):1149–1158). Science Press. https://doi.org/10.1007/s11802-018-3463-4

Thakur, R. K., Jindal, R., Singh, U. B. and Ahluwalia, A. S. 2013. Plankton diversity and water quality assessment of three freshwater lakes of Mandi (Himachal Pradesh, India) with special reference to planktonic indicators. Environmental Monitoring and Assessment, 185(10), 8355–8373. https:// doi.org /10. 1007/ s10661-013-3178-3

Wadjikar, B., Masram, A., Rai, M., & Darvekar, A. 2017. Monitoring Water Quality Using Zooplankton Organisms As Bioindicators In The Well Water Of Nagpur City (M.H.) India, International Journal of Researches in Biosciences,Agriculture and Technology, Monitoring Water Quality, Issue 2). www.vmsindia.org

Witty, L. M. (n.d.). Practical Guide to Identifying Freshwater Crustacean Zooplankton. http://coopunit.laurentian.ca

16

Estimation of Physico-chemical Parameters of a Freshwater Perennial Lake: Satara Tukum

A.K. Pimpalshende and S.R. Sitre

ABSTRACT

Satara tukum is very small village located in Pombhurna tehsil of Chandrapur District of vidarbha region in Maharashtra state. The freshwater lake under study in this research work is located in the rural village of Satara Tukum. All village people depends on this lake water and use it for cloth washing, irrigation, farming and for use of domestic animals. The studies on physico-chemical characteristics of this fresh water lake was undertaken for one year period in during 2019-2020. During study period physical and chemical parameters viz. atmospheric temperature, water temperature, turbidity, conductivity, total dissolved solids, free CO_2, total alkalinity, pH, dissolved oxygen, total hardness, Calcium, chlorides, phosphate, nitrate, sulphate, Phosphate, COD and BOD were recorded. In all 17 different physico-chemical parameters were analyzed during our study on this freshwater ecosystem of perennial nature.

***Keywords**: Satara Tukum, Physico-chemical parameters, Pombhurna tehsil, Chandrapur district.*

INTRODUCTION

In nature water is considered as the most important natural resource which is utilized in large quantity throughout the world for drinking, washing, bathing, recreation, irrigation, aquaculture and many industries. Water also plays a fundamental role in regulation of water cycle in nature. India is a agriculture based country totally based on different water sources. The quality of water plays a important role to mankind because it is directly linked with human health. Lake water's are mainly used for farming and domestic activities like cloth washing. Increasing human activities continuously in lake basins is posing a great threat to all forms of life in aquatic environment. Most of the people who live in villages use irrigation facilities for farms and use excessive use fertilizers and pesticides for agriculture purposes and subsequently and degrade the aquatic ecosystems of lakes. We need to conserve water for future generations and safeguard our natural resources for sustainable development and biodiversity conservation.

Many researchers investigated the freshwater lakes and ponds in the Indian subcontinent *viz.* Bagde and Verma (1985), Desai (2014), Patel and Patel (2012), Nirbhavanae and Khobragade (2017), Mehta *et* al (2016), Gupta (2018), Bastola (2013), Raj and Sevankodiyane (2018), Rana Phul Kunwar Singh (2016), Shyamala and Hemvathy (2018), Meshram *et* al (2014), Fule and Nimgare (2018), Chaudhary and Sitre (2020), Pimpalshende and Sitre (2021).

The analysis of physico-chemical parameters points out towards existing water quality conditions of a particular lake water in a particular district and a real picture emerge from it about its status. In India a number of ponds, lakes and reservoirs are naturally found but they are not being utilized fully due to lack of insufficient knowledge about them.

Looking from this point present research is an attempt to bridge the gap between it. The physico-chemical conditions vary according to different climatic conditions and the surrounding factors which are recorded during present analysis of a perennial freshwater lake of Pombhurna tehsil of Chandrapur district in Maharashtra state during a year span on this beautiful lake.

MATERIALS AND METHODS

Satara Tukum is very small village located 12 km away from pombhurna tehsil and 25 km away from district head quarter in Maharashtra state in which the perennial freshwater body is situated. The Satara Tukum lake is about 194 mts above mean sea level and is at 19°89'56.63' N latitude and 79°62'98.79' E longitude. The water depth of Satara Tukum lake is near about 7 feet during the monsoon season and 5 feet during the summer season. Satara Tukum lake is perennial freshwater ecosystem which attracts bird in winter season due to abundance of flora and fauna present in it.

The study on this aquatic ecosystem was done during the year 2019-2020. Water samples were collected from 3 different sampling spots following standard protocol (APHA,1989). Collected water samples were immediately brought to Department of Zoology, N.S.College, Bhadrawati laboratory (IHLR & SS) every month and total 17 different physico-chemical parameters were analyzed. The physical parameters, air temperature, water temperature and pH of lake water were analyzed on the sampling spot itself while the other parameters turbidity, Dissolved Oxygen (D.O.), Conductivity, total alkalinity, total Hardness, calcium, chlorides, sulphate, total dissolved solids, nitrate, Phosphate, COD and BOD were determined using standard literature i.e. APHA (2005), Kodarkar (1992), Trivedi and Goel (1984) in the college laboratory.

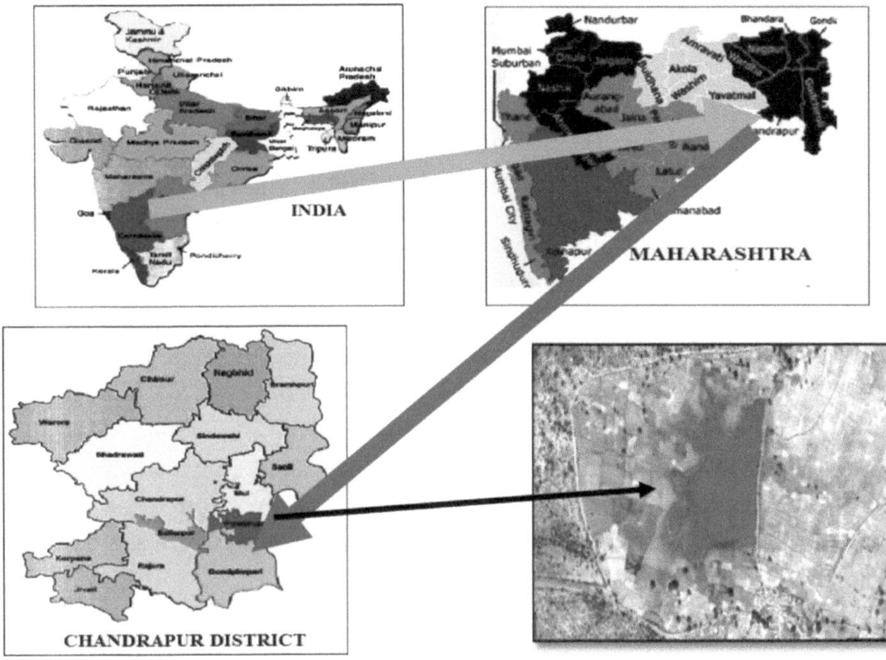

Satara Tukum lake (Aerial View)

Fig. 1: Study Area- Map of Chandrapur district & lake under investigation
(See colour version on page 352)

RESULTS AND DISCUSSION

The analyzed monthly values of all the parameters of satara Tukum lake water are are shown in table no 1. Air temperature affects the growth and reproduction of all algae, plants,vertebrates and invertebrates. Rising temperature reduces oxyen solubility in water.Highest temperature is recorded in May month while minimum temperature recorded in December month. Similar observations were recorded by Tichkule and bakre (2017), Khan *et al.* (2012) Supporting our work. The water temperature was clearly affected by seasonal changes of ambient atmospheric temperature.As temperature drops in winter months. Similar findings reported by Rahane (2015) who worked on Girna reservoir, supporting our findings. Turbidity was high in rains and low in summer seasons. Turbidity occurs due to presence of suspended particles in monsoon season in lakes. Turbidity makes water unfit for domestic purposes and our observations were supported by works of Nirbhavane and Khobragade (2017) who worked on Sion lake Mumbai. Water conductivity was high in summer and low in winter months. Similar observations were recorded by Sharma *et al* (2015) supporting our work. pH is a measure of acidic or alkaline nature of water and measures the concentration of H+ ions in water. The pH is very important for phytoplankton, zooplankton, benthos in water.The pH changes affect the flora and fauna .The toal hardness fluctuate between 120-150 mg/litre.Similar observations recorded by Chittora *et al* (2017) supporting our findings. BOD of lake ranged between 7-12.06 mg/litre.Similar observations recorded by Verma *et al* (2012) supporting our work. COD is a measure of oxygen utilized during the oxidation of organic matter by very strong chemical oxidizing agent like potassium dichromate.High COD is found in summer and minimum in winter months. Similar observations are recorded by Chaudhary and Pillai (2008) in Sasthamcottah lake supporting our work. Nitrate ranges from 0.2 to 0.42 mg/litre in lake water, which is maximum during summer. Similar observations were recorded by Arafat et al (2008) in Ullal Lake supporting our findings.

In water quality assessment dissolved oxygen is the most important parameter which reflects physical and biological processes prevailing in water. The supply of oxygen in aquatic ecosystems are mainly from physical and biological processes to maintain aquatic life.Very low oxygen can kill fish and other organisms.

Alkalinity is one of the important parameter of natural water.The carbonate and bicarbonate form the alkalinity. Nitrate is mostly dissolved in water and is highly soluble. Nitrates come from run off from agricultural fields. In present study nitrate values ranges from 0.2 to 0.5 mg/litre in a lake water.

In aquatic environment physico-chemical parameters exert a great impact on the biotic components.It controls biodiversity and distribution of aquatic comunities. The physical and chemical parameters exert their influence singally as well as in combination and their interaction creates abioti environment.In aquatic ecocsytems the balance get upset by man made activities, resulting in pollution. Thje adverse impact of pollutionis felt on the unique physical and chemical properties of water.

The physico-chemical and biological characters of lakes provide clear idea about the tropic status of the inland water bodies.The lakes provide abundent water for drinking and other household uses to mankind. The examination of water quality from a particular region is very important for developemental acivities of that region, because lakes and reservoirs are used for supplying water for all human needs as well as for comercial activities.

Table 1: Monthly Readings of Physicochemical Parameters : Satara Tukum Lake (2019-2020)

S. No.	Parameter	Feb.	Mar	Apr	May	Jun	July	Aug	Sep.	Oct.	Nov	Dec	Jan	Min. Value	Max. Value
1	Atmos. Temp.(0c)	30	34	42	43	40	33	29.2	28.9	29	28	24	28.2	**24**	**42**
2	Water Temp.(0c)	24.2	25.3	27.1	29.5	28.5	25.5	25.6	24.6	24.2	24.1	22	22.2	**22**	**29.5**
3	Turbidity (NTU)	9.8	10.1	12.5	13.05	13.5	13.4	25.1	24	22	18	17.1	16.2	**9.8**	**25.1**
4	pH	7.9	8.1	8	8.21	8.2	7.9	7.7	7.8	7.8	8	8.1	8.11	**7.7**	**8.21**
5	Cond. (MS/Cm)	256	289	304	305	298	250	220	225	235	248	252	256	**220**	**305**
6	TDS	160	185	191	198	201	198	210	208	201	199	195	165	**160**	**210**
7	DissolvedOxygen	6.2	6.0	5.5	5.2	5.6	5.7	6.3	6.7	6.8	6.9	6.8	6.4	**5.2**	**6.9**
8	Free CO_2	2.9	3.0	3.3	3.6	3.1	3.0	2.8	2.5	2.4	2.4	2.5	2.2	**2.2**	**3.6**
9	Total Alkalinity	105	110	114	114	118	113	136	140	125	115	109	106	**105**	**140**
10	Total. Hardness	128	135	144	150	148	144	132	128	126	129	120	121	**120**	**150**
11	Calcium	0.42	0.41	0.52	0.6	0.51	0.5	0.4	0.3	0.31	0.2	0.3	0.4	**0.2**	**0.6**
12	Chloride	7.8	9.8	9.8	10	10.1	10.1	10.5	9.7	9.7	9.5	9.7	9.6	**7.8**	**10.5**
13	Phosphate	3.8	4.0	4.3	5.0	4.9	4.7	4.5	4.8	4.8	4.9	4.6	4.7	**3.8**	**4.9**
14	Sulphate	3.6	3.8	4.5	4.9	4.9	4.8	4.3	3.9	3.6	3.8	3.1	3.2	**3.1**	**4.9**
15	Nitrate	0.2	0.25	0.3	0.5	0.4	0.42	0.42	0.26	0.33	0.35	0.4	0.3	**0.2**	**0.50**
16	BOD	7.0	10	11	12	11.8	12.06	11.8	11.5	11.1	10.1	9.9	9.85	**7.00**	**12.6**
17	COD	39.8	39.8	40.2	40	39.5	38.5	37.6	36	35.5	36	37	38.9	**35.5**	**40.2**

Note: All the parameters are in mg/litre (Except Sr.No. 1-5)

ACKNOWLEDGEMENT

The authors wish to express their sincere thanks to the Principal and Head, Institution of Higher Learning, Research and Specialized Studies (IHLR& SS) N.S. Science and Arts College Bhadrawati, District Chandrapur to carry out this research work at above mentioned research center.

REFERENCES

Agale M.C, Patil J.V. and Patel N.G.2013. Study of seasonal variations of phytoplankton and their correlation with physic-chemical parameters of Budakki Medium Irrigation tank, Shirpur Dist.Dhule,India. European J. of Zoology Research,2(3): 8-16.

APHA.2005.Standard Methods for the Examination of Water and Waste Water, 21st Edition, Washington, D.C.

Bagde, U.S. and Verma, A.K.1985. Physicochemical characteristics of water of J.N.U. lakes at New Delhi, India. Journal of Ecology. 12:151-156.

Bastola S. C.2013. Study of Physico-chemical parameter of Deepang Lake in Pokhara valley, Nepal. Janapriya Journal of Interdisciplinary Studies, 2(1):90-95.

Chaudhary G. and Site S. 2020. Physico-Chemical analysis of lake water quality: A study with special reference to Naleshwar Lake of Mul Tehsil, Maharashtra. Parishod Journal, IX:6478-6480.

Chaudhary Ruchira and Pillai R.S.2008. Studies of the physic-chemical parameters of Sasthamcottah lake (India) with reference to suitability for aquaculture. J.of Env. Res. and Devpt. 2(3): 402-405.

Chittora A.K., Chandrashekhar Kapoor and Vidhya Kapasya. 2017. Comparative Assessment of Physico-chemical Parameters of Udaipur City (Rajasthan) India. International J. of Env. Sci. & Nat.Resources. Vol 1(2): PP. 1-12.

Choyal, P.C., Chourey Vijay R., Dwivedi D., Choubey, V. 2021. Seasonal Variation of physicochemical parameters in water stream of Mahi river in M.P. region, India. JETIR, 8 (10): 509-524.

Desai, Pooja.2014. Water quality assessment of Lakes in Vashi, Navi, Mumbai, Maharashtra, India., Int. J. of Sci. Eng. and Res. Vol. 3 Issue7, 66-69.

Devi, A, V.Bashkarna and Sureshkumar R.2013. Physico-chemical parameter and zooplanktons diversity of a temple pond in Virudhnagar,tamil Nadu. Intl.J.of Sci.Env.& Technol. 2(2): 250-257.

Edmondson, W. T.1959.Freshwater Biology, 2nd Edition. John Willey and Sons, Inc. New York London:1148

Fule, C.W. and Nimgare, S.S.2018.Seasonal changes in Physico-chemical characteristics of various water sources in Ashti Tehsil of District Wardha (M.S.). Aayushi International Interdisci.Res. Journal,25:658-661.

Gupta, S.S. 2018. Physico-chemical status of water quality from water sources of Drug Dam, Taluka Kalamb, district Yavatmal, Maharashtra, India., Aayushi Int. Interdisciplinary Research Journal, Special Issue No. 25. Pp. 396-398.

Khan Rafiullah M, Jadhav M.J. and Ustad R.K.2012. Physicochemical analysis of Triveni Lake Waters of District Amravati In India. Bioscience Discovery, 3(1): 64-66.

Kodarkar M. S.1992.Methodology for Water analysis (Physicochemical, Biological & Microbiological)., Indian Association of Aquatic Biologists, Publication No.2.

Kudesia, V. P.1998.Water Pollution, Fourth Edition, Pragati Prakashan, Merut (India).

Mehta G., Deshbhratar S., Raut S. and Mahaley J.2016. Assessment of certain physicochemical parameters of Satpala Lake, virar, palghar, Maharashtra, India, Inter. Jour. of Innovative Research in Sci. Eng. and Tech..5 (8): 14598-14605.

Meshram, M. P., Cherian, K. J. and Bodkhe, S. S.2014. Studies on the Physico-chemical Status of Two Water Bodies at Nagpur City Under Anthropogenic Influences. International Journal of Researches in Bioscience, Agricultue and Technology. Vol II, pp.130-144.

Murlidharan, Patel A.C. and Patel,R.S.(2012): Comparison of the physicochemical parameter of two lakes at Lodra & Nardipur under Biotic stress. Int. Jour. of Sci. & Research Publication Vol. 2, Issue 9, pp. 01-07

Nirbhavane, Gangotri and Khobragade,Kshama.2017. Study of water quality of Sion lake Mumbai, Maharashtra. Scholors J. of Eng. & Tech. 5(8); 413-415.

Patel, A.C and R.S. Patel,R.S.2012.Comparison of the physicochemical parameter of two lakes at Lodra and Nardipur under Biotic Stress. Int. Jour. of Sci. & Research ,2,(9): 01-07.

Pimpalshende A.K. and Sitre, S.R.2021. Assessment of Physico-chemical parameters of Satara Bhosale lake of Pombhurna Tehsil of Chandrapur District in Maharashtra State. Journal of Emerging Technologies and Innovative Research, Vol.8. pp. d540-d545.

Rahane Balasaheb, Waykar Bhalchandra and Bhalla Resham.2015. Assessment of seasonal variations in water quality of Girna Reservoir in Nashik District (M.S.). Intl. Journal of Recent Sci.Res., 6(10):6570-6573.

Raj Arun J and SP Sevarkodiyone.2018. A study on physico-chemical parameters of Urinjikulam Pond, Thiruthangal (Virudhunagar District, Tamil Nadu), International Journal of Aquaculture and Fishery Sciences. Vol. 4(1): pp.010-012.

Rana Phul Kunwar Singh .2016. Physico-chemical study of Mohan Ram lake Shahdol (M.P). International Journal of Applied Research. Pp. 239-241.

Sharma J.N., Konkiya Shailesh Raj and Singh S.K.2015.Limnological study of water quality parameters of Dal lake India. International J. of Innovative Res.in Sci Eng. & Technol. 4(2): 380-386.

Shyamala, R and Hemavathy. E.2018. Physico-chemical parameters and land use patterns of Pulicat lake, Tamil Nadu, India., International Journal of Advanced Scientific and Technical Research,8 (6):10-37.

Solanki Meenakshi and Saraswat Heena.2020. analysis of water quality using physico-chemical parameters of river Narmada, Madhya Pradesh, India. Int. J. Adv. Res. 9(01), 754-757.

Tandale M.R. and D.S. Dabhade 2014. Study of some physico-chemical parameter of Lonar crater, India., Journal of Global Biosciences, 3: 941-950.

Tichkule, G.C. and Bakare, S.S.2017. Physicochemical analysis of Two fresh water lakes near Lakhani, Dist.-Bhandara(M.S.)., Asian Journal of Multidisciplinary Studies,5 (9):31-35.

Trivedi,R. K. and Goel, P. K.1984. Chemical and Biological Methods for Water Pollution Studies. Environmental Publication, Karad, India. pp. 1-251.

Verma S.R., Chaudhari P.R. Singh R.K and Wate S.R.2011. Studies on the ecology and trophic status of an urban lake of Nagpur city,India. Rasayan J.Chem., 4(3): 652-659.

17

Aquatic Insects Diversity in Lentic Waterbodies: Baseline Study

Sanyogita Verma and Sanyogta Meshram

INTRODUCTION

India is rich in biodiversity and possesses about 108276 species of insects. Indian subcontinent is one of the mega biodiversity countries of the world occupying ninth position in terms of freshwater mega biodiversity (Mittermeier *et al.,* 1997) comprising 30 order (Dijkstra *et al.* 2013). Inland wetlands of India serve as the habitat for more than 500 species of aquatic insects which are represented predominantly by Ephemeroptera (mayflies), Odonata (dragonflies) and Trichoptera (caddisflies), (Subramanian and Sivaramakrishnan, 2007). Aquatic insects play important ecological roles in keeping freshwater ecosystems functioning properly (Choudhary and Janak, 2015). The aquatic insects are important part of aquatic ecosystem which play vital role in ecosystem functioning and are known to be representative of structure and function of freshwater ecosystem due to their high abundance, high birthrate with short generation time, large and rapid colonization (Choudhary and Ahi 2015). Aquatic insect function on multiple trophic level such as shredders, collectors, scraper and predators on animal debris, wood and leaf litter reaching the wetland from the surrounding landscape. Nutrients processed by aquatic insects are further degraded into absorbable form by fungal and bacterial action. Plants in the riparian zone absorb this nutrient soup transported through the wetlands. In addition to this significant ecosystem function, aquatic insects are also a primary source of food for fishes and amphibians (Tachet *et al.*, 2003 ; Kumar, 2014).They interact in nutrient cycling within lentic and lotic systems and are regarded as an effective bioindicator (Lundquist and Zhu 2018) and serve as an important component of natural foodweb in aquatic ecosystem. However, the aquatic insect fauna of this part in India is inadequately documented. Therefore, present study carried out to delineate the baseline status of aquatic insects of Anandwan lakes at Warora Dist: Chandrapur.

MATERIALS AND METHODS

Aquatic insects were collected in post monsoon i.e October 2023-February 2024 by using "kick-net method" and "all out search method" from the midstream and downstream The kick-net is held against water current and an area of ($1m^2$) in front of the net is disturbed for one minute (Subramanian and Sivaramakrishnan, 2007). All out search method is used when water flows through boulder and cobble with high turbulence. Aquatic insects were collected from 10 square meter area for one hour. Within the sampling area, aquatic insects were searched in all the possible substrata such as boulders, cobbles, leaf litter and dead wood. A sable hairbrush or forcep was used for collection of all samples (Subramanian and Sivaramakrishnan, 2007). After collection, insects were preserved in 70% ethanol. Collected samples were examined under a motic stereozoom microscope and identified upto family using standard taxonomic literature (Merritt and Cummins, 1996; Subramanian and Sivaramakrishnan, 2007; Bouchard *et al* 2004; Winterbourn and Gregson, 1981; Morse *et al.*, 1994; Cummins and Merritt, 1996; Epler, 2001)

Study area

Anandwan was established with a few leprosy patients in 1949 by Shraddhey Baba Amte with a hope of giving marginalised people a dignified life through self-reliance and restoring a sense of belonging which they had lost due to the ill treatment at the hands of society. Anandwan is located in Chandrapur district of Maharashtra having co-ordinates 20.2604° N, 79.0235° E. Number of manmade ponds are constructed for irrigation , pisiculture and duck farming. These lakes play a keyrole to *replenish groundwater* and enhancing aesthetic value. The present study carried out at two lakes, because of their different surroundings and easily accessiblility for sampling. The details are described below :

Study sites	Geo coordinate position	Details of study sites
Lake 1	Lat 20.25802 Long. 79.02632	Surrounded by human settlement, receive sewage waste and the dominant vegetations
Lake 2	Lat 20.28505 Long. 79.06234	Surrounded by scrub forest, Inhabiting aquatic birds, limited number of aquatic vegetations

Lake 1: Near Human Settlement

Lake 2: Surrounded by Sparse Forest

Table 1: Checklist of Aquatic Insects in Study Areas

Order	Family	Scientific name
Hemiptera	Girridae	Gerris remigis (Water strider)
	Belostomatidae	Belostoma flumineum
	Nepidae	Ranatra elongate
	Corixidae	Corisella sp.
	Notonectidae	Notonecta glauca
	Nipidae	Ranatra filiformis
	Corixidae	Micronecta ludibunda
	Hydrometridae	Hydrometra martini
Diptera	Culicidae	Anopheles culifibacies
		Aedes sp.
		Culex sp.
	Chironomidae	Cladopelma sp.
		Chironomous sp.
Odonata	Aeshnidae	Anax immaculifrons
	Coenagrionidae	Ischnura posita
		Ceriagioncoro mandelianum
		Pseudagrian decorum
		Pseudagrion microcephalum
		Ischnura aurora (Brauer)
	Lestidae	Lestesthoracicus Laidlaw
	Gomphidae	Ictinogomphus rapax
	Libellulidae	Crocothemis servillia
		Orthetrum neglectum
		Zyxomma petiolatum
		Urothermis signata

Coleoptera	Dystiscidae	Predaceous Diving Bettle
		Laccophilus sp.
		Eretes sticticus
	Hydrophilidae	Hydraena quadricollis
		Hydeophilus olivaceous
		Hydrophillus piceus (Water scavenger bettles)
	Psephenidae	Water penny bettles
	Scirtidae	Scirtes sp.
Ephemeroptera	Heptageniidae	Rhithrogena sp.
Trichoptera	Philopotamidae	*Dolophilodes sp.*
Plecoptera	Pteronarcyidae	Pteronarcys sp.
Araneae	Argyronetidae	*Argyroneta aquatica (Water spider)*
Isopoda	Armadillidiidae	*Pillbugs*

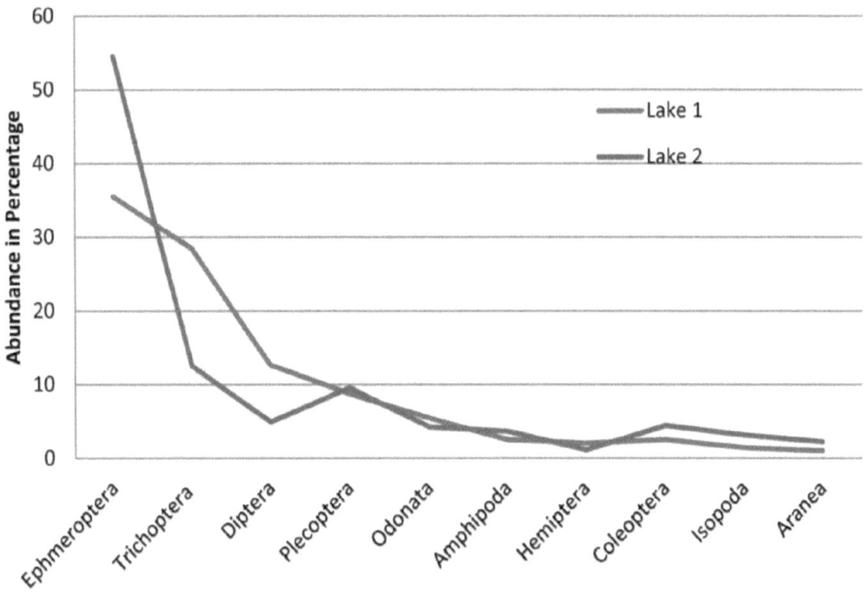

Fig. 1: Adundance (%)of Aquatic Insects Order in Sampling Locations

RESULT AND DISCUSSION

The Checklist of recorded aquatic insects is shown with their taxa, order, family and scientific name and common name given in Table 1. During the present study a total of 39 Genus of aquatic insects belonging to 10 orders and 24 families have been recorded from the two sampling sites of the study area. About 29 genus were recorded from the lake 1 and from lake 2 total 36 genus were recorded. Occurrence of diverse insect species in relation to luxuriant aquatic vegetation on account of shelter, breeding sites and food which has also been recorded at lake 2.

Family wise recorded species from study area showed highest number for hemiptera (08), followed by Odonata(05), Coleoptera (04) and only two from Diptera while only one from remaining families (Table 1). Details of abundance with reference to order in both lakes are depicted as Figure 1. While overall species diversity revealed that the insects from the order Odonata were the most dominant having distinguished 12 genus followed by Hemiptera and Coleoptera sharing equal number i.e. 08 genus and Diptera only comprising of 05 genus. While all others comprising only one genus from each family.

The order Odonata is regarded as a monophyletic group, divided into three suborders Zygoptera (damselflies), Anisoptera (true dragonflies) and a small suborder Anisozygoptera (damsel dragons). The adult of Odonata is terrestrial and larva is aquatic. Morphological variation occurs in all three suborders larvae, Damselflies larva have three or sometimes two caudal gills for respiration (also used as flippers for swimming) While Dragonfly larvae have lack of caudal gills and Damsel dragons also have no caudal gills, like dragonfly absorb oxygen through gills in the rectum (Thorp and Covich 2015). Mainly the Odonata larvae inhabit in freshwater but some species tolerate moderate pollution of water system. Coenagrionidaes and Libellulidae family is show the slightly highest abundance in lake 1 than lake 2 indicating eutrophic condition of Lake 1.

Coleoptera commonly called as order of beetle, large group of holometabolous insects constitute almost one–fourth of all known life-forms on earth. Beetles are aquatic or semiaquatic in nature sometimes larvae or adult or both are spending their part of lifecycle in water bodies. Typically the adults have strongly sclerotized body with the forewings hardened into elytra while larvae have sclerotized head, three pair of segmented thoracic legs and have no wing pad (Bouchard *et al;* 2004). Mostly Coleoptera are inhabits to pools and riffles area of rivers. Some aquatic beetle acts as good biomonitoring agent such as family Hydrophilidae, good biomonitor for heavy element and their presence indicates low dissolved oxygen content. Dystiscidae are predatory beetles, mostly inhabit shallow area present in high abundance in Lake 2. Prevalence of Dytiscidae is polluted. Indicative of the ecological health of studied lakes. Dytiscidae insects generally prefer leaves of submerged aquatic vegetations in clean freshwater lakes and are predacious in nature. In contrast, Hydrophilidae beetles inhabit shallower regions of water bodies with abundant macrophytes and feed on detritus, algae, decaying .

The aquatic Hemiptera are composed of semi-aquatic bugs (Gerromorpha) that live primarily on the water's surface and aquatic bugs (Nepomorpha) that live submerged beneath the water. About 97% of this diversity is associated with freshwater habitats (Thorp and Covich, 2015). Aquatic bugs are characterized by their half sclerotized elytra, piercing and sucking mouth parts, segmented antenna, some without wings have wing pad. Most of the aquatic bugs are carnivorous in nature and have modified leg for seizing and holding the prey. They are important as fish food, bioindicators and predators. Gerridae abundtly present in lake 1, confirming presence of organic pollution in Lake 1 due to runoff from human settlement

The order Diptera includes "true" or two-winged flies, divided into three suborders Nematocera, Brachycera and Cylorrhapha. Larval stage of Diptera inhabits water bodies and adults are terrestrial. Morphologically the larvae are distinguished by lacking jointed thoracic legs. On the basis of presence of their cranial structure larval dipterans show tremendous variation in morphological character. Nematocera larvae have well develop head with stout and toothed mandible, moving horizontally or obliquely, Brachycerans larvae are more claw-like, has fewer inner teeth, and moves vertically and Cyclorrhaphan larvae lack sclerotized head, and mouth hooked move vertically. Aquatic dipteran larvae play significant role in bio-monitoring of water quality, chironomids are good indicators rich source of organic material. Chironomus is abundantly high at lake 1. The genus Chironomus is recognized as a taxon characteristic of organic enrichment and has been widely used as an indicator of aquatic ecosystem health .

Ephemeroptera (Mayflies) is the oldest order of insects and they are widely distributed in all type of running and standing freshwater habitats. Typically, the nymphs have tracheal gills on their abdomen, unpaired tarsal claws, an enlarged mesothorax, usually three or sometime two caudal filaments. The adult stage of Mayfly having two wings called subimago and imago which are unique among the insects. Mayflies are pollution sensitive insects and they were used in the bioassessment and monitoring of freshwater bodies worldwide because of their relative abundance in a wide variety of substrates and their increasing chances of detecting pollution impacts (Arimoro and Muller 2010).

Plecoptera is commonly called as stonefly, small order of hemimetabolous insects. Nymph of stonefly are aquatic whereas adult are terrestrials. Morphologically stonefly have soft body, three segmented tarsi, elongate filiform antennae, mandibulated mouthparts, two compound eyes, two or three ocelli, two usually long cerci, 10- segmented abdomen with vestiges of the eleventh segment. They are used extensively as indicators of pollution and environmental change. Mostly, greatest diversity of stonefly nymphs in running water point out well-oxygen level of water and they are also called cold water specialists.

Trichoptera is an order of holometabolous insects, closely related to the Lepidoptera, the moths and butterflies. Which are commonly known as Caddisfly and their larval and pupal stages are aquatic. Morphologically the larva is easily recognized by their segmented thoracic legs, a pair of anal prologs with a single curved terminal claw (very short), sometimes almost invisible, single segment antennae and hair like abdominal gills. On the basis of case making behaviour the larvae are divided into five groups (Ward and Whipple 1966) namely purse-case makers (Hydroptilidae), Saddle-case makers (Glossosomatidae), tube-case makers (Limnephilidae), retreat makers (Philopotamidae, Hydropsychidae, Polycentropidae) and free-living (Rhyachophilidae). They are important in aquatic ecosystem because they process organic material and important food source for fish. Like Ephemeroptera, Plecoptera and many Tricoptera species are sensitive to pollution and use as water quality indicator (Diantari *et al.* 2017).

CONCLUSION

Prior to this study, the aquatic fauna of Anandwan was poorly known Current study outlined the patterns of composition and distribution of different aquatic insects. Occurrence of more diverse and abundant insects in lake-2 than lake-1 suggested the presence of luxuriant aquatic vegetation which is necessary for shelters, oviposition sites and foods. Another may be the larger size and greater depth than the other. Low abundance of hemipteran insects in lake-2 suggested that this Lake was less polluted. However the findings of this study are not conclusive and will contribute in future in-depth studies. Further studies are needed to identify the seasonal pattern, preference of habitat and positive or negative impact of environmental changes on entomofauna.

ACKNOWLEDGEMENT

Authors are grateful to the Principal Anand Niketan College Anandwan, Warora for extending support and providing laboratory facilities.

REFERENCES

Bouchard, R.W., Ferrington, L.C. and Karius, M.L., 2004. Guide to aquatic invertebrates of the Upper Midwest. St. Paul: Water Resources Centre, University of Minnesota, 208 p.

Choudhary, A. and Ahi, J. 2015. Biodiversity of freshwater insect: A review. The International Journal of Engineering and Science, 4(10): 25-31

Choudhary, A. and Janakahi.2015. Diversity and distribution of aquatic insect population in Lakha banjara Lake, Sagar (MP), India. Journal of International Academic Research for Multidisciplinary, 3(5): 367.

Cummins, K. and Merritt, R.W., 1996. An introduction to the aquatic insects of North America. Dubuque: Kendall Hunt Publishing Company

Diantari N.P.R, Ahyadi H, Rohyani I.S, Suana I.W.2017. Diversity of Ephemeroptera, Plecoptera and tricoptera as bioindicator of water quality in Jangkok River, West Nusa Tenggara. Indonesian Journal of Entomology, 14(3): 135-42

Epler, J. H., 2001. Identification manual for the larval Chironomidae (Diptera) of north and south Carolina. Mooresville: North Carolina Department of Environment and Natural Resources.

Francis O Arimoro 1, Wilhelmine J Muller. Mayfly (Insecta: Ephemeroptera) community structure as an indicator of the ecological status of a stream in the Niger Delta area of Nigeria. Environ Monit Assess. 2010 Jul;166(1-4):581-94. DOI: 10.1007/s10661-009-1025-3

Kumar, A. 2014. Studies on qualitative and quantitative abundance of aquatic entomofauna in glacial fed mountainous Goriganga river of Kumaun Himalaya Uttrakhand, India. International Research Journal of Environment Science, 3(4): 51-63.

Lundquist, M.J. and Zhu, W. 2018. Aquatic insect functional diversity and nutrient content in urban streams in a medium-sized city. Ecosphere, 9(5): 1-11.

Merrit, R.W., Cummins, K.W., Berg, M.B. 2008. An introduction to the aquatic insect of Rhgimhart, M. 1889. Contributions a la faunc indo-chinoisc. 2c memoire. Hydrocantharcs. - Annls Soc. ent. Fr., (6)9: 147-156.

Merritt, R.W., K.W. Cummins and M.B. Berg, 2008. 15. McAleece, N., 1998. Biodiversity Professional Beta. An Introduction to the Aquatic Insect of North The Natural History Museum and The Scottish America. 4 ed., pp: 1158..Mittermeier P. Robles Gil and C.G. Mittermeier, (Eds.) Megadiversity: Earth's Biologically

Morse, J.C. and Yang, L. and Tian, L., 1994. Aquatic insects of China useful for monitoring water quality. Nanjing: Hohai University Press.

Sivaramakrishnan, K. G., Subramanian, K. A. and Ramamurthy, V. V. 2009. Annotated checklist of ephemeroptera of the Indian subregion. Oriental Insects, 43(1): 315– 339.

Tachet, H. Richoux, P., Bourneau, M., Usseliopolatera, P. 2003. Invertebratesd eaudouce, systematique, biologie, ecologie. Fresh water Invertebrates: Systematic, Int.J.Curr. Microbiol.App.Sci. 5(9): 273-281 281

Thorp and Covich's (2020).Freshwater Invertebrates, Volume 5: Keys to Neotropical and Antarctic Fauna, Fourth Edition, ISBN 978-0-12-804225-0 DOI: .

Wahizatul, A.A., Long, S.H. and Ahmad, A. 2011. Composition and Distribution of Aquatic Iinsect communities in relation to Water quality in two Freshwater streams of Hulu Terengganu, Terengganu. Journal of Sustainability Science and Management, 6(1): 148-155.

Ward, H.B. and Whipple, G.C. 1996. Freshwater Biology. Ed. W.T. Edmondson. University of Washington Wealthiest Nations, CEMEX, Monterrey, Mexico. Pp 39-49

Westfall, M.J. Jr. and K.J. Tennessen.1996. Odonata. possible cause of insect abundance and richness in Lake In: R.W. Merrit and K.W. Cummins, (eds). An 1 Introduction to the Aquatic Insects of North America two lakes. 3rd Edition. Kendell/Hunt Publishing Company.

Winterbourn, M.J. and Gregson, K.L.D., 1981. Guide to the aquatic insects of New Zealand. Bulletin of the Entomological Society of New Zealand, 5: 1-80.

18

Rotifers Deteriorating Significantly The Water of Madhuganga Lake at Kalmeshwar, District Nagpur

Aasidhara Darvekar

ABSTRACT

Analysis of the community structure, zooplankton diversity and density of rotifers in lentic dam water of Madhuganga lake at Kalmeshwar in Nagpur District, was carried out during January to December, 2023. During analysis 10 species of rotifer came across belonging to 5 genera, 4 families and 2 orders. Order Ploima has maximum species richness (8) around, 87% belonging to two families,viz. Asplanchnidae & Brachionidae. The genera Brachionus was the most diversified with five species, namely, Brachionus plicatilis, Brachionus falcatus, Brachionus calyciflorus, Brachionus diversicornis and Brachionus caudatus . Testudinella sp belonhing to order Floscularinaceae was seen only in summer season was rarest zooplankton. These zooplankton play a vital role in Eutrophication of the dam water hence detoriating the pristine waters in rural regions.

***Keywords**: Brachionidae, Floscularinaceae, Madhuganga lake, Rotifera.*

INTRODUCTION

Wheeler bearers commonly referred to as 'Rotifers 'are significantly they are often taken as reliable indicators of mesotrophic characteristics in the lake ecosystems (Gannon and Stemberger, 1978). Because of their wide tolerances to varied dissolved oxygen levels and temperature distributions, they are even found to inhabit the deeper waters of aquatic ecosystems (Esparcia *et al.*, 1989). They belong to the Kingdom Animalia, Subkingdom Eumetazoa, Phylum Rotifera (Cuvier, 1798) having three classes Monogononta, Digononta and Seisonidea (Pennak 1953). In other words these wheel animalcules named, because of the presence of their corona, which is composed of several ciliated tufts around their mouth region. This ciliated tuft helps in locomotion and resembles a wheel. These create a current that sweeps food into the mouth, where it is chewed up by a characteristic pharynx (called the *mastax*) containing a tiny, calcified, jaw-like structure called the *trophi*. The cilia functions to pull the animal, when unattached, through the water. Most free-living forms have pair of posterior toes to anchor themselves while feeding. Rotifers are bilaterally symmetrical found in various shapes. There is a well-developed cuticle, thick and rigid, providing a box-like shape, or flexible, giving the animal a worm-like shape; such rotifers are respectively called *loricate* and *illoricate*.

Habitats of Rotifers are relatively easy to find. Many live in ponds, moist soil, or any stagnant water. Rotifers can be free swimming or sessile. Rotifers are mostly omnivorous and some have been observed to be cannibalistic. They normally eat algae or decomposing organic material. As they are a major food source for Cladoceran , Copepodans , juvenile fishes (fry) and fishes, therefore , an important food link of water ecosystem (Arora and Mehra 2003). Presence of abundant Rotifers in freshwaters determines them to be one of the main groups of zooplankton in limnological studies. They are permanently and obligatorily connected to aquatic habitats in all active stages and only their resting stages are drought-resistant (Chopra *et al* 2012). Various researchers have conducted pioneer study on the diversity and habitat of rotifers such as Anderson 1889; Sharma 1987; Sharma 1998a; Gilbert and Schreiber 1998; Arora and Mehra 2003; George *et al.* 2011; Tyor and Chawala 2012. However such studies on Madhuganga dam water has not been conducted, hence for assessment of Madhuganga Lentic water has been conducted for the seasonal variation in diversity and density of rotifers.

MATERIALS AND METHODS

Madhuganga Lake: The lake is situated around 15 km from Kalmeshwar in Nagpur District of Maharashtra. **Coordinates:** 21°18'51"N 78°47'43"E. Human interference is seen in most of the parts of the banks of the lake. Fishing is done by local fishermen on daily basis. The locations of sampling stations were selected as per the accessibility. Six sampling stations (M1-M6, Fig:1) were selected for collecting water samples for Zooplankton analysis (Figure 1). Monthly water samples were collected during morning hours (between 7:00 am & 8:00 am) for analysis of Zooplankton population during January 2023 to December 2023 period. An amount of 50 litres water was filtered through plankton net of mesh size 62 μ to concentrate the Zooplankton in the filtrate collected in the bottle attached to the plankton net. The filtrate containing Zooplankton was preserved by adding 4 ml commercial Formalin to 96 ml filtrate to make the final concentration of 4% Formalin in filtrate.

The monthly collected samples were tagged as (M1-M6) and conserve samples at room temperature. Sedgwick Rafter counting chamber was used for Zooplankton analysis. The qualitative and quantitative analysis of Zooplankton were done following standard methods (APHA, 1976). Zooplankton were identified by using standard keys (Ward and Whipple, 1959 and Needham, 1972). Species diversity index of zooplankton was determined using Shannon and Wiener Diversity Index (SWDI) method (Shannon, 1948).

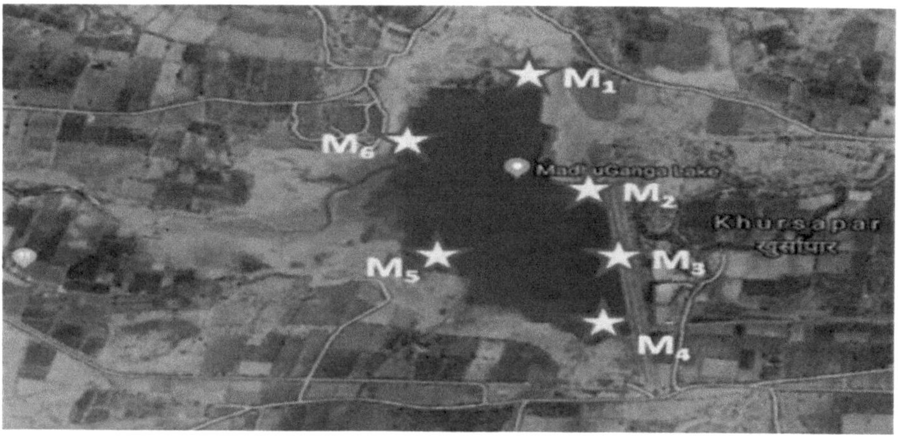

Fig. 1: Location of Madhuganga Lake site , Kalmeshwar, District- Nagpur (Sampling Site) (See colour version on page 352)

RESULTS AND DISCUSSION

In lake Madhuganga, near Kalmeshwar study area, a total of 10 species of rotifers belonging to 5 genera, 4 families and 2 orders were observed (Table 1; Plate 1). Order Ploima had maximum species richness (8) around, 87 % belonging to two families,viz. Asplanchnidae & Brachionidae. The genera *Brachionus* was the most diversified with five species, namely, *Brachionus plicatilis, Brachionus falcatus, Brachionus calyciflorus, Brachionus diversicornis and Brachionus caudatus*. Another rotifer, *Testudinella* sp belonging to order Flosculariaceae was seen only in summer season was rarest zooplankton. Most dominant species *Brachionus diversicornis (52) was observed at* M_2 *sampling location during Winter season* (Table 2 & 3).These species have been reported to be the best indicators of water pollution. Rotifer is popularly used as indicator by other workers (Sladecek 1983). Species of Brachionus, especially is a superior indicator of water pollution as it is least affected by the algal blooms in eutrophic water (Ceirans, 2007; Dadhich *et al.,* 1999). Attayde and Bozelli (1998) reported that Asplanchna sp., Brachionus sp., and Filinia sp. were superior indicators of eutrophic waters and probably Brachinus sp. alone can determine the eutrophic status of water body. Rotifera are most significant group in the Zooplankton community (Saler 2004, Barrabin 2000, Aboul-Ezz *et al.* 1996). Rotifers are reported as the faster growth in eutrophic water bodies (Cajander, 1983). Rotifers especially Brachionus sp. were recorded at higher density in two eutrophic lakes (Ismail and Adnan, 2016). In the view of all authors Rotifers are not visible to predatory fishes due to turbid and low transparency water and small size of Rotifers in the eutrophic waters. Similar results were also observed by Barrabin (2000) and Saler (2004) as well.

The Table 3 describes the Seasonal variation where Total numbers of individuals were 365 in summer, 322 in Winter and only 60 species in Rainy season with lowest population size. Indicating that Summers being most productive season for Rotifers. Shannon Weiner Index showing mesotrophic nature of water with 1.76 in summer, 1.6 in winter and 1.72 during rainy season respectively. Hence, signifying dam water is declining towards Eutrophic condition. For diversity, Species Evenness was analyzed that showed similar value of 0.9 in all seasons justifying similar species occurred. Species Richness of Zooplankton were 10 in summer , 7 in Winter and only 6 species were observed in Rainy season respectively showing lower diversity during rainy season.

Here at Lake Madhuganga, Kalmeshwar , District-Nagpu maximum diversity of rotifers is analysed in the months of winter and summers that coincided with Sharma (2005) and Sharma (2011) noted maximum diversity in winter and Kumar and Nandini (2012) in summer season. Whereas on the contrary Anitha and George 2006; Tidame and Shinde 2012 showed more diversity during the monsoon months probably due to addition of more nutrients due to rainfall.

CONCLUSION

Thus on conclusion we utilize the work done by Sharma and Srivastav (1986), Ahmed (1990) and Shyeshefer *et al.* (2008) which refer that summer months have high temperature, photoperiod is increased and higher light intensity result in phytoplankton and algae growth to the highest that influences the abundance, adaptation and reproduction of the rotifer population.

Table 1: Systematic position of rotifers composition in Lake Madhuganga, Kalmeshwar, District-Nagpur during January-December 2023

Phylum- Rotifera Class-Monogononta Order- Ploima (Hudson & Gosse, 1886	Family	Species
	Asplanchnidae (Harring & Myres,1933)	*Asplanchana brightwelli* (Gosse, 1850)
		Asplanchana herricki (de Guerne, 1888)
	Brachionidae (Ehrenberg 1838)	*Brachionus plicatilis* (Muller 1786)
		Brachionus falcatus(Zacharias,1898)
		Brachionus calyciflorus (Pallas 1776)
		Brachionus diversicornis (Pallas 1776)
		Brachionus caudatus (Barrois and Daday 1894)
		Keratella tropicana (Apstein 1907)
Order-Flosculariaceae (Harring, 1913)	Filinidae (Bartos,1959)	*Filinia longiseta* (Ehrenberg 1834)
	Testudinellidae (Segers, H. 2012)	*Testudinella sp* (Bory de Saint-Vincent, 1822)

Table 2: Seasonal Density of Rotifers in Lake Madhuganga, Kalmeshwar, District-Nagpur during January-December 2023

S.no.	SUMMER	M1	M2	M3	M4	M5	M6	TOTAL
1	*Asplanchana brightwelli*	6	4	0	2	3	6	21
2	*Asplanchana herricki*	7	6	1	3	1	1	19
3	*Brachionus plicatilis*	2	5	1	6	4	1	19
4	*Brachionus falcatus*	13	9	30	2	4	3	61
5	*Brachionus calyciflorus*	10	23	14	9	7	7	70
6	*Brachionus diversicornis*	27	24	12	14	0	8	85
7	*Brachionus caudatus*	5	0	2	3	9	4	23
8	*Filinia sp*	7	9	3	5	2	7	33
9	*Keratella tropicana*	1	0	5	5	8	12	31
10	*Testiculla sp*	1	0	2	0	0	0	3
	Total Zooplankton/site	**79**	**80**	**70**	**49**	38	49	365
	No. of Species identified	**10**	**7**	**9**	**9**	**8**	**9**	
	WINTER	**M1**	**M2**	**M3**	**M4**	**M5**	**M6**	**TOTAL**

1	*Asplanchana brightwelli*	0	2	1	0	4	0	7
2	*Asplanchana herricki*	0	0	0	0	0	0	0
3	*Brachionus plicatilis*	10	15	7	0	0	0	32
4	*Brachionus falcatus*	8	0	12	1	7	11	39
5	*Brachionus calyciflorus*	13	17	5	3	0	0	38
6	*Brachionus diversicornis*	4	52	37	23	12	0	128
7	*Brachionus caudatus*	4	0	0	0	1	3	8
8	*Filinia sp*	3	7	6	0	1	1	18
9	*Keratella tropicana*	7	15	20	8	2	0	52
10	*Testiculla sp*	0	0	0	0	0	0	0
	Total Zooplankton/site	**49**	**108**	**88**	**35**	**27**	**15**	322
	No. of Species identified	**7**	**7**	**7**	**4**	**7**	**3**	
	Rainy	**M1**	**M2**	**M3**	**M4**	**M5**	**M6**	**TOTAL**
1	*Asplanchana brightwelli*	0	0	1	0	1	1	3
2	*Asplanchana herricki*	2	0	1	1	1	0	5
3	*Brachionus plicatilis*	0	0	0	0	0	0	0
4	*Brachionus falcatus*	0	2	0	0	0	2	4
5	*Brachionus calyciflorus*	7	3	5	0	1	1	17
6	*Brachionus diversicornis*	1	1	3	1	1	2	9
7	*Brachionus caudatus*	0	0	0	0	0	0	0
8	*Filinia sp*	1	0	0	0	1	1	3
9	*Keratella tropicana*	6	5	2	4	1	1	19
10	*Testiculla sp*	0	0	0	0	0	0	0

	Total Zooplankton/site	**17**	**11**	**12**	**6**	**6**	**8**	60
	No. of Species identified	**5**	**5**	**5**	3	6	6	

Table 3: Seasonal variation, Shannon Weiner Index , Species Evenness , Species Richness of Zooplankton belonging to Phylum –Rotifera in Lake Madhuganga, Kalmeshwar , District-Nagpur during January-December 2023

Sampling location	**Total No. of Zooplankton (Phylum-Rotifera) (organisms/ml)**		
	Summer	**Winter**	**Rainy**
M1	79	49	17
M2	80	108	11
M3	70	88	12
M4	49	35	6
M5	38	27	6
M6	49	15	8
Shannon diversity index	1.755	1.6	1.72
Evenness	0.98	0.89	0.95
Richness (number of species)	10	7	6
Total number of individuals	365	322	60
Average population size	60.83%	53.67%	10%

Plate 1: Some common species of Phylum –Rotifera in Lake Madhuganga, Kalmeshwar , District-Nagpur during January-December 2023

ACKNOWLEDGEMENT

The author is grateful to Dr. Anand Tikhe, Principal K.Z.S. Science College, Bramhni, Kalmeshwar for permission to carry out work in college premises.

REFERENCES

Aboul-Ezz, S.M., Salem, S.A., Samaan, A.A., Latif, A.F.A. and Soliman AA.1996. Distribution of rotifers in the Rosetta Nile branch (Egypt), Journal of Egyptian and German Social Zoology, 20(D): 85– 123. [Google Scholar].

Ahmed, M. 1990. Hydrobiological studies of whorl reservoir Aurangabad (Maharashtra), India. Journal of Environment Biology, 11: 335–343.

Anderson, H.H. 1889. Notes on Indian rotifers. Journal of Asiatic Society Bengal, 58: 345–358.

Anitha, P.S., and R.M. George. 2006. The taxonomy of Brachnious palicatilies species complex (Rotifera: Monogononta) from the Southern Kerela (India), with a note on their reproduction preference. Journal of Marine Biological Association, India 48(1): 28–32.

APHA Standard methods for the examination of water and waste water. 4th edition, American Public Health Association, Washington, 1976; 1-1193.

Arora, J., and Mehra, N.K. 2003. Species diversity of planktonic and epiphytic rotifers in the back waters of Delhi segment of Yamuna River, with remarks on new records from India. Zoological Studies, 42(2): 239–247.

Attayde JL, Bozelli RL.1998.Assessing the indicator properties of zooplankton assemblages to disturbance gradients by canonical correspondence analysis. Canadian Journal of Fisheries and Aquatic Sciences,55(8):1789–1797. http://dx.doi.org/10.1139/f98-033. [Google Scholar].

Barrabin, J.M.2000.The rotifers of Spanish reservoirs: Ecological, systematical and zoogeographical remarks. Limnetica, 19: 91–144. [Google Scholar].

Cajander V.1983. Production of planktonic Rotatoria in Ormajarvi, an eutrophicated lake in southern Finland. Hydrobiologia,104(1): 329–333. http://dx. doi. org/10. 1007/ BF00045986. [Google Scholar]

Ceirans, A.2007. Zooplankton indicators of trophy in Latvian lakes. Acta Universitatis Latviensis, 723: 61–69 [Google Scholar].

Chopra, G., A.K. Tyor and S. Kumari. 2015. Diversity of Rotifers in Shallow Lake of Sultanpur National Park, Gurgaon (Haryana). Proc Zool Soc .68(1):58–63. DOI 10.1007/s12595-013-0094-z .

Dadhich, N. and Saxena, M.M. 1999.Zooplankton as indicators of trophical status of some desert aters near Bikaner. Journal Environment and Pollution,6(4): 251–254. [Google Scholar].

Esparcia,A. *et al.* 1989.Brachionus plicatilis tolerance to low oxygen concentrations Hydrobiologia ,186:331–337.

GannonJ. E. *et al.* 1978.Zooplankton (especially crustaceans and rotifers) as indicators of water quality,Trans. Am. Microscopical Soc.

George, M.G.1966. Comparative plankton ecology of five fish tanks in Delhi. Hydrobiologia, 27: 81- 108

Gilbert, J.J., and D.K. Schreiber. 1998. Asexual diapause induced by food limitation in the rotifer Synchaeta pectinata. Ecology,79: 1371–1381.

https://www.wikidoc.org/index.php/Rotifer

Ismail AH and Adnan AAM, Zooplankton composition and abundance as indicators of eutrophication in two small man-made lakes. Trop. Life Sciu. Res., 2016; 27 (suppl): 31-38. doi: 10.21315/tlsr2016.27.3.5

J. E. Gannon and R. S. Stemberger, R.S.1978. Zooplankton (Especially Crustaceans and Rotifers) as Indicators of Water Quality, Transactions of American Microscopy Society, 97(1):16-35. http://dx.doi.org/10.2307/3225681.

Kumar, A.T., and C. Nandini. 2012. Rotifer diversity in a semi urban shallow perennial water body. A case study of Jalaser tank Mandalgarh. Indian Journal of Environment Science 1(1): 25–29.

Needham J.G. and Needham P.R.1972. A Guide to the Study of Freshwater Biology. 5th edition. HoldenDay Inc., California,108.

Pennak, R.W.1953. Freshwater invertebrates of United States. The Ronald Press Company, New York.

Saler, S.2004. Observation on the seasonal variation of rotifer fauna of Keban Dam Lake (Cemisgezek Region) Science and Engineering Journal of Firat University, 16(4): 695–701. [Google Scholar].

Shannon, C.E. 1948.A mathematical theory of communication. The Bell System Technical Journal, 27: 379–423 and 623–656.

Sharma, B.K. 2005. Rotifer communities of floodplain lake of Brahmaputra basin of lower Assam (N.E. India): Biodiversity, distribution and ecology. Hydrobiologia, 533: 209–221.

Sharma, J.P., and J.B. Srivastav. 1986. Ecological observation on rotifer fauna of some fresh water ponds of Jammu (J&K) India. Geobios New Report, 5(10): 715–720.

Shyeshefer, S., M. Mousavi, and F. Shirazi. 2008. Ecological study of Rotifera from Kor river,

Fars, Iran. Journal of Environmental Biology,5: 29–37.

Sladecek V.1983. Rotifers as indicators of water quality. Hydrobiologia,100(1): 169– 201. http://dx.doi.org/10.1007/BF00027429. [Google Scholar]

Tidame, S.K., and S.S. Shinde. 2012. Seasonal variation in rotifer diversity of temple ponds of Nashik district (M.S.) India. International Multidisciplinary Research Journal, 2(5): 19–22.

Tyor, A.K., and D. Chawala. 2012. Survey and study of phytoplankton ecology in Sukhna Lake, Chandigarh (India). International Journal of Applied Biology and Pharmaceutical Technology, 3(2): 229–238.

Ward, H.B. and Whipple, G.C.1959. Fresh water Biology, 2nd Edition, John Wiley and Sons, New York.

19

Indigenous Technical Knowledge in Aquaculture

Hardik Sikotariya, T.H. Dave, Mayur Bhadarka, Ketan Makwana and Mohit Bamaniya

ABSTRACT

Indigenous Technical Knowledge (ITK) stands as a reservoir of ancient wisdom, often underestimated but profoundly impactful across diverse realms. Within aquaculture, it emerges as a cornerstone, shaping sustainable practices and bolstering ecosystem health. Through the lens of indigenous communities, aquaculture transcends mere production, embodying a holistic ethos of stewardship and symbiosis with nature. Rooted in generations of wisdom, indigenous practices offer a tapestry of innovative solutions, from periphyton-based aquaculture to integrated farming systems seamlessly blending tradition with modernity. Success stories from Thailand to Assam underscore the transformative potential of ITK, catalyzing sustainable aquaculture revolutions. However, challenges loom, from water quality concerns to climate change threats, necessitating a harmonious fusion of indigenous wisdom with cutting-edge technologies like IoT and AI. As we chart the future course, preserving and honoring ITK becomes paramount, not merely as a repository of knowledge, but as a testament to cultural heritage and a beacon guiding humanity towards a more sustainable coexistence with our aquatic ecosystems.

Keywords: *Indigenous Technical Knowledge (ITK), Aquaculture Sustainability, Indigenous Knowledge Integration, Traditional Aquaculture Practices.*

INTRODUCTION

Indigenous Technical Knowledge (ITK) represents the treasure trove of wisdom held by indigenous communities, often overlooked yet immensely valuable. Its impact spans various domains, notably aquaculture, where it has been pivotal. Local fisher communities, drawing from their indigenous knowledge, have been crucial in safeguarding wild fish genetics and improving aquaculture yields. Moreover, their Indigenous and local knowledge (ILK) contributes multifaceted insights to fisheries and aquatic ecosystems science, underscoring the need to preserve and honor this invaluable heritage. Additionally, traditional indigenous systems like aub-anbars have historically demonstrated their effectiveness in practical applications such as passive cooling and water storage, further highlighting the relevance and efficacy of ITK.

Importance of ITK in Aquaculture

Indigenous Technical Knowledge (ITK) is essential in aquaculture, providing valuable methods for pond preparation, water quality management, disease control, and post-harvest operations. These traditional practices, backed by scientific validation, can greatly improve fish production sustainability while being environmentally friendly. Moreover, modern biotechnological tools are increasingly used in aquaculture to enhance fish quality, disease resistance, growth rates, and genetic diversity. Additionally, the concept of intelligent fish farms, employing technologies like IoT and artificial intelligence, aims to optimize aquaculture processes for precise feeding, disease control, and harvesting, promoting green and sustainable practices. Combining ITK with modern biotechnology and intelligent technologies is crucial for advancing aquaculture practices towards efficiency, sustainability, and increased productivity.

Historical Perspective

Traditional Aquaculture Practices of Indigenous Peoples

The traditional aquaculture methods of indigenous peoples are deeply connected to nature and sustainable resource management. Passed down through generations, these practices often involve selective fishing, cultural beliefs, and stewardship to conserve and use aquatic resources wisely. For example, some indigenous communities use natural substances to stun fish or plants that affect fish for fishing. These practices not only help livelihoods but also foster harmony among communities and support ecosystem health. Bringing back these traditional methods can lead to sustainable fisheries, strong fishing communities, and better conservation of biodiversity.

Evolution and Adaptation of ITK over Time

Over time, Indigenous Traditional Knowledge (ITK) has changed and adjusted to new situations. In a study by Sarkar *et al.*, practices like 'mind' cultivation and 'chal' for water harvesting in Himachal Pradesh, as well as 'Khadin' farming and 'tanka' for water harvesting in Rajasthan, demonstrate how ITK has been used to tackle climate change challenges. Similarly, Public Key Infrastructure (PKI) has developed over two decades to become a trustworthy system for building trust in digital platforms, showing how technologies improve and strengthen over time. Additionally, in the world of cloud-based systems, iObserve introduces a method for ongoing quality analysis and adaptation through real-time observation, illustrating how systems evolve and adjust to changes in their surroundings.

Key Components of Indigenous Technical Knowledge in Aquaculture

Sustainable Resource Management Techniques

In aquaculture, sustainable resource management techniques rely heavily on indigenous technical knowledge (IK) for effectively conserving and using aquatic genetic resources. Indigenous communities hold valuable knowledge and practices that contribute to sustainable management of marine and aquaculture resources. By integrating IK with modern approaches, genetic management of farmed aquatic species can be enhanced, ensuring their conservation and sustainable use. Research highlights the importance of blending indigenous knowledge systems with modern practices for successful conservation of natural resources. Passed down through generations, indigenous knowledge plays a crucial role in ensuring sustainable food security, healthcare, and ecological protection. By tapping into IK, aquaculture can benefit from traditional wisdom and practices to develop effective and sustainable resource management strategies.

Traditional breeding and cultivation methods

Traditional breeding and cultivation methods of fish encompass a variety of approaches aimed at enhancing output and quality while promoting ecological benefits and genetic diversity. One common practice involves stocking fish fries in rice fields, where they benefit from organic fertilizers and duckweed, resulting in improved growth and yield. Another method is the use of hollow feed in breeding ponds, which consists of layers of various raw materials and facilitates efficient feeding while reducing waste compared to traditional sinking feeds. Additionally, traditional rice-fish coculture systems have been found to maintain high genetic diversity in common carp populations, highlighting the importance of such practices in preserving genetic variability in aquaculture settings. These traditional methods not only enhance production but also contribute to ecological sustainability and the conservation of genetic diversity in fish farming practices.

Aquaculture practices based on indigenous knowledge

Sticks in the mud- the periphyton-based aquaculture

The acadja practice, initially documented by Welcomme, involves trapping fish using periphyton-based food production systems established in West African countries. Similar practices, such as the kathafishery in Bangladesh and the samarah fishery in Cambodia, independently developed worldwide, following a similar strategy. The concept of utilizing periphyton techniques in ponds, inspired by traditional farming methods, has garnered significant research attention. The results unequivocally show the potential to boost fish production through the utilization of the periphyton system. The practice stemmed from indigenous wisdom aimed at attracting fish, with fish farmers finding simple and practical ways to grasp its principles and implement it in aquaculture. In Bangladesh, where substrate-based fish culture is common, farmers believe that periphyton can grow on substrates, serving as both fish feed and a deterrent to fish poachers, as it's challenging to use nets in ponds with substrate. These substrates are readily available within their farming systems.

Integrated aquaculture

Chinese farmers pioneered integrated fish farming systems like crop-livestock-fish culture integration millennia ago, and these systems continue to serve as vital nutrient-recycling strategies in many developing countries today.

Paddy-cum-fish culture

Paddy-cum-fish culture is recognized as one of the most ancient and straightforward forms of integrated fish farming practiced globally. Archaeological evidence indicates that agriculture and integrated aquaculture systems may have developed simultaneously in China over 8000 years ago, These systems underwent various designs and experiments, contributing to their long history of evolution and implementation. In the integrated fish-in-paddy field system, typically, a small portion (5-20%) of the rice field area is transformed into a trench, refuge pond, or both. The layout of trenches varies within rice fields. This system operates by allowing fish to feed on organisms, including insects and potential rice pests, as well as weeds. Additionally, the stirring action of fish foraging leads to the resuspension of nutrients in the sediment. It's commonly observed that incorporating fish into rice fields often leads to increased rice yields. In India, traditional aquaculture was historically practiced along the coastline by fisheries communities. Some of the oldest traditional fish farming systems include the bheri system in West Bengal, the gheri system in Orissa, the pokkali system in Kerala, the khazan system of Goa, and the khar lands or gazani (coastal khar lands) in Karnataka. In West Bengal, where salinity is either low or reduced by freshwater discharge diluting tidal water, fish cultivation occurs in paddy fields. The bheri system is commonly used for rice-fish culture or fish monoculture. Most bheries are utilized for fish culture, with Kolkata city domestic sewage serving as the primary feeding source. The sewage-fed system is considered unique and is the largest of its kind globally. In Kerala's pokkali fields, covering approximately 1,250,000 hectares, the summer fallow months are utilized for brackish water aquaculture. These fields, influenced by the Vembanad backwaters and tidal control, are flooded during the southwest monsoon for rice cultivation (June-September). Fish and prawns are cultured during other periods, with small fishes and prawns entering the fields with high tides from nearby shore waters. Fish and shrimp production in this culture varies from 500 to 1,200 kg/ha. After prawn harvest, water is drained off, and the fields regain their saline nature due to monsoon rains, making them suitable for rice cultivation again. The traditional pokkali paddy variety, known for its saline, flood, and acid resistance, withstood the 2004 tsunami. This organically grown variety is esteemed for its unique taste, high protein content, and medicinal properties. In North East India, the Apatani tribe practices paddy-cum-fish farming alongside shifting cultivation (Jhum), reflecting their progressive agricultural methods. The system incorporates a blend of paddy, fish, and finger millet on the bunds separating each plot. In each terrace where paddy is grown, a small pit is dug and stocked with fingerlings. During the monsoon season, when water supply is sufficient, the entire paddy field is submerged shallowly (5-10 cm), allowing fish to move around the terrace. In times of water scarcity, fish retreat to the pits to grow. Fish obtain nutrition from manuring of the paddy field, runoff from hill slopes, and larger surface areas for grazing. Studies indicate the potential for up to 5 tons of rice/ha and an average of 500 kg of fish/ha. Along the coast of Goa, the khazan system of paddy-cum-fish culture is practiced, exemplifying a community-managed agriculture-aquaculture integrated ecosystem with roots dating back to the sixth century. The khazan system was developed by local farmers drawing upon their traditional knowledge of climate, tidal cycles, geomorphology, monsoon precipitation, runoff, sediment dynamics, soil properties, and drainage characteristics of estuarine lands. This expertise enabled them to create a suitable practice tailored to their environment. The production system is situated within reclaimed mangroves, utilizing a network of dykes, canals, and gates. The traditional and highly adapted khazan technology relies on principles of salinity regulation and the tidal clock. However, urban growth poses a threat to the system's sustainability, prompting efforts to preserve this traditional fish farming technology. In Guinea Bissau, a country in West Africa, a similar traditional paddy-cum-fish culture system has been practiced in mangrove areas since ancient times. This system integrates the culture of Tilapia spp. and Clarias spp., both capable of tolerating high salinity rates, with rice production. The system involves constructing a main dam and secondary dykes to control the influx of seawater and enable the storage of rainwater into the rice field. This creates a brackish environment suitable for both rice and fish culture. By mixing rainfall water with seawater, an "artificial" ecosystem is formed, reducing the presence of predatory species intolerant to low salinity. The integration of paddy cultivation with fish culture is also a traditional practice in other Asian countries. In Bangladesh, a notable accomplishment is the control of the golden apple snail, a rice pest, through the introduction of common carp. Smallholders benefit from using rice fields as nurseries to rear fish fry into fingerlings during the 3-month rice-cropping season. In Indonesia, old-fashioned methods merged rice and fish farming, with leftovers often heading downstream into brackish water fish farms (tambaks). These tambaks were multi-purpose ponds, mixing fish, veggies, and trees.

Integration of Makhana cultivation with fish culture

In Manipur, there are reports of combining Makhana (Euryale ferox) farming with fish cultivation. In India, both native carps and exotic species, along with air-breathing fishes like Channa sp. and Anabas testudineus, are utilized in aquaculture. Makhana fruits are not only delicious but also have medicinal properties; both ripe and unripe fruits are used as vegetables, while tender leaves and petioles are consumed after removing spines. The plant's rhizome is employed as a diuretic and in treating various ailments like dropsy, jaundice, scabies, and gonorrhea. Ripe seeds are used to address chronic diarrhea. Propagation is done from mature seeds, sown in separate ponds by the end of January. Seedlings are later transplanted into culture ponds with a spacing of 5-6 ft, accommodating around 700-800 plants per hectare. This integrated approach reportedly yields a net profit of approximately Rs. 1,78,525 per hectare per year for farmers.

Other integrations

In the Mekong Delta of southern Vietnam, farmers utilize trenches within their fruit orchards, typically bordered by a lateral trench and connected to nearby rice fields [47]. Fish and freshwater prawns can freely move between these sub-systems, benefiting from decomposing rice straw, fallen fruit, and insects that fall into the water. In Malaysia, the Philippines, and Vietnam, the term 'forestry-fish' co-culture refers to cultivating brackish water fish and shrimps within fenced-off mangrove forests. In China, bamboo-fish culture is practiced, where mud from fish ponds fertilizes bamboo plantations surrounding the ponds. Waste from processing bamboo shoots is also fed into the fishponds. In China, another widely used system combines aquaculture with mulberry trees grown near ponds. Silkworm droppings, waste pupae, and washings from silkworm trays are all fed into fishponds in this integrated approach.

Case Studies

Indigenous Aquaculture Practices in Specific Regions

Indigenous aquaculture practices play a vital role in sustainable resource management across diverse regions. In Hawaiian fishponds, Indigenous knowledge guides the sustainable harvest of economically important species like the mud crab, Scylla serrata, ensuring long-term viability. Similarly, in the Amazon region, fish farming focuses on native species, particularly tambaqui or tambatinga hybrids, in semi-intensive systems. Indigenous communities contribute significantly to fish conservation and aquaculture development, as seen in Kenya, where local fisherfolk use traditional strategies for sustainable resource utilization. Moreover, globally, aquafarmers adopt indigenous practices across various aquaculture stages, including site selection, periphyton-based aquaculture, and diverse aquaculture integrations. These examples underscore the importance of integrating Indigenous knowledge into aquaculture practices to enhance sustainability and resource management.

Success Stories and Challenges Faced

Success stories of Indigenous Technical Knowledge (ITK) in aquaculture are evident in various regions, demonstrating its significant impact on enhancing practices and sustainability. In Thailand, the development of practical techniques for Nile tilapia fry production and grow-out sparked increased interest among farmers, leading to tilapia becoming the most widely cultured fish in the country. In Assam, fish farmers effectively apply Indigenous Technical Knowledge in pond preparation, water quality management, and disease control, showcasing its rationality and effectiveness. Similarly, in West Bengal, documented indigenous technological knowledge in fish farming has resulted in socially and ecologically acceptable innovations that are economically viable and low-risk, promoting higher and sustainable fish production. These success stories highlight the crucial role of ITK in improving aquaculture practices and ensuring sustainability. Information Technology in Aquaculture (ITK) faces several challenges, including poor water quality, feeding-related issues, disease outbreaks, and the need for precise monitoring and control. Additionally, the sector struggles with overproduction, leading to imbalanced water environments, increased disease occurrences, and reduced product quality, necessitating innovative solutions such as IoT adoption. Furthermore, the sustainability and resilience of aquaculture livelihood projects are threatened by factors like limited knowledge, climate change, and technological advancements, highlighting the importance of continuous adaptation and readiness. These challenges emphasize the significance of leveraging modern technologies like IoT, artificial intelligence, and edge computing to improve productivity, optimize feeding practices, and ensure the sustainability of aquaculture operations.

Integration of Indigenous Knowledge with Modern Aquaculture

The integration of Indigenous Knowledge (IK) with modern aquaculture practices is essential for sustainable fish conservation and improved aquaculture production. Indigenous fisher communities hold valuable knowledge and practices for sustainable aquatic resource management. This knowledge, which includes nature observations, spiritual beliefs, and cultural values, offers a holistic approach that can complement conventional fisheries management techniques. By incorporating IK into aquaculture research and management, we can enhance the conservation of fish genetic resources and develop more effective strategies for sustainable fish production. Recognizing the diverse contributions of fishers' Indigenous and local knowledge (ILK) is crucial for advancing fisheries science, conservation efforts, and ecosystem management. It underscores the importance of preserving this knowledge and involving ILK holders in decision-making processes.

Future Directions and Recommendations

Promoting the Preservation and Recognition of ITK

Preserving and promoting Indigenous Traditional Knowledge (ITK) involves safeguarding valuable wisdom passed down through generations. Globally, various initiatives aim to document and protect ITK, empowering communities to prevent misuse and promote development. Recognizing the cultural significance of heritage sites is vital for preservation efforts, emphasizing the importance of context, relationships, and identity. Additionally, raising awareness and acknowledging traditional games contribute to preserving national culture and heritage. Implementing preservation activities like preventive, curative, and restorative measures in libraries is essential to combat damage and ensure the longevity of collections. By appropriately designating cultural sites and introducing educational programs, the intrinsic value of ITK and heritage can be effectively conserved and celebrated for future generations.

Supporting Indigenous Communities in Aquaculture Development

Supporting Indigenous communities in aquaculture development involves recognizing the value of indigenous knowledge in fish conservation and aquaculture improvement. Engaging Indigenous community members in monitoring harvested populations enhances sustainable resource use and creates employment opportunities. Despite evidence showing that culture-aligned economies effectively engage Indigenous Australians in work, there is resistance to substantial investment in these economies. Establishing an Indigenous Australian business development framework is crucial for driving business development in remote Indigenous communities and ensuring key success factors for viable fisheries-based businesses. Aquaculture in rural areas contributes to livelihoods, food security, and improved purchasing power, particularly through small indigenous fish farming systems. By incorporating indigenous knowledge, training, and business support frameworks, aquaculture development can empower Indigenous communities sustainably.

CONCLUSION

The studies on Indigenous Technical Knowledge (ITK) in aquaculture highlight its importance in fish conservation and aquaculture development. Indigenous knowledge plays a crucial role in sustainably managing fish genetic resources. Various indigenous practices, such as periphyton-based aquaculture and integrated aquaculture systems, have been identified as effective methods in aquaculture. Research in Assam emphasizes the applicability and limitations of ITKs in aquaculture, highlighting the need for validating and popularizing these practices. Additionally, ITKs have played a vital role in disaster management in flood-prone regions, demonstrating innovative approaches in pond construction, fish preservation, and modified fishing techniques. Overall, these studies underscore the importance of integrating indigenous knowledge with scientific research to enhance fisheries management and promote sustainable aquaculture practices.

REFERENCES

Adiati, M. P.2016. Recognition of traditional games in Indonesia as cultural preservation efforts through special event. In International Conference on Tourism, Gastronomy, and Tourist Destination (ICTGTD 2016) (pp. 216-221). Atlantis Press.

Akiona, A. K., Popp, B. N., Toonen, R. J., Siple, M. C., Kotubetey, K., Kawelo, H., and Franklin, E. C. 2022. Predatory fish diets shift towards an invasive mullet in a traditional Hawaiian aquaculture system. Aquaculture, Fish and Fisheries, 2(5):307-320.

Anon. Report of the Agricultural Land Development Panel. Report submitted to the Government of Goa, Panjim, Goa, 1992.

Atlas, W. I., Ban, N. C., Moore, J. W., Tuohy, A. M., Greening, S., Reid, A. J., ... and Connors, K. 2021. Indigenous systems of management for culturally and ecologically resilient Pacific salmon (Oncorhynchus spp.) fisheries. BioScience, 71(2):186-204.

Ayaa, D. D., and Kipterer, J. K.2018. The role of indigenous technical knowledge and geographical information systems (GIS) in sustainable natural resource management around the Teso Community in Kenya. Journal of Ecology and The Natural Environment, 10(6): 97-107.

Azim, M. E., Verdegem, M. C., van Dam, A. A., and Beveridge, M. C. (Eds.).2005. Periphyton: ecology, exploitation and management. CABI.

Berio, M. J. M., and Salugsugan, J. C. P.2022. Analysis of the Sustainability, Vulnerability, and Challenges of Aquaculture Livelihood Projects. Technium Business and Management, 2(3):38-62.

Bhujei, R. C. 2008. The role of the Asian Institute of Technology in the promotion of tilapia for aquaculture.

Biswajit Goswami, B. G., and Gangopadhyay, P. K. 2017. Controlling fish diseases through the use of ITK by the fish farmers of West Bengal, India.

Cagauan, A. G., Symoens, J. J., and Micha, J. C. 1994. Overview of the potential roles of pisciculture on pest and disease control and nutrient management in rice fields.

Cahyani, R. G., & Khadijah, U. L. S. 2023. Kegiatan preservasi koleksi di Perpustakaan Institut Teknologi Nasional. Informatio: Journal of Library and Information Science, 3(2):139-158.

Coope,J.2020.Indigenous knowledge and techno-scientific modernity:"Hierarchical integration" reconsidered. Ecopsychology, 12(2):151-157.

Costa-Pierce, B. A. 1988. Traditional fisheries and dualism in Indonesia. Naga, 11(2): 3-4.

Das, B. K., Meena, D. K., Das, A., and Sahoo, A. K. 2022. Prospects of Smart Aquaculture in Indian Scenario: A New Horizon in the Management of Aquaculture Production Potential. In Smart and Sustainable Food Technologies (pp. 59-85). Singapore: Springer Nature Singapore.

Das, S. C. S., Joshi, K. D., Alam, A., and Jha, D. N. 2017. Some Indigenous Traditional Fishing Methods of Tharu Tribals of Lakhimpur kheri District, Uttar Pradesh, India. Journal of Scientific Achievements, 2(6): 1-5.

De Silva, S. S., and Davy, F. B.2010. Success stories in Asian aquaculture. IDRC, Ottawa, ON, CA.

Dehghani-Sanij, A., and Sayigh, A.2022. An Introduction to Indigenous Knowledge and Water Storage Systems. In Cisterns (pp. 3-21). River Publishers.

Devi, R., Saha, B., Pandit, A.,and Kashyap, D.2014. Assessment of applicability of Indigenous Technical Knowledge (ITK) in aquaculture as perceived by fish farmers in Assam. Indian J. Fish, 61(3):104-110.

Edwards, P. J. W. A. 200. Traditional Chinese aquaculture and its impact outside China. World Aquaculture, 35(1).

Elena, Vital'evna, Pishchenko., Irina, Moruzi. 2023. Historical aspects of carp domestication and breeding. Rybovodstvo i rybnoe hozâjstvo, doi: 10.33920/sel-09-2304-05

Fleming, A. E. 2015. Improving business investment confidence in culture-aligned Indigenous economies in remote Australian communities: a business support framework to better inform government programs. The International Indigenous Policy Journal, 6(3).

Folorunso, T. A., Bala, J. A., Adedigba, A. P., Olatunji, B. E., and Mingyi, C. 2022. Internet of Things-based surveillance and feeding system for aquaculture applications.

Gangadhar, B., and Keshavanath, P. 2012. Growth performance of rohu, Labeo rohita (Ham.) in tanks provided with different levels of sugarcane bagasse as periphyton substrate. Indian Journal of Fisheries, 59(3):77-82.

Gangadhar, B., Sridhar, N., Hemaprasanth, K., Raghunath, M., and Jayasankar, P. 2016. Indigenous technical knowledge in aquaculture sector: A literature review. International Journal of Fisheries and Aquatic Studies, 4(1):373-378.

Gongfu, Z., Symoens, J. J. and Micha, J. C. 1995. Integration of agriculture and fish-farming (Dyke-Ponds) in Canton region (China).

Guan, R. J., and De Chen, Y.1989. Reform and development of Chinas fisheries. Food and Agriculture Organization of the United Nations.

Halwart, M. 1994. Fish as biocontrol agents in rice: the potential of common carp Cyprinus carpio (L.) and Nile tilapia Oreochromis niloticus (L.) (pp. iii+-169).

Halwart, M. 1998. Trends in rice-fish farming. FAO Aquaculture Newsletter, 18(3), C11.

Hien, T. T. T., Minh, T. H., Phuong, N. T., and Wilder, M. N. 1998. Current status of freshwater prawn culture in the Mekong River Delta of Vietnam. JIRCAS Journal, 6: 89-100.

Ho-Cheol, Ryu.2023. Expanding the recognition and promoting the preservation and valuing of Buddhist temples as cultural properties through designation as historic sites - The example of Bulgapsa Temple in Yeonggwang -. The East Asian Ancient Studies, doi: 10.17070/aeaas.2023.3.69.257

Howes, M., and Chambers, R. 1979. Indigenous technical knowledge: Analysis, implications and issues. The IDS Bulletin, 10(2):5-11.

Hurley, K. K., Kapur, M. S., Siple, M., Kotubetey, K. I., Kawelo, A. H. I. and Toonen, R. J.2020. A codeveloped management tool to determine harvest limits of introduced mud crabs, Scylla serrata (Forskal, 1775), within a Native Hawaiian fishpond. Pacific Conservation Biology, 27(4): 418-431.

IIRR, I.1992. Farmer-Proven Integrated Agriculture–Aquaculture: A Technology Information Kit. ICLARM, Manila, Philippines and IIRR, Silang, Cavite, Philippines.

International Center for Living Aquatic Resources Management, & International Institute of Rural Reconstruction.2001. Integrated agriculture-aquaculture: a primer (No. 407). Food & Agriculture Organization.

Jabali, W., Wamukota, A., and Fulanda, B.2020. The role of indigenous knowledge in the management of marine resources: A case study of Kuruwitu and Mkunguni fishing areas in Kenya. Western Indian Ocean Journal of Marine Science, 19(1):19-31.

Johnston, D.1999. Mixed shrimp-mangrove forestry farming systems in Ca Mau Province, Vietnam. Aquaculture Asia, 6-12.

Lightfoot, C., Costa-Pierce, B. A., Carangal, V. R., and Bimbao, M. P. 1992. Rice-fish research and development in Asia. Monographs.

Lightfoot, C., Roger, P. A., Cagauan, A. G., and Dela Cruz, C. R.1993. Preliminary steady-state nitrogen models of a wetland rice field ecosystem with and without fish. In Trophic models of aquatic ecosystems. ICLARM Conf. Proc. 26 (pp. 56-64).

Little, D. C., Innes-Taylor, N. L., Turongruang, D., and Komolmarl, S. 1991. Large fish seed for small-scale aquaculture.

Nandeesha, M. C., Alvarez, C., Atanda, T., Bhujel, R., Bosma, R. H., Giri, N. A., ... & Yuan, D. 2010. Expert Panel Presentation VI. 4. Supporting farmer innovations, recognizing indigenous knowledge and disseminating success stories.

Nandeesha, M. C.2002. Farmers as Scientists: Sewage-fed aquaculture systems of Kolkata: A centuryold innovation of farmers. Aquaculture Asia, 7(2): 28-32.

Obiero, K. O., Mboya, J. B., Ouko, K. O., Kembenya, E. M., Nyauchi, E. A., Munguti, J. M., ... and Githukia, C. M. 2023. The role of indigenous knowledge in fisheries resource management for aquaculture development: A case study of the Kenyan Lake Victoria region. Aquaculture, Fish and Fisheries, 3(2):175-183.

Olaganathan, R., and Kar Mun, A. T.2017. Impact of aquaculture on the livelihoods and food security of rural communities. International Journal of Fisheries and Aquatic Studies, 5(2):278.

Ovung, E. Y., Kithan, L. M., Brearley, F. Q., and Tripathi, S. K. 2022. Indigenous Community Fishing Practices in Nagaland, Eastern Indian Himalayas. Sustainability, 14(12):7049.

Pam, Perkins. 2022. Domestication of aquaculture species. doi: 10.1093/oso/9780198850229.003.0006

Poorna, R. L., Mymoon, M., and Hariharan, A.2014. Preservation and protection of traditional knowledge–diverse documentation initiatives across the globe. Current Science, 1240-1246.

Prasetyo, N., Carr, A., and Filep, S. 2020). Indigenous knowledge in marine ecotourism development: The case of Sasi Laut, Misool, Indonesia. Tourism Planning & Development, 17(1): 46-61.

Previous column'Peri-urban, A. (2008). An increasingly secure future for wastewater-fed aquaculture in Kolkata, India? Aquaculture Asia.

Rai, S. C. 2005. Apatani paddy-cum-fish cultivation: An indigenous hill farming system of North East India.

Rajendran, B.2017. Evolution of PKI ecosystem. In 2017 International Conference on Public Key Infrastructure and its Applications (PKIA) (pp. 9-10). IEEE.

Ranga, M. R.2006.Transformation of coastal wetland agriculture and livelihoods in Kerala, India.

Renkui, C., Dashu, N., and Jianguo, W. 1995. Rice fish culture in China: the past, present, and future. In Rice fish culture in China. IDRC, Ottawa, ON, CA.

Ruddle, K., and Zhong, G.1988. Integrated agriculture-aquaculture in South China: the dike-pond system of the Zhujiang Delta. CUP Archive.

Sahoo, P. K., and Paul, A. 2023. Opportunities and challenges in aquaculture biotechnology. Frontiers in Aquaculture Biotechnology, 15-23.

Saunders, T., and Xuereb, S. 2016. Optimising the monitoring of tropical aquatic resources through the development of Indigenous scientific capability. Reviews in Fish Biology and Fisheries, 26(4): 727-736.

Sebestianus Lakra, S. J.2019. Sustainable resource management through indigenous knowledge and practices–a case of food security among the Baiga Tribe in India. European Journal of Sustainable Development, 8(4): 233-233.

Sihoe, A. D. 2019. The evolution of video assisted thoracic surgery. Atlas of Uniportal Video Assisted Thoracic Surgery, 3-15.

Silvano, R. A., Baird, I. G., Begossi, A., Hallwass, G., Huntington, H. P., Lopes, P. F., ... and Berkes, F. 2023. Fishers' multidimensional knowledge advances fisheries and aquatic science. Trends in Ecology & Evolution, 38(1): 8-12.

Singh, A., Maurya, S. P.,and Singh, V.2022. Indian Research Journal of Extension Education.

Singh, K. B. 2011. Eurale Ferox Salisb Integration into Fish Ponds in Manipur, India: A Sustainable Approach* Kumnam Bedajit Singh,"Soibam Khogen Singh,"Himadri Saha. African Journal of Basic & Applied Sciences, 3(5):198-200.

Sonak, S., Kazi, S., and Abraham, M. 2005. Khazans in troubled waters (p. 54). New Delhi: TERI Press.

Sonesson, A. K., Hallerman, E., Humphries, F., Hilsdorf, A. W. S., Leskien, D., Rosendal, K., ... and Mair, G. C.2023. Sustainable management and improvement of genetic resources for aquaculture. Journal of the World Aquaculture Society, 54(2): 364-396.

Sujit Sarkar, S. S., Padaria, R. N., Vijayragavan, K., Himanshu Pathak, H. P., Pramod Kumar, P. K., & Jha, G. K. 2015. Assessing the potential of indigenous technological knowledge (ITK) for adaptation to climate change in the Himalayan and arid ecosystems.

Vitiello, V., Castelluccio, R., and Trampetti, S. 2022. The Recognition of Cultural Value as an Element for the Preservation of the 20th-Century Heritage: Application of the ICOMOS Multidisciplinary Approach to the ex SMOM of Pozzuoli. Heritage, 6(1), 284-299.

Wang, C., Li, Z., Wang, T., Xu, X., Zhang, X., and Li, D. 2021. Intelligent fish farm—the future of aquaculture. Aquaculture International, 1-31.

Wang, Guangjun., Xie, Jun., Zhang, Kai., Yu, Ermeng., Li, Zhifei., Gong, Wangbao., Wu, Jingrong., Jingjing, Tian., Peng, Huihui.2018. Breeding method for improving fish cultivation output and quality in rice fields.

Welcomme, R. L. 1972. An evaluation of the acadja method of fishing as practised in the coastal lagoons of Dahomey (West Africa). Journal of fish Biology, 4(1): 39-55.

Yap, F. 2001. Culture of Fish in Rice fields. FAO, Rome, Italy and ICLARM, Penang, Malaysia.

Yiu, E.2022.Traditional knowledge in the management of seascapes. In Routledge Handbook of Seascapes (pp. 369-384). Routledge.

20

Aquaponics: In a Nutshell

Sayan Biswas, Ketan V. Tank, Mayur Bhadarka, Ketan Makwana and Mohit Bamaniya

ABSTRACT

Aquaculture in India is intensifying to meet the rising demand for fish protein amidst shrinking agricultural land and mounting pressure on natural resources due to land use conversion. In response, aquaponics emerges as a transformative solution, seamlessly integrating recirculating aquaculture and hydroponics. Despite scientific exploration, there remains a dearth of emphasis on its commercial implementation. However, aquaponics offers significant advantages, notably reducing water usage while boosting fish and vegetable yields per unit area compared to traditional systems. This efficiency has sparked increasing interest, particularly for urban food production. This innovative system fosters symbiotic relationships between aquatic animals and plants, highlighting global strides toward sustainable food production. Embracing aquaponics not only aligns with sustainability principles but also enhances economic efficiency and promotes human health through eco-friendly agricultural practices. Aquaponics combines hydroponics, cultivating plants in nutrient-rich water without soil, with aquaculture, the farming of aquatic organisms. Fish waste serves as nutrients, undergoing transformation by nitrifying bacteria into forms usable by plants. Within the system, water rich in nutrients cycles between fish tanks and grow beds, where plants absorb nutrients and purify water, ensuring fish welfare. This closed-loop system provides a model for environmentally conscious food production, emphasizing the urgent need to integrate such technologies into agricultural frameworks for a healthier, sustainable future.

Keywords**:** *Urban; Agriculture; Aquaponics; Hydroponics; Culture; Nutrient; Ammonia; Nitrate; Nitrite; Bacteria.*

INTRODUCTION

The expansion of aquaculture, a vital source of income for a large portion of the global population (Salam *et al.*, 2014), has created an unforeseen challenge: increased competition for land use with traditional agriculture. This land squeeze is further intensified by a multitude of factors. Shrinking arable land due to urbanization and desertification, coupled with the relentless growth of the human population, puts immense strain on existing agricultural resources. Man-made environmental pollution from industrial waste and agricultural runoff, along with the far-reaching effects of climate change, add another layer of complexity. These combined pressures necessitate a paradigm shift towards innovative and sustainable agricultural practices.

One promising solution gaining traction is aquaponics, a system designed to cultivate plants and aquatic animals together in a carefully balanced, recirculating environment (Salam *et al.,* 2014). This approach leverages a symbiotic relationship between fish and plants. Fish waste, rich in nitrogenous compounds, acts as a natural fertilizer for the plants growing in the system. These plants, in turn, function as biofilters, removing excess nutrients and purifying the water for the fish. In essence, aquaponics integrates two established agricultural techniques: intensive aquaculture, the practice of raising fish in controlled tank environments, and hydroponics, the cultivation of plants without soil (FAO, 2012). The aquaculture component of aquaponics utilizes eco-friendly, indoor recirculating systems that enable meticulous control over water quality and temperature year-round. This controlled environment offers significant advantages over traditional methods, minimizing dependence on external factors like weather patterns.

The rise of aquaponics coincides with a growing public interest in bio-integrated food production systems. The "grow your own organic food" movement resonates strongly with a population increasingly concerned about food safety and environmental impact. Recent advancements by researchers and practitioners have further bolstered the viability of aquaponics. Through continuous innovation in areas like system design, nutrient management, and optimal plant and fish selection, aquaponics is rapidly transforming from a niche concept into a robust model for sustainable food production. With its potential to address land scarcity, reduce water usage, and minimize environmental footprint, aquaponics offers a promising path towards a more secure and sustainable future for our food systems.

Advantages of Aquaponics

- Highly Water-Efficient: Aquaponics uses significantly less water compared to traditional agriculture. The recirculating system cleans and reuses water, minimizing waste.
- Eliminates Chemical Dependency: There's no need for artificial fertilizers. Fish waste provides a natural, nutrient-rich source of fertilizer for plants.

- Built-in Water Filtration: The plants act as natural biofilters, removing harmful compounds from the water and keeping it clean for the fish.
- Location Flexibility: Unlike traditional farming, aquaponics doesn't require fertile soil. It can be set up almost anywhere, even in urban areas or locations with poor soil quality.
- Naturally Organic Produce: Since no chemicals are used, the plants grown in aquaponics are naturally organic.
- Potential for Agritourism: Unique and innovative aquaponics systems can be a draw for tourists interested in sustainable agriculture, benefiting rural communities.
- Reduced Pest and Disease Problems: The controlled environment of aquaponics minimizes damage from pests and diseases often encountered in outdoor farming.
- Weed-Free Cultivation: Aquaponics eliminates the need for weeding, saving time and effort.
- Faster Plant Growth: Plants in aquaponics systems often experience accelerated growth due to the readily available nutrients and controlled environment.
- Sustainable Food Production: Aquaponics promotes a closed-loop system that minimizes waste and maximizes resource efficiency, making it a sustainable approach to food production.
- Clean and Efficient Gardening: Compared to traditional soil-based methods, aquaponics offers a cleaner way to grow food, minimizing mess and potential contamination.
- Relatively Easy Setup: While some knowledge is needed, aquaponics systems can be set up at various scales, making them accessible to both hobbyists and commercial growers.

Working of Aquaponics

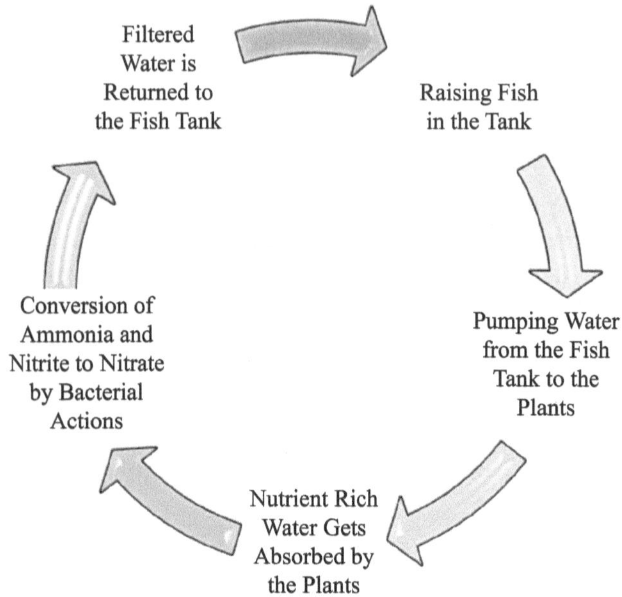

Components

As described by Rakocy and Diver (2006), aquaponics relies on two fundamental components: aquaculture and hydroponics. The aquaculture component, typically a fish tank, serves as the rearing environment for fish or other aquatic organisms. Here, fish produce waste rich in ammonia, a key nutrient for plants. The hydroponics component, often a designated grow bed, provides a platform for cultivating plants without traditional soil.

Optimizing the System: Additional Subsystems

While the aquaculture and hydroponics sections form the foundation of an aquaponic system, additional components can significantly enhance its efficiency and overall health (Rakocy, 2006). These subsystems address various aspects of water quality management and environmental control.

Water Movement

Water Pump(s): These pumps are the lifeblood of the system, ensuring continuous water circulation between the fish tank, grow beds, and other components. Proper pump selection considers factors like system size, water flow rate requirements, and head pressure needed to overcome elevation changes.

Oxygenation

Air Pump and Air Stones: Maintaining adequate dissolved oxygen levels in the water is crucial for both fish and beneficial bacteria. Air pumps push air through diffusers or air stones, creating tiny bubbles that oxygenate the water.

Water Quality Management

Settling Basin (Optional): This basin allows uneaten fish food, detached biofilms, and fine particulates to settle at the bottom. This captured material can be periodically removed to prevent water quality degradation.

Filtration System (Optional): While the plants act as natural filters, additional filtration systems can be employed for further water purification. Options include mechanical filters to remove physical debris and biological filters to promote the growth of beneficial bacteria that break down harmful ammonia into usable nitrates for plants.

Environmental Control (Optional)

Water Heater: In cooler climates, a water heater can be used to maintain optimal water temperature for fish and plant growth.

Grow Lights: For indoor aquaponics systems or situations with limited natural light, grow lights can supplement or replace sunlight, ensuring sufficient light for healthy plant growth.

Live Components

Fish: The choice of fish species depends on factors like water temperature tolerance, ammonia tolerance, and market demand. Common choices include tilapia, catfish, and koi.

Plants: A wide variety of plants can thrive in aquaponics systems. Popular choices include leafy greens like lettuce and kale, herbs like basil, and fruiting vegetables like tomatoes and peppers.

Sump: The Collection Point

The sump serves as the lowest point in the aquaponics system. Water from the grow beds drains into the sump, where it's collected before being pumped back to the fish tank, completing the recirculating loop.

By incorporating these core and additional components, aquaponic systems can be tailored to specific needs and environmental conditions, fostering a thriving ecosystem for both fish and plants.

Plants in the system

Aquaponics emerges as a game-changer in sustainable food production. It cleverly merges aquaculture, the practice of raising fish, with hydroponics, where plants flourish in water-based solutions. This ingenious system operates in a closed loop, fostering a mutually beneficial relationship between plants and fish. Imagine plant roots nestled in nutrient-rich wastewater from the fish tank. These plants act as nature's water filters, removing ammonia, a toxin harmful to fish, while utilizing the dissolved nutrients to fuel their own growth. The cleansed water then makes its way back to the fish tank, completing the cycle. This symbiotic dance between plants and fish ensures the success of the entire system.

Unlike traditional soil-based methods that require significant space and constant attention to moisture levels, aquaponics offers a space-saving and low-maintenance solution. However, ensuring consistent water flow is critical, as power outages or pump failures can threaten the fish within hours. While the plants might survive for days depending on the temperature, prioritizing a reliable power source is crucial.

The beauty of aquaponics lies in its diverse crop compatibility. Plants with high nitrogen demands, like tomatoes, can flourish alongside their low-maintenance counterparts like lettuce. This is because the nutrient-rich water reaches all the plants, but due to the single-pass system, they only absorb what they need. Even with this efficient plant filtration, there might still be excess nutrients flowing back to the fish tank. While this isn't necessarily a problem unless the fish tank water becomes cloudy, it might require monitoring or the potential to support another set of grow beds.

The possibilities with aquaponics are truly exciting. This method boasts the potential to cultivate a vast array of vegetables, from leafy greens like lettuce and spinach to fruiting favorites like tomatoes, peppers, and even strawberries. Herbs like basil and cilantro thrive alongside zucchini, squash, and broccoli, creating a diverse and productive food source. Even some flowers can be grown using this method, adding a touch of beauty to your sustainable bounty. So, if you're looking for a way to grow fresh, healthy food in a space-efficient and eco-friendly manner, then look no further than aquaponics.

Why plants like aquaponics?

- **Nutrients constantly provided:** Plants in aquaponics enjoy a veritable smorgasbord of nutrients delivered directly to their roots. No more waiting for slow-release fertilizers or struggling to find sustenance in depleted soil.
- **Warm water bathing the roots:** Imagine basking in a warm, nutrient-rich bath! That's what aquaponics offers plant roots, promoting optimal growth and nutrient uptake.

- **Don't have to search for water or food:** Aquaponics eliminates the stressful hunt for water and nutrients. Plants can simply focus on thriving in a self-contained paradise.
- **Less effort needed in putting out roots:** Forget extensive root systems searching for resources. In aquaponics, plants can put less energy into root development and more into producing delicious fruits and vegetables.
- **All the energy goes into growing UP not DOWN:** Aquaponics empowers plants to channel their energy skyward. Instead of expending resources on a sprawling root system, they can focus on producing lush foliage and bountiful harvests.
- **No weed competition:** Say goodbye to pesky weeds stealing valuable resources! Aquaponics creates a controlled environment where plants can flourish without competition.

Fish Selection

In aquaponics, a diverse range of freshwater fish can flourish, although water temperature is a determining factor. While certain cold-water species such as trout may not be optimal, there are numerous alternatives available. Popular choices encompass tilapia, catfish, yellow perch, bass, bluegill, carp, koi, goldfish, and even freshwater prawns.

Fish Care

Ensuring the well-being of fish within your aquaponics setup is vital. Here's a breakdown of essential practices

1. Feeding: Offer fish 2-3 meals daily, being cautious not to overfeed. Targeting 1.5-2% of their body weight per day ensures they consume the food within 5-10 minutes.
2. Temperature Regulation: Fish thrive within consistent and suitable temperature ranges, avoiding extremes or stressful conditions.
3. Water Quality Monitoring: Regularly assess the water quality within the fish tank, including parameters like ammonia, pH, and oxygen levels.
4. Fish Observation: Regularly observe fish behavior and appearance for early indications of stress or disease.

Fish Health

Maintaining fish health is crucial for the success of your aquaponics system. Here are some strategies for promoting fish health:

1. Biosecurity: Employ proper hygiene and biosecurity practices, including quarantining new fish and monitoring their health for several days before introduction.
2. Stress Reduction: Foster a low-stress environment by maintaining optimal water parameters, providing adequate hiding spots, and preventing overcrowding.
3. Immune System Support: Supporting a healthy immune system is paramount in disease prevention. Strive to create a stress-free environment to bolster your fish's natural defences.

When constructing your aquaponics system, choosing the right materials is crucial for the health of your fish, plants, and ultimately, yourself! Here's a breakdown of safe and effective options:

Fish-Friendly and Food-Grade Materials:

Polypropylene (PP): Look for this label to ensure safe plastic components.

High-Density Polyethylene (HDPE): This sturdy plastic is a great choice for tanks and other containers.

High Impact ABS (Hydroponic Grow Trays): These durable trays are ideal for supporting your plants.

Stainless Steel Barrels: A long-lasting and safe option for fish tanks.

EPDM or UV-resistant PVC Pond Liner: Line your system with these materials for water containment. Avoid fire retardant materials as they can be harmful.

Additional Options:

Fiberglass Tanks and Grow Beds: Another safe and long-lasting choice for your aquaponics setup.

Rigid White PVC Pipe and Fittings: Perfect for plumbing your system for efficient water flow.

Black Flexible PVC Tubing: This flexible option allows for easy connection points.

Some ABS: While generally safe, double-check the specific type of ABS before using it in your system.

Always Say No to Copper: Copper is highly toxic to fish, so avoid using it in any part of your aquaponics system.

We should remember, choosing safe materials lays the foundation for a thriving and healthy aquaponics system!

CONCLUSION

Aquaponics, a system blending fish farming (aquaculture) with plant cultivation (hydroponics), offers a solution to environmental regulations limiting water use. This innovative approach recycles nutrient-rich wastewater from fish tanks, nourishing plants and creating a closed-loop system. This not only benefits the environment but also presents a profitable alternative to traditional fish farming, boosting farm income. Particularly valuable in regions facing water scarcity and poor soil quality, aquaponics ensures a year-round supply of fresh vegetables and fish, even in markets with fluctuating demand. While solar power can further reduce operational costs, success hinges on a deep understanding of how the system functions. In ecologically sensitive areas like the Sundarbans, mangrove-friendly aquaponics systems can empower local communities by providing alternative income sources, while also protecting vital mangrove forests and improving water management. Although West Bengal's coastal regions hold potential for aquaponics, a lack of production data highlights the need for locally-adapted technologies. By embracing and developing aquaponics, we can cultivate self-sufficient local food systems that prioritize both environmental sustainability and food security. Integrating these ecological technologies has the potential to transform wetland ecosystems positively, emphasizing the importance of collaboration in building a sustainable future.

REFERENCES

Chakravartty, D., Mondal, A., Raychowdhury, P., Bhattacharya, S. B. and Mitra, A. 2017. Role of aquaponics in the sustenance of coastal India-Aquaponics is a solution for modern agriculture in ecologically sensitive Indian mangrove Sundarbans: A review. International Journal of Fisheries and Aquatic Studies, 5(2): 441-448.

Datta, S., Mahapatra, B. K., Bhakta, J. N., Bag, S. K., Lahiri, S., Mandal, R. N., and Jana, B. B. 2018. Aquaponics: a green and sustainable eco-tech for environmental cum economic benefits through integration of fish and edible crop cultivation. Wastewater Management Through Aquaculture, 207-224.

Jena, A. K., Biswas, P. and Saha, H. 2016. Aquaponics-a step towards urban agriculture. Innovative Farming, 1(4):163-167.

Mishra, V.2018. 'Aquaponics'-Self Sustainable Ecosystem'. International Journal of Science and Research.(https://doi. org/10.21275/ART20197410).

Panigrahi, G. K., Panda, S. and Padhi, S. N.2016.Aquaponics: An innovative approach of symbiotic farming. International Journal of Bioassays, 5(9): 4808-4814.

Roy, M., Salam, M. A., Hossain, M. B. and Shamsuddin, M. 2013. Feasibility study of aquaponics in polyculture pond. World Applied Sciences Journal, 23(5): 588-592.

Sharma, P. K., Kumar, J. S. S. and Anand, S. 2018. Aquaponics: A boon for income generation in water deficient areas of India like Rajasthan. Int J Fisheries Aquatic Stud., 6:170-173.

21

Advances in Aquaculture Waste Management Strategies for Sustainability and Compliance

Mayur Bhadarka, D.T. Vaghela, Hardik Sikotariya, Pinak Bamaniya, Ketan Makwana and Neha Kharadi

ABSTRACT

The aquaculture sector expands globally, the necessity for efficient waste management methods grows more pressing to ensure sustainability and compliance with environmental regulations. Aquaculture waste, primarily in the form of effluent or sludge, poses significant challenges for farm managers, particularly in meeting discharge standards. The worldwide proliferation of aquaculture has raised concerns about environmental harm from the release of pollutants such as uneaten feed and fecal matter into natural water systems. To address these issues, novel approaches to wastewater treatment have been developed, including environmentally friendly methods like aquaponics and integrated multi-trophic systems, as well as biofilm technologies utilizing algae and fungi. Additionally, recycling wastewater through oxidation ponds and the adoption of constructed wetlands and recirculation systems have proven effective in removing nutrients and reusing water. Adherence to regulatory frameworks is essential, with international, regional, and national agreements playing a crucial role in addressing waste management challenges, including marine plastic debris. Overall, sustainable waste management strategies are vital for reducing environmental impacts and promoting the long-term sustainability of aquaculture practices worldwide.

Keywords: *Aquaculture, Aquaculture Waste, Waste Management, Management Techniques.*

INTRODUCTION

The aquaculture sector expands and environmental regulations become stricter, waste management has emerged as a significant responsibility for managers of aquaculture farms. Waste generated by aquaculture facilities usually comes in the form of either effluent or sludge. Effective management and disposal of wastewater in aquaculture are becoming progressively crucial due to the strict regulations concerning the discharge of waste into natural water systems. The worldwide expansion of aquaculture industries has led to a rise in adverse environmental effects due to the release of significant quantities of polluting effluents containing uneaten feed and fecal matter (Chávez-Crooker and Obreque-Contreras, 2010) Efficient management of wastewater in aquaculture is crucial for fostering sustainable practices. A variety of innovative methods have been developed to treat aquaculture wastewater effectively. These encompass eco-sensitive systems such as aquaponics and integrated multi-trophic systems, along with biofilm technologies that exploit the mutual growth of algae and fungi to remove nutrients. Furthermore, the reuse of wastewater via oxidation ponds can both provide natural sustenance for fish and treat household sewage effectively. Technologies like constructed wetlands and recirculation systems have proven highly effective in removing nutrients and recycling water, thus aiding in sustainable aquaculture practices. Additionally, specialized continuous automatic treatment equipment has been devised specifically for aquaculture wastewater, ensuring efficient purification processes. Together, these diverse strategies collectively strive to address the challenges posed by nutrient pollution and resource depletion in managing aquaculture wastewater.

Types of Waste in Aquaculture Systems

Solid Waste: Solid waste within aquaculture systems encompasses uneaten feed, fecal matter, and additional organic substances that settle at the bottom of the aquatic environment. As time progresses, the buildup of such waste can deteriorate the quality of water and lead to a reduction in oxygen levels within the aquatic ecosystem.

Dissolved Waste: Dissolved waste comprises nitrogenous compounds (like ammonia and nitrate) and phosphorus, which are discharged into the water as metabolic residues of aquatic organisms. Elevated concentrations of dissolved waste can trigger eutrophication, characterized by the overabundance of nutrients in water bodies, leading to harmful algal blooms, oxygen depletion, and biodiversity decline. Strategies such as biofiltration, denitrification, and water exchange are frequently utilized to address the buildup of dissolved waste in aquaculture setups.

Uneaten Feed and Excretion: Leftover feed and waste produced by cultured organisms add to both solid and dissolved waste in aquaculture settings. This waste influx can introduce surplus nutrients into the water, stimulating microbial proliferation and changing nutrient patterns in the aquatic milieu. Moreover, deploying effective waste elimination systems and biological filtration techniques can help diminish the influence of leftover feed and waste on water quality.

Environmental Impacts of Aquaculture Waste

Pollution Concerns: The waste produced by aquaculture has a considerable environmental impact, prompting concerns about pollution. Waste generated by aquaculture encompasses both organic and inorganic pollutants such as fish feed, disinfectants, antibiotics, hormones, and plastics. If not recycled properly, these pollutants can have detrimental effects on the environment. Components of aquaculture waste, like nutrients from fish food metabolism, leftover food, and chemicals from disease treatments, contribute to eutrophication and genetic impacts on wild fisheries. To address these concerns, various waste management techniques have been proposed, including biofloc technology, recycling methods, and advanced treatment systems such as recirculation aquaculture systems (RAS) and biological, physical, and chemical treatments. It's essential to implement sustainable waste management strategies to minimize the environmental harm caused by aquaculture waste.

Nutrient Enrichment and Eutrophication: Aquaculture waste, especially from intensive operations, plays a significant role in enriching nutrients and causing eutrophication in aquatic environments. (Ajie and Prihatiningtyas, 2022). The waste, consisting of uneaten feed and excreted nutrients, results in declining water quality, increased turbidity, and decreased levels of dissolved oxygen. (Elvines *et al.,* 2024). Phosphorus, a significant constituent of aquaculture waste, worsens eutrophication on a global scale. Different technologies, such as adsorption and biological removal methods, are employed to minimize phosphorus discharge into the environment. The composition of aquafeed directly influences the quality of waste, which in turn affects parameters like water pH, turbidity, and oxygen levels. Furthermore, the particulate waste from fish farms affects seabed habitats and food webs, highlighting the need for biochemical tools to track waste dispersion and its ecological effects. Implementing proper waste management strategies is essential to alleviate the environmental impacts of aquaculture waste.

Habitat Degradation: Waste generated by aquaculture, which includes organic materials from fish farming, can degrade habitats by inducing oxygen-depleted sediments, altering the quality of water and sediment, and influencing benthic macrofauna. This waste, consisting of feces, leftover feed, and various chemicals, adds to environmental decline, impacting the health of farmed fish as well as the profitability of aquaculture operations. (Rebeca, 2022; Elvines *et al*., 2024). The accumulation of organic material beneath fish cages can result in a decrease in dissolved oxygen levels and the production of hydrogen sulfide, making the sediment unsuitable for benthic organisms to inhabit (Shih, 2018). Moreover, aquaculture waste has the potential to enter the food web, leading to possible broader ecosystem alterations and impacting biogeochemistry and benthic ecology (Paraska *et al.,* 2015). Effective management strategies, such as employing biochemical tools to track waste dispersion and comprehend its environmental impact, are essential for mitigating these environmental effects and promoting the sustainable expansion of aquaculture (Mungati *et al.,* 2020).

Impact on Biodiversity: The impact of aquaculture waste on biodiversity is indeed significant, given its environmental repercussions. Waste from aquaculture systems, comprising uneaten feed and aquafeed residues, can result in degradation of water quality, eutrophication, harmful algal blooms, and elevated fish mortality rates (Rebeca, 2022; Elvines *et al.,* 2024). Moreover, particulate waste from fish farms can induce organic enrichment in seabed habitats, influencing benthic ecology and integrating into the food web, thereby potentially instigating broader ecosystem shifts (Mungati *et al.*, 2020). Globally, the expansion of the aquaculture industry exacerbates these challenges, as the disregard for fish industry waste adds to environmental pollution, fosters disease outbreaks, and perpetuates unsustainable practices. Employing strategies such as controlled formulation of aquafeeds, enzyme-based feeds, and precision feeding can aid in reducing aquafeed waste and minimizing its impact on biodiversity (Alvarado-Ramírez *et al.,* 2023). In conclusion, tackling aquaculture waste is vital for preserving biodiversity and fostering sustainable aquaculture practices.

Traditional Waste Management Techniques

Pond-Based Systems: In conventional pond-based wastewater management for aquaculture, natural processes are harnessed to treat wastewater and improve fish production. For instance, in regions like the mountainous areas of Northern Vietnam, traditional pond systems often adopt a polyculture strategy, combining grass carp with other fish species (Sharique and Ali, 2022). These ponds typically depend on continuous water flow and natural food supplies, yet encounter issues such as deteriorating water quality and mortality rates specific to certain species. Moreover, innovative methods such as employing floating beds with aquatic plants can notably improve wastewater purification by combining plant absorption, matrix adsorption, and microbial activity synergistically (Xu *et al.,* 2015).

Sedimentation Ponds: Sedimentation ponds are pivotal for wastewater treatment in aquaculture setups, serving to facilitate the exchange of nutrients between sediment and water, thereby impacting the overall water quality (Vitoria and Philips, 2023). Furthermore, sedimentation ponds are integral components of a holistic treatment system, which encompasses ecological ponds, artificial wetlands, and filtration mechanisms. This comprehensive approach aims to guarantee that the treated water meets the required discharge standards (Sharique and Ali, 2022). These ponds aid in the reduction of nutrients such as nitrogen and phosphorus, thereby enhancing water quality parameters such as dissolved oxygen levels and turbidity (Ajie and Prihatiningtyas, 2022). Through harnessing natural processes like photosynthesis and symbiotic relationships between algae and bacteria, sedimentation ponds actively contribute to the purification and recycling of wastewater, thereby promoting sustainable aquaculture practices (Qin *et al.*, 2019). Incorporating sedimentation ponds into aquaculture systems is imperative for upholding environmental sustainability and fostering effective wastewater treatment.

Effluent Treatment Ponds: Effluent treatment ponds play a pivotal role in treating wastewater in aquaculture by harnessing solar energy for photosynthesis. This process aids in nutrient removal and water purification, ensuring a suitable environment for fish culture (Sharique and Ali, 2022). Effluent treatment ponds are instrumental in removing pollutants such as suspended solids, nitrogenous wastes, and phosphates, thereby fostering sustainable development in aquaculture (Julia and Subbotina, 2022). In summary, effluent treatment ponds alongside advanced technologies play a crucial role in upholding environmental sustainability and facilitating resource recovery in aquaculture wastewater management.

Waste Stabilization Ponds: Waste stabilization ponds (WSPs) prove to be efficient in the treatment of wastewater for use in aquaculture. Research indicates that Waste Stabilization Ponds (WSPs) can be enhanced through the integration of mechanical and anaerobic pre-treatment methods (Sinh *et al.,* 2023). This augmentation significantly enhances treatment efficiency, particularly in the reduction of pathogens such as E. coli, thus meeting the required reuse standards (Sharique and Ali, 2022). Moreover, incorporating baffles into WSPs can boost hydraulic efficiency, prolong mean residence time, and diminish stagnant areas, consequently enhancing treatment efficacy (Lian *et al.*, 2022). This enhancement is achieved by offering additional surface area for microorganism attachment and augmenting nutrient removal. In Ghana, the Chirapatre wastewater treatment plant showcased notable stabilization proficiency across multiple parameters, rendering the effluents suitable for aquaculture reliant on wastewater With fish survival rates surpassing 65% and displaying satisfactory growth, the plant demonstrated its effectiveness in supporting wastewater-fed aquaculture (Nsiah- Gyambibi, 2022).

Use of Natural Filters and Vegetation: The evolution of aquaculture wastewater treatment has transcended conventional methods like natural filters and vegetation techniques. Modern approaches prioritize sustainability, embracing innovative methods such as Biological Aerated Filters (BAFs) enhanced with biological filter media like Marchantia polymorpha (Ding *et al*., 2023). Efficient treatment is imperative due to the surplus of organic pollutants produced by aquaculture activities, underscoring the significance of implementing suitable wastewater treatment processes for fostering sustainable development (Chatla *et al*., 2020). Constructed wetlands primarily serve the purpose of water purification, but additional advantages can be integrated into their design. Marsh vegetation not only enhances the aesthetic appeal and plant diversity of the landscape but also provides habitat for fauna and recreational areas. Various strategies can be employed to utilize plants, including halophytes, macro-algae, and micro-algae, in treating saline aquaculture effluent (De Lange *et al.,* 2013).

Modern Waste Management Techniques

Recirculating Aquaculture Systems (RAS): Recirculating Aquaculture Systems (RAS) are employed in aquaculture management for diverse reasons. RAS present a sustainable solution by efficiently managing wastewater and preserving water quality to ensure the well-being of fish. Electrochemical water treatment within RAS substitutes conventional biological techniques, delivering cost efficiency, disinfection, and organic matter removal (Ben-Asher *et al*., 2024). Furthermore, marine anammox bacteria such as Candidatus Scalindua demonstrate potential in addressing nitrogen-enriched RAS wastewater, achieving notable rates of ammonia and nitrite removal (Micolucci *et al.,* 2023). The utilization of microalgae-based technology in RAS plays a crucial role in wastewater remediation. This involves cultivating microalgae with the nutrient-rich water from fish, generating biomass suitable for fish feed while simultaneously purifying the water for reuse (Villanova *et al.,* 2023). Moreover, nitrifying biofilters are essential components of RAS, playing a critical role in converting harmful ammonia into less toxic nitrate. Additionally, water disinfectants such as peracetic acid hold promise for pathogen control within RAS without interfering with the biofiltration process (Lepine *et al.,* 2023). In summary, RAS exhibit effectiveness in managing aquaculture wastewater and upholding water quality standards, thereby promoting sustainable practices in fish farming.

Biofilters and Bioreactors: Modern biofilters and bioreactors are pivotal in efficiently treating aquaculture wastewater. Numerous studies underscore the effectiveness of diverse systems. For instance, Biological Aerated Filters (BAFs) enhanced with Marchantia polymorpha have demonstrated consistent treatment performance, attaining impressive removal rates for NH4+-N and algae cells (Ding *et al*., 2023). Nature-based systems employing anaerobic bioreactors with either horizontal or vertical flow, integrating materials such as coarse sand and almond pruning, have showcased notable decreases in water quality parameters, particularly nitrogen content (Rodríguez-Espinosa *et al.*, 2023). A dual-stage biofiltration system employing zeolite and pyrite has demonstrated efficacy in nutrient removal from brackish aquaculture wastewater, facilitated by specific microbial communities that enhance denitrification processes (Fel *et al.,* 2020). Moreover, Chlorella sp. strains isolated from shrimp aquaculture wastewater have displayed swift nutrient removal in tubular photo-bioreactors, indicating potential for widespread implementation in large-scale wastewater treatment (Anh, 2020). The innovative biofiltration and bioreactor systems provide sustainable solutions for aquaculture waste treatment.

Aquaponics Systems: Aquaponics systems are vital for sustainable aquaculture waste treatment. These setups combine aquaculture with hydroponics, leveraging fish wastewater to feed plants. In return, the plants filter out dissolved nutrients such as nitrogen and phosphorus from the water. Aquaponics not only assists in waste management but also supports intensive food production, requiring minimal water and eliminating the need for soil, fertilizers, or pesticides (Cifuentes- Torreset *et al.,* 2021; Lad, 2022). Studies in aquaponics have concentrated on diverse areas, including practical techniques, modelling, system variations, and species interactions, indicating a rising interest in this domain (Vargas *et al.,* 2022). Moreover, aquaponics is acknowledged as a versatile technology that contributes to wastewater treatment, bio-filtration, and nutrient supply, aligning with the aim of sustainable waste management and ecosystem equilibrium for a cleaner future (Mohanan *et al.,* 2022).

Integrated Multi-Trophic Aquaculture (IMTA): IMTA systems integrate species from different trophic levels, converting uneaten feed and waste into valuable resources, promoting ecologically sustainable aquaculture (Aguilo-Arce *et al.*, 2023). IMTA seeks to enhance water quality by decreasing nutrient levels such as nitrogen and phosphorus, employing inventive techniques like constructed wetlands. This approach boosts the environmental sustainability of aquaculture endeavours. Studies indicate that IMTA systems notably influence the growth, survival, and biomass output of aquatic species, underscoring their efficacy in aquaculture treatment. However, despite its promise, the adoption of IMTA faces obstacles due to governance challenges and insufficient policies. This highlights the necessity of addressing regulatory frameworks to facilitate the broad application of IMTA in aquaculture treatment.

Waste-to-Energy Technologies: Waste-to-energy technology is pivotal in treating aquaculture waste. Numerous inventive methods have emerged to transform waste into energy sources. Biotechnological interventions, like genetic engineering and bioinformatics, are harnessed to convert agricultural waste, including aquaculture byproducts, into eco-friendly bioethanol and biogas. Moreover, microbial fuel cell methods have been investigated for fish waste treatment, repurposing it into a valuable reservoir of both nutrients and energy (Imam *et al.*, 2022). Furthermore, electrocoagulation, a physical method, has been scrutinized for its efficacy in managing organic waste generated by aquaculture. It has demonstrated substantial enhancements in water quality parameters when implemented with precise voltage and duration settings (Syukor *et al.*, 2021). Collectively, these technologies contribute to sustainable waste management practices in aquaculture.

Nutrient Recovery and Recycling Systems: Nutrient recovery and recycling technologies play a crucial role in fostering sustainable practices in aquaculture waste treatment. A range of innovative approaches has been proposed, including integrated multi-trophic systems (IMTA), which aim to recycle nutrients from aquaculture wastewater to cultivate diverse organisms. Biofloc technology (BFT) stands out as another significant method that improves feed conversion, enhances biosecurity, and promotes wastewater recycling by perpetually recycling and reusing nutrients within the system (Sharique and Ali, 2022). Hybrid membrane technologies have emerged as promising solutions for efficiently concentrating nutrients in aquaculture effluent for recovery. However, further research is necessary to enhance their performance in nutrient recovery (Luo, 2022). These technologies, spanning from IMTA to BFT and hybrid membrane systems, play a vital role in mitigating environmental impact and advancing sustainable aquaculture practices.

Challenges and Limitations

Economic Constraints: The aquaculture industry encounters economic constraints in waste treatment due to the imperative of pursuing sustainable development amid escalating pollution concerns. Efficient waste management poses challenges for the industry (Chatla *et al.,* 2020). Numerous technologies exist for treating aquaculture effluents, yet their implementation can entail significant costs (Brinker and Bergheim, 2005). To overcome economic limitations, approaches like integrated multitrophic aquaculture are proposed, leveraging ecological engineering to harmonize ecosystem functions and minimize environmental impacts. Moreover, biological treatments, particularly those employing secondary filtering organisms, are underscored as effective and sustainable methods for waste treatment in aquaculture (Buschmann *et al.,* 2008). In essence, economic constraints in aquaculture waste treatment emphasize the significance of devising cost-effective solutions that prioritize both environmental sustainability and resource efficiency.

Technical Challenges: A notable technical hurdle in aquaculture waste treatment involves effectively removing pollutants from wastewater to avert environmental harm. Numerous treatment technologies have been investigated, including natural coagulants such as chitin and Moringa oleifera seeds, which have exhibited promising outcomes in decreasing physicochemical parameters by 30-50% (Zhang, 2019). Furthermore, the adoption of recirculation aquaculture systems (RAS) has proven pivotal in efficiently treating aquaculture effluents. These systems, in conjunction with biological, physical, and chemical treatments, play a crucial role in mitigating the impact of aquaculture wastewater on the environment. The advancement of innovative systems such as aquaculture farm sewage treatment systems and wastewater treatment equipment designed for aquaculture seawater underscores the continuous efforts to improve the efficiency of aquaculture waste treatment.

Environmental Considerations: Environmental considerations in aquaculture waste treatment are paramount for fostering sustainable practices. The aquaculture industry grapples with challenges stemming from the generation of substantial volumes of wastewater (Ahmad *et al.,* 2022). To tackle this issue, innovative approaches such as integrated multi-trophic systems (IMTA) and recirculation aquaculture systems (RAS) are being embraced to ensure effective treatment (Qiao *et al.*, 2022). Employing aquaculture wastewater in constructed wetlands and plant lagoons can contribute to sustainable reduction of nutrient levels (Tom *et al.*, 2021). Furthermore, converting aquaculture waste into poly-β-hydroxybutyrate (PHB) for fish feed offers an eco-friendly solution, fostering fish growth and immune response. Implementing technologies such as bioreactors, biofloc, wetlands, and phytoremediation can enhance treatment effectiveness while mitigating environmental impact. Collectively, these approaches strive to enhance water quality, diminish pollution, and uphold the long-term sustainability of aquaculture systems.

Regulatory Compliance The aquaculture industry is subject to stringent environmental legislation, requiring efficient treatment methods for aquaculture effluents. Extensively researched for their efficacy in treating aquaculture wastewater, technologies such as recirculation aquaculture systems (RAS), biological, physical, and chemical treatments are prominently featured (Ahmad *et al.*, 2022). Specialized methods like the Sequencing Biofilm Batch Reactor (SBBR) have been devised to fulfil effluent quality standards, guaranteeing minimal levels of CODcr (Chemical oxygen demand chromium) NH4-N, and maintaining a pH range of 6-9 (Acierno *et al.,* 2006). Furthermore, immediate biological treatment methods that involve fermenting organic waste materials are suggested to produce organic fertilizers that meet regulatory standards (Gao and Wang, 2019). Adhering to regulations is vital for fostering sustainable aquaculture development in the face of evolving environmental concerns.

Regulatory Frameworks and Compliance

Regulatory frameworks are pivotal in effectively managing aquaculture waste. Research underscores the influence of regulatory costs on U.S. tilapia farms, where compliance expenses accounted for 15% of total production costs, significantly impacting revenue and constraining industry growth (Engle, 2023). Furthermore, compliance frameworks offer a structured approach to monitor and measure the effectiveness of regulations, assisting organizations in aligning their business practices with mandated requirements (Radhakrishnan and Dineshbabu, 2012). Emphasizing transparent and efficient decision-making processes in mariculture policies is crucial to minimize environmental impacts and integrate operations sustainably into ecosystems (Niaounakis, 2017). Management of aquaculture waste is pivotal for ensuring sustainability and adherence to regulatory frameworks. European legislation places a strong emphasis on implementing reutilization and reduction strategies for managing aquaculture waste, thereby promoting circular bio-economy principles aimed at safeguarding human health and the environment (Fraga- corral *et al.,* 2022). Global strategies for solid waste management, integrating concepts like the Waste Hierarchy and Circular Economy, underscore the importance of enhanced compliance, stakeholder engagement, and awareness to tackle waste generation and environmental challenges on a worldwide scale (Awino and Apitz 2023) .Moreover, the significance of international, regional, and national agreements in tackling marine plastic debris, including the implementation of specific measures for different sources of plastic, highlights the broader regulatory landscape's role in waste management (Haber *et al.,* 2018).

CONCLUSION

In Summary, as the aquaculture sector expands worldwide, effective waste management becomes increasingly crucial for both sustainability and compliance with regulations. Aquaculture waste, primarily effluent or sludge, poses challenges for farm managers, especially regarding discharge standards. The global growth of aquaculture raises concerns about environmental damage from pollutants entering natural water systems. Innovative wastewater treatment methods, such as aquaponics and biofilm technologies, alongside recycling approaches like oxidation ponds and constructed wetlands, play key roles in nutrient removal and water reuse. Adherence to regulatory frameworks, backed by international agreements, is essential for addressing waste management challenges. Ultimately, sustainable waste management practices are vital for minimizing environmental harm and ensuring the ongoing viability of aquaculture worldwide.

REFERENCES

Acierno, R., Blancheton, J. P., Bressani, G., Ceruti, L., Chadwick, D., Roque D'Orbcastel, E., ... and Donaldson, G. 2006. Manual on effluent treatment in aquaculture: Science and Practice. Aquaetreat.

Aguilo-Arce, J., Ferriol, P., Trani, R., Puthod, P., Pierri, C., and Longo, C. 2023. Sponges as emerging by-product of Integrated Multitrophic Aquaculture (IMTA). Journal of Marine Science and Engineering, 11(1): 80.

Ahmad, A. L., Chin, J. Y., Harun, M. H. Z. M., and Low, S. C. 2022. Environmental impacts and imperative technologies towards sustainable treatment of aquaculture wastewater: A review. Journal of Water Process Engineering, 46, 102553.

Ajie, G. S., and Prihatiningtyas, E. 2022. Nutrients removal from integrated multi-trophic aquaculture (IMTA) water using waste stabilization ponds (WSP). In IOP Conference Series: Earth and Environmental Science,976(1):012029. IOP Publishing. stabilization ponds (WSP). IOP Conference Series: Earth and Environmental Science, doi: 10.1088/1755-1315/976/1/012029.

Ajie, G. S., and Prihatiningtyas, E. 2022. Nutrients removal from integrated multi-trophic aquaculture (IMTA) water using waste stabilization ponds (WSP). In IOP Conference Series: Earth and Environmental Science (Vol. 976, No. 1, p. 012029). IOP Publishing.

Alvarado-Ramírez, L., Santiesteban-Romero, B., Poss, G., Sosa-Hernández, J. E., Iqbal, H. M., Parra-Saldívar, R., ... and Melchor-Martínez, E. M. 2023. Sustainable production of biofuels and bioderivatives from aquaculture and marine waste. Frontiers in Chemical Engineering, 4, 1072761.

Anh, L. H. 2020. Khuyen VTK Biotreatment of aquaculture wastewater with Chlorella sp. in tubular photo-bioreactors. International Journal of Scientific and Technology Research, (9):4.

Awino, F. B., and Apitz, S. E. 2024. Solid waste management in the context of the waste hierarchy and circular economy frameworks: An international critical review. Integrated Environmental Assessment and Management, 20(1):9-35.

Ben-Asher, R., Gendel, Y., and Lahav, O. 2024. Electrochemical applications in RAS: A review. Reviews in Aquaculture, 16(1): 86-105.

Brinker, A., and Bergheim, A. 2005. Waste treatment in fish farms. Water encyclopedia, 1:681-684.

Buschmann, A. H., Hernández-González, M. C., Aranda, C., Chopin, T., Neori, A., Halling, C., and Troell, M. 2008. Mariculture waste management.

Chatla, D., Padmavathi, P. and Srinu, G.2020. Wastewater treatment techniques for sustainable aquaculture. Waste management as economic industry towards circular economy, 159-166.

Chávez-Crooker, P., and Obreque-Contreras, J. 2010. Bioremediation of aquaculture wastes. Current Opinion in Biotechnology, 21(3): 313-317.

Cifuentes-Torres, L., Correa-Reyes, G., and Mendoza-Espinosa, L. G.2021. Can Reclaimed Water Be Used for Sustainable Food Production in Aquaponics?. Frontiers in Plant Science, 12, 669984.

De Lange, H. J., Paulissen, M. P. C. P., and Slim, P. A. 2013. 'Halophyte filters': the potential of constructed wetlands for application in saline aquaculture. International Journal of Phytoremediation, 15(4):352-364.

Ding, J., Meng, Y., Lu, S., Peng, Y., Yan, W., Li, W., ... and Zhang, H. 2023. The Treatment of Aquaculture Wastewater with Biological Aerated Filters: From the Treatment Process to the Microbial Mechanism. Toxics, 11(6): 478.

Ding, J., Meng, Y., Lu, S., Peng, Y., Yan, W., Li, W., ... and Zhang, H. 2023. The Treatment of Aquaculture Wastewater with Biological Aerated Filters: From the Treatment Process to the Microbial Mechanism. Toxics, 11(6): 478.

Elvines, D. M., MacLeod, C. K., Ross, D. J., Hopkins, G. A., and White, C. A. 2024. Fate and effects of fish farm organic waste in marine systems: Advances in understanding using biochemical approaches with implications for environmental management. Reviews in Aquaculture, 16(1):66-85.

Engle, C. R., van Senten, J., Clark, C., and Boldt, N. 2023. Has the regulatory compliance burden reduced competitiveness of the US tilapia industry?. Fishes, 8(3): 151.

Fei, X., Sun, S., He, S., Huang, J., and Zhou, W. 2020. Application of a novel two-stage biofiltration system for simulated brackish aquaculture wastewater treatment. Environmental Science and Pollution Research, 27: 636-646.

Fraga-Corral, M., Ronza, P., Garcia-Oliveira, P., Pereira, A. G., Losada, A. P., Prieto, M. A., ... and Simal-Gandara, J. 2022. Aquaculture as a circular bio-economy model with Galicia as a study case: How to transform waste into revalorized by-products. Trends in Food Science & Technology, 119, 23-35.

Gao, Yinhua and Wang, Youming. 2017. Aquaculture wastewater purification treatment method and device.

Haber, M. J., Hibbert, B., Haber, M. J., and Hibbert, B.2018. Risk Management Frameworks. Asset Attack Vectors: Building Effective Vulnerability Management Strategies to Protect Organizations, 241-247.

Imam, T. S., Ari, H. A., and Ugya, A. Y. 2022. Biotechnological Intervention for the Management of Agricultural Waste. In Environmental Biotechnology (pp. 229-243). Apple Academic Press.

Julia, M., Subbotina. 2022. Process diagram of pig-breeding effluents treatment in classic fish-water-biological ponds. Проблемы ветеринарной санитарии, гигиены и экологии, doi: 10.36871/vet.san.hyg.ecol.202203012

Lad, D. 2022. Aquaponics: A Way of Green Technology.

Lepine, C., Redman, N., Murray, M., Lazado, C. C., Johansen, L. H., Espmark, Å. M., ... & Good, C. 2023. Assessing peracetic acid application methodology and impacts on fluidized sand biofilter performance. Aquaculture Research, 2023.

Lian, Y., Coggins, L. X., Hay, J., van de Ven, A., and Ghadouani, A. 2022. Effect of attached growth on treatment performance in waste stabilization ponds. Water, 14(20): 3245.

Luo, G. 2023. Review of waste phosphorus from aquaculture: Source, removal and recovery. Reviews in Aquaculture, 15(3):1058-1082.

Micolucci, F., Roques, J. A., Ziccardi, G. S., Fujii, N., Sundell, K., and Kindaichi, T. 2023. Candidatus scalindua, a biological solution to treat saline recirculating aquaculture system wastewater. Processes, 11(3): 690.

Mohanan, V. P., Mahanta, N. R., Kushwaha, A., and Goswami, L.2022. Leveraging the potential of aquaponics for urban sustainability. In Emerging Trends to Approaching Zero Waste (pp. 59-78). Elsevier.

Niaounakis, M. (2017). Regulatory Framework. Management of Marine Plastic Debris, 361–413. doi:10.1016/b978-0-323-44354-8.00007-0

Nsiah-Gyambibi, R. 2022. Municipal wastewater stabilization and the potential reuse for African Catfish (Clarias gariepinus) culture in waste-fed earthen ponds in the tropics.

Paraska, D., Bruce, L., Shiell, G., and Hipsey, M. 2015. Predicting critical thresholds of aquaculture waste loading to coastal sediment. In Predicting critical thresholds of aquaculture waste loading to coastal sediment (pp. 536-542). Modelling and Simulation Society of Australia and New Zealand Inc..

Qiao, G., Li, X., Li, J., Zhang, M., Shen, Y., Zhao, Z., ... and Wang, Z. 2022. An Eco-Friendly Conversion of Aquaculture Suspended Solid Wastes Into High-Quality Fish Food by Improving Poly-β-Hydroxybutyrate Production. Frontiers in Physiology, 13, 797625.

Qin, Zhiqing., Lin, Jianbin., Fan, Haiping., Liang, Ping., Qiu, Manli., Li, Xuegui. 2019. Freshwater aquaculture tail water treatment system.

Radhakrishnan, E. V., and Dineshbabu, A. P. 201. Regulatory framework for mariculture development and management.

Rebeca, Osorio.2022. Impact of Aquaculture Solid Waste on Environment in the Delta Region of Andhra Pradesh: A Case Study. doi: 10.1007/978-981-19-6774-0_35

Rodríguez-Espinosa, T., Pérez Gimeno, A., Almendro Candel, M. B., Gómez Lucas, I., & Navarro-Pedreño, J. 2023. Low-quality irrigation water treated using waste biofilters. Water, 15(13), 2464.

Sharique, A., Ali. 2022. Fish Poly Culture in Domestic Wastewater Ponds: A Step Towards Protein Recovery and Pollution Reduction. Bioscience Biotechnology Research Communications, doi: 10.21786/bbrc/15.3.3

Shih, Y. C.2018. Established the Environmental Monitoring Program Indicators to Prevent Disease Infection and Promote Sustainable Development.

Sinn, J., Agrawal, S., Orschler, L., Schubert, S., and Lackner, S. 2023. Upgrade of waste stabilization ponds to improve effluents for reuse purposes. H2Open Journal, 6(2): 315-330.

Syukor, A. A., Sulaiman, S., Chandrakant, J. P., Mishra, P., Nasrullah, M., Singh, L., and Zularism, A. W. 2021. Energy generation from fish-processing waste using microbial fuel cells. In Delivering Low-Carbon Biofuels with Bioproduct Recovery (pp. 101-121). Elsevier.

Tom, A. P., Jayakumar, J. S., Biju, M., Somarajan, J., and Ibrahim, M. A. 2021. Aquaculture wastewater treatment technologies and their sustainability: A review. Energy Nexus, 4, 100022.

Vargas, S. A. V., Navarrete, Ó. L. G., and Gómez, M. A. C. 2022. Automated, technified and traditional aquaponic systems: a systematic review. Ingeniería Solidaria, 18(2):1-39.

Victoria, M, Philips.2023.Sediment and Remediation of Aquaculture Ponds. doi: 10.1007/978-981-19-5486-3_8

Villanova, V., Roques, J. A. C., Forghani, B., Shaikh, K. M., Undeland, I., and Spetea, C.2023. Two-phase microalgae cultivation for RAS water remediation and high-value biomass production. Frontiers in Plant Science, 14, 1186537.

Xu, G., Duan, D., Tian, G., Zhang, M., and Liu, F. 2015. Aquaculture waste water treatment by an enhanced ecological floating bed based on synergistic effect. Journal of Dalian Ocean University, 30(1): 62-67.

Zhang, Xuechao. 2019. Aquaculture farm sewage treatment system.

22

Sustainable Phenol Removal from Aqueous Solution Through Natural Polymeric Substances

P.U. Meshram, V.P. Meshram and S.D. Indurkar

ABSTRACT

Phenol and its derivatives are compounds that are found in many industrial wastewaters such as in paints, paper, plastics, oil and gasoline, steel textile and wood industries. They are one of the largest group of water pollutant owing to their presence in many industrial waste water because of their wide uses. In addition, leakage accidents that took place during the transportation and storage of phenol causes contamination of water resources. It is in the list of priority pollutants of US Environmental Protection Agency (EPA) and it has toxic, carcinogenic effects on human and aquatic organisms. Therefore, waste water containing phenol should be treated before discharge to its high toxicity even at low concentration. Over the last few years, the treatment of wastewater contaminated with phenol and its compound has attracted great attention due to their toxicity and low bio-degradability properties. Several treatment processes are used to remove or recover phenol and its compound from waste water in the literature and many of them have some advantages and drawbacks.

The present study reveals the liquid phase adsorption of phenol from water and wastewater by tree barks of different species (Ficus racemosa) and (Artocarpus heterophyllus). Kinetic and equilibrium experiments have been performed for the above low cost absorbents and effects of pH, time and dose are investigated. The data generated under effective optimization conditions are used to fit in the adsorption model (Freundlich Adsorption) and Adsorption Isotherms generated through which the nature and efficiency of the absorbents have been worked out.

The studies show more than 90% removal of Phenol from aqueous solution and isotherms indicate good agreement for the validity of the model.

***Keywords**: Adsorption, Phenol, Artocarpus Heterophyllus (Jack Fruit), Ficus Racemosa (Umber).*

INTRODUCTION

Phenolic compounds in the wastewater are harmful pollutants which originate as byproducts from several industries like pulp and paper mills, coaking plants, petroleum refineries, tanning industries, phenolic resin manufacturing units etc. Phenolic compounds impart colour and odour to water and are toxic even at a very low level of concentration. The various processes for the removal of phenolics from the wastewater can be grouped into two broad categories. In one category are those processes like microbiological degradation, chemical treatment, incineration etc. which aim at destroying the phenolics, while the other category has processes involving the recovery of phenolics e.g. solvent extraction, activated carbon treatment, uptake by porous polymeric adsorbents etc. Among these various methods polymeric adsorbents are very promising for the removal of phenolic compounds due to following reasons. The efficiency of removal of phenolics by porous polymeric adsorbents and their subsequent elution from the polymer matrix is exceptionally good. The requirement equipment is very simple and the operation is usually trouble free. The operation can be carried out at ambient temperature and the space requirements are comparatively small. Very stringent specifications regarding the quality of the effluent can be met by the use of polymeric adsorbents.

Porous polymeric adsorbents show quite a good sorption of phenolic compounds. Kennedy has reported a series of macroreticular adsorbents such as Amberlite XAD-2, XAD-4 for removing chlorinated phenols from pesticide industrial wastewater (1). Crook et.al. (2) reported the removal of phenols from industrial waste effluents using Amberlite XAD-7 porous polymeric adsorbent. Farrier et.al. (3) reported the uptake of phenol and benzoic acid by macroreticular adsorbent, Amberlite XAD-8. Dobrevsky *et. al.* (4) carried out a series of experiments for the removal and recovery of phenolics from wastewater using Amberlite XAD-2, XAD-4 and XAD-7, Duolite AIOID, Warion ATM, Wofatit SZ-30 and Wafatit EA-60 adsorbents/ resins. Schaeffer *et. al.* (5) reported the efficiency of Amberlite XAD-4 and XAD-8 polymeric absorbents for removing organic compounds from coke plants effluent. Wojaczynska *et. al.* (6) studied the sorption of various aromatic compounds on porous polymeric adsorbents.

Morris and Weber (7-8) studied adsorption of phenol on activated carbon. Belford (9) studied the adsorption of organics on to activated carbon from dilute aqueous solutions and explained the correlation of molar adsorbability with physico-chemical parameters.

Randtke (10), Boening (11) studied the adsorption on bench scale with different types of activated carbons for the adsorption of humic and fulvic acid in water Zogorski (12) and Snoeyink (13) evaluated the efficiency of different commercial activated carbons for the adsorption of phenol at different pH values. Lawrence K. Wang

(14) experimented the adsorption of organics from industrial effluents on to activated carbon and suggested that the change in adsorption is affected by lowering the pH. They studied the adsorption of organic compounds which included anionic and cationic surfactants, low molecular weight organics and industrial wastes from paper mills and glue factories.

Digiano and Weber and others (15-17) studied the adsorption of phenol, pnitrophenol, 2-4 dinitro phenol and anions of these species. He observed that the anions of weak electrolytes show much lower adsorption affinity for activated carbon.

In view of the high cost and tedious procedure for the preparation and regeneration of activated carbon, and other porous polymeric substances there is still continuous search for low cost adsorbents. Recently, some cheap and efficient adsorbents have been developed from the natural wastes. In continuation of our interest in this area, an attempt was made to use natural polymeric substances, such as commonly available Indian tree barks namely *Artocarpus heterophyllus* and *Ficus racemosa* for the removal and recovery of phenol from aqueous solution.

MATERIALS AND METHODS

Glassware

All glasswares used in the study were of Borosil make. Before using, the glasswares were carefully washed with chromic acid then with tap water and thoroughly rinsed with single and then 2-3 times with double distilled water immediately prior to use.

Distilled Water

Double distilled water was used in the entire study.

Chemical & Reagents

All chemicals and reagents used in the present study were of A.R. Grade.

Adsorbate (Phenol Solution)

Approximately 1.0 gm of A.R. grade phenol was dissolved in freshly boiled and cooled double distilled water and diluted to 1000 ml in one litre volumetric flask. The solution was standardized as per standard method (18).

Adsorbent (Modified bark substrate)

Modified bark substrate *(Arcorpus heterophyllus* and *Ficus racemosa)* were used in the adsorption studies.

Instruments

pH Meter

pH meter (control dynamics) was used for adjusting the pH after calibration with standard buffers.

Spectrophotometer

The U.V. spectrophotometer (systronics) was used for the estimation of phenol at λmax 500 nm.

Experimental

Batch Studies

i) ***Optimisation of pH*** 100 ml of desired concentration of phenol solution prepared from standardized stock solution was taken in a series of BOD bottles and the pH of the solution was adjusted from 3 to 10 with 0.1 N NaOH and 0.1 N HCI. A definite quantity of treated bark (1 g) was added in each bottle. The solutions were kept in the shaking machine for 30 minutes duration and then filtered. The filtrates were estimated for residual phenol concentration by spectrophotometrically as per standard method (18).

ii) ***Optimisation of Time*** To find out the effect of time on phenol adsorption with barks, desired concentration of phenol at optimized pH were taken in 100 ml capacity BOD bottles in which desired quantity of barks were added in each bottle. The solutions were stirred on shaking machine from 10,20,.....,120 minutes to attain adsorption equilibrium and filtered with whatman no.42 filter paper and estimated spectrophotometerically as per standard method. iv.)

iii) ***Optimisation of Dosage*** 100 ml of desired concentration of phenol solutions were taken in 300 ml capacity BOD bottles. 1 to 10 g. of treated barks were added in each bottle and shaken on shaking machine at definite rpm after adjusting optimum pH and time. The solution were filtered through whatman No 42. Filter papers and filtrtaes were estimated spectrophotometerically for their residual concentration at λ max 500 nm.

Results and Discussion

i) **Effect of pH on Phenol Adsorption by Barks** The data generated to observe the pH effect on phenol removal by barks are presented in Fig. 1. It was observed that there is maximum removal of phenol at pH 5 to 6 for both the barks and decline with increase in pH. This indicates that acidic pH is responsible for the maximum adsorption of phenol due to protonation effect.

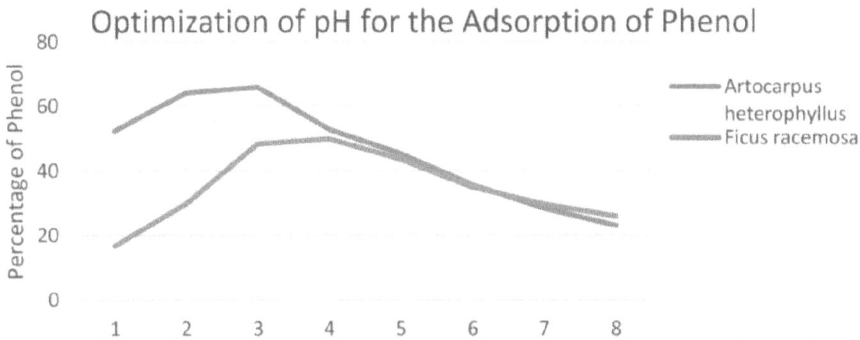

Fig. 1: Optimization of pH for the Adsorption of Phenol in *Artcarpus heterophyllus* and *Ficus racemosa*

Table 1: Optimization of pH for the Adsorption of Phenol with *Artcarpus heterophyllus* Bark

Sr. No.	pH	Original Phenol Concentration (mg/l)	Residual Phenol Concentration (mg/l)	Phenol Adsorbed (mg/l)	Percentage of phenol
1.	3	57	27.19	29.81	52.3
2.	4	57	20.46	36.54	64.1
3.	5	57	19.44	37.56	65.9
4.	6	57	26.96	30.04	52.7
5.	7	57	31.18	25.82	45.3
6.	8	57	36.71	20.29	35.6
7.	9	57	40.81	16.19	28.4
8.	10	57	43.95	13.05	22.9

ii) ***Effect of Time on Phenol Adsorption by Barks*** The data generated to observe the effect of time at optimized pH on phenol adsorption is shown in Fig. 2. It was observed that instant adsorption was observed and increased with increase in time up to 30 minutes for *Artocarpus heterophyllus* bark and remained constant with increase in time. This indicates that adsorption on this bark is a instant phenomeno whereas adsorption and desorption phenomenon was observed with the *ficus recemosa* bark.

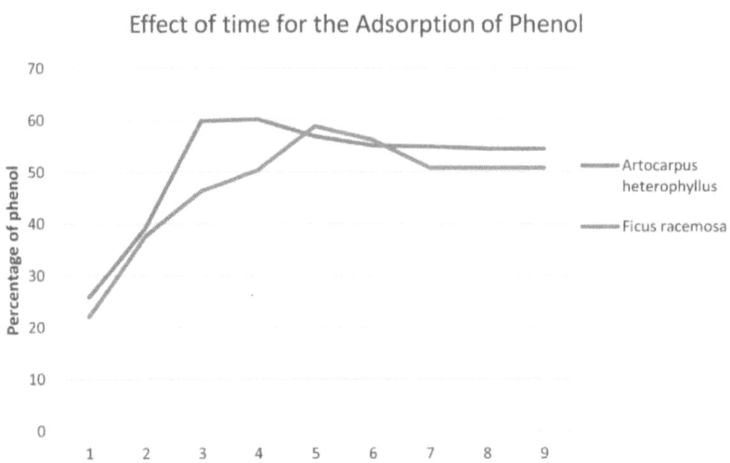

Fig. 2: Effect of time for the Adsorption of Phenol in *Artcarpus heterophyllus* and *Ficus racemose*

Table 2: Effect of time for the Adsorption of Phenol with *Artcarpus heterophyllus* Bark

Sr. No.	Time (min.)	Original Phenol Concentration (mg/l)	Residual Phenol Concentration (mg/l)	Phenol Adsorbed (mg/l)	Percentage of phenol
1.	10	57	42.27	14.73	25.85
2.	20	57	34.66	22.34	39.20
3.	30	57	29.64	27.36	59.80
4.	40	57	26.28	30.72	60.20
5.	50	57	24.58	32.42	56.88
6.	60	57	25.59	31.41	55.10
7.	90	57	25.71	31.29	54.90
8.	100	57	25.94	31.06	54.50
9.	120	57	25.94	31.06	54.50

iii) ***Effect of Dose on Phenol Adsorption*** The results obtained to observe the effect of barks dosage on phenol removal are presented in Table 3. It was observed that phenol adsorption is increased with increase in doses for both the barks up to certain extent and further remained constant with increase in doses. This indicates that, since the surface area of both the barks is less, there is a definite role of chemical adsorption rather than the surface area (physical and Adsorption).

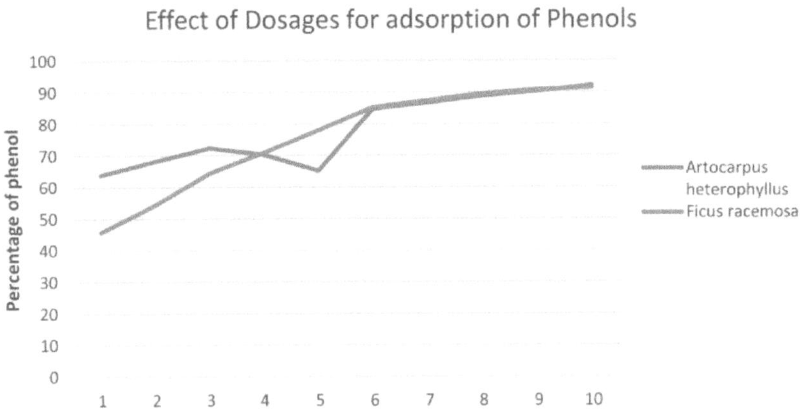

Fig. 3: Effect of Dosages for adsorption of Phenols in *Artcarpus heterophyllus* and *Ficus racemosa*

Table 3: Effect of Dosages (*Artcarpus heterophyllus)* for adsorption of Phenols.

Sr. No.	Dose of bark (gm/l)	Original Phenol Concentration (mg/l)	Residual Phenol Concentration (mg/l)	Phenol Adsorbed (mg/l)	Percentage of phenol
1.	10	57	20.58	36.42	63.90
2.	20	57	18.13	38.87	68.20
3.	30	57	15.91	41.09	72.40
4.	40	57	16.93	40.07	70.30
5.	50	57	19.84	37.16	65.20
6.	60	57	08.72	48.28	84.70
7.	70	57	07.59	49.41	86.68
8.	80	57	06.43	50.57	88.72

Adsorption Mechanism

The data generated was fit in the adsorption model (Freundlich Adsorption Model). Adsorption isotherms were drawn by plotting log versus log Ce where is the concentration of adsorbate adsorbed per gm of adsorbent and Ce is the equilibrium concentration of phenol. The isotherms drawn are shown in Fig. 4. It was observed that the adsorbate undergone chemical interaction with the adsorbent and referred as "chemical adsorption" or chemisorption. The adsorbate molecules are strongly held on the surface of the adsorbent, because of the formation of strong localized bond at the active sites of the adsorbent as explained by the other researchers in case of activated carbon (19-20). This type of interaction between adsorbate and adsorbent is favoured according to the surface functional groups or surface characteristics of the adsorbents (21-23). However, several factors influence the different types of adsorption often interacting to cause concentration of a particular solute at the interface. The adsorption of organic compound from water is influenced by factors associated with adsorbent, the adsorbate and the aqueous system (23). The nature of isotherms for both the barks indicates that there is a linear relationship between the adsorbent and adsorbate. The steep curve indicates that the adsorption is significant at a wide range of concentration, independent of the quantity of the adsorbent and applicable for higher concentration.

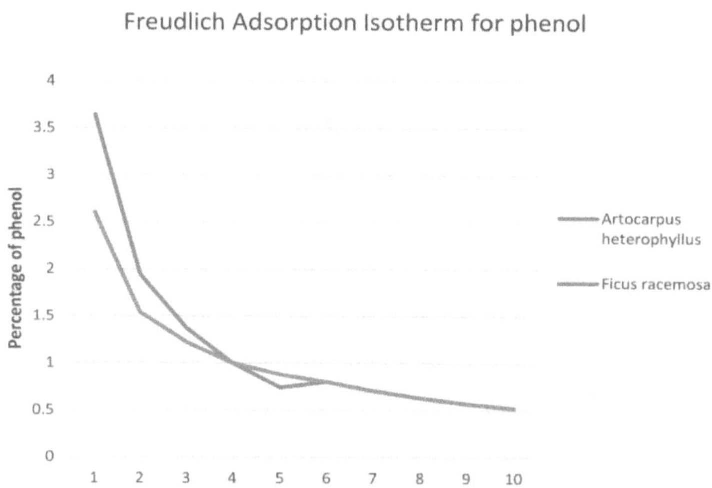

Fig. 4: Freudlich Adsorption Isotherm for phenol in *Artcarpus heterophyllus* and *Ficus racemosa*

Table 4: Freudlich Adsorption Isotherm for phenol with *Artocarpus heterophyllus* Bark.

Sr. No.	Dose of bark (gm/l) (M)	Original Phenol Concentration (mg/l) (C)	Residual Phenol Concentration (mg/l) (Ce)	Phenol Adsorbed (mg/l) (C-Ce)=M	Percentage of phenol (X/M)
	10	57	20.58	36.42	3.642
	20	57	18.13	38.87	1.943
	30	57	15.91	41.09	1.369
	40	57	16.93	40.07	1.001
	50	57	19.84	37.16	0.743
	60	57	08.72	48.28	0.804
	70	57	07.59	49.41	0.705
	80	57	06.43	50.57	0.632
	90	57	05.56	51.44	0.571
	100	57	04.56	52.44	0.524

Table 5: Optimization of pH for the Adsorption of Phenol with *Ficus racemosa* Bark

Sr. No.	pH	Original Phenol Concentration (mg/l)	Residual Phenol Concentration (mg/l)	Phenol Adsorbed (mg/l)	Percentage of phenol
1.	3	57	19.01	37.99	16.66
2.	4	57	40.01	16.99	29.82
3.	5	57	29.53	27.47	48.20
4.	6	57	28.62	28.38	49.80
5.	7	57	32.20	24.8-	43.50
6.	8	57	37.11	19.89	34.90
7.	9	57	40.11	16.89	29.64
8.	10	57	42.29	14.71	25.82

Table 6: Effect of time for the Adsorption of Phenol with *Ficus racemosa* Bark

Sr. No.	Time (min.)	Original Phenol Concentration (mg/l)	Residual Phenol Concentration (mg/l)	Phenol Adsorbed (mg/l)	Percentage of phenol
1.	10	57	44.40	12.60	22.1
2.	20	57	35.57	21.43	37.6
3.	30	57	30.61	26.39	46.3
4.	40	57	28.27	28.73	50.4
5.	50	57	23.49	33.51	58.8
6.	60	57	24.97	32.03	56.2
7.	90	57	28.04	28.96	50.8
8.	100	57	28.04	28.96	50.8
9.	120	57	28.04	28.96	50.8

Table 7: Effect of Dosages (*Ficus racemosa)* for adsorption of Phenols.

Sr. No.	Dose of bark (gm/l)	Original Phenol Concentration (mg/l)	Residual Phenol Concentration (mg/l)	Phenol Adsorbed (mg/l)	Percentage of phenol
1.	10	57	30.95	26.05	45.70
2.	20	57	26.05	30.95	54.30
3.	30	57	20.29	36.71	64.40
4.	40	57	16.64	40.36	70.80
5.	50	57	12.60	44.40	77.90
6.	60	57	08.44	48.56	85.20
7.	70	57	07.20	49.80	87.38
8.	80	57	06.04	50.96	89.42
9.	90	57	05.46	51.54	90.68
10.	100	57	04.93	52.07	91.36

Table 8: Freudlich Adsorption Isotherm for phenol with ***Ficus racemosa*** Bark.

Sr. No.	Dose of bark (gm/l) (M)	Original Phenol Concentration (mg/l) (C)	Residual Phenol Concentration (mg/l) (Ce)	Phenol Adsorbed (mg/l) (C-Ce)=M	Percentage of phenol (X/M)
1.	10	57	30.95	26.05	2.60
2.	20	57	26.05	30.95	1.54
3.	30	57	20.29	36.71	1.22
4.	40	57	16.64	40.36	1.00
5.	50	57	12.60	44.40	0.88
6.	60	57	08.44	48.56	0.80
7.	70	57	07.20	49.80	0.71
8.	80	57	06.04	50.96	0.63
9.	90	57	05.46	51.54	0.57
10.	100	57	04.93	52.07	0.52
11.					

CONCLUSION

Investigations have been made to determine whether or not cheaply available materials could be used in the treatment of wastewater in place of other commercially available adsorbents. In view of this, bark materials such as *Artocarpus heterophyllus* and *Ficus recemosa* were used as adsorbent to remove phenol or phenolic compounds. Batch adsorption studies were conducted for the adsorption of phenol by modified bark substrate. It was observed that the rate uptake of phenol is higher at a wide range of concentration.

The Freundlich adsorption isotherms show a good agreement for the validity of the equation. The steepness of the isotherm indicates that there is a uniform distribution of the solute (phenol) on the surface of the modified bark (adsorbent). It is concluded that products prepared from the tree bark materials could be used as adsorbent in the field of wastewater treatment processes. Further it was also observed that the product is efficient for adsorption of dissolved organics and could be useful in tertiary treatment of potable water as well as industrial effluent.

Thus, overall objective of this paper was to develop inexpensive, easily available, effective metal ion adsorbent from natural polymeric substances as alternative to existing commercial absorbent. The product prepared if used in waste water system in a huge scale, may become economical and sustainable.

ACKNOWLEDGEMENT

The authors are thankful to the Dr. P. N. Charde, Principal, Sevadal Mahila Mahavidyalaya for granting permission to attend and present the paper.

REFERENCES

Belfort, S.1979.Reaction Adsorption of Organic Homologous into Activated Carbon from Dilute Aqueous Solutions Solvophobic Interaction Approach & Corraction of Molar Adsorbility with Physico-chemical Parameters, Env. Sci. & Technol., 13(8):956-945.

Boening P.H., Beckmann D.D., Snoeyink, V.L.1980.Activated Carbon Versus Resin Adsorption of Phenol Humic Substances, J. AWWA, 54-59.

Crook, E.H., McDonnell, R.P. and McNultry, J.T.1975.Removal and Recovery of phenols from Industrial Waste Effluents with Amberlite XAD Polymeric Adsorbents", Ind. Eng. Chem., Prod. Res. Dev., 14, 113.

Digiano, F.A. and Weber W.J. (Jr)/1973. Sorption Kinetics in Infinite Batch Experiments, J. WPCF, 45 (713).

Digiano, F.J and Weber (Jr.) W.J.1969.Technical Publication", T-69-1 Deptt. of Civil Engg. Univ. of Michigan, Ann Arbor, Michigan.

Dobrovsky, J. and Zuezdov, A.1979.Dynamic Adsorption Method with Macroreticular and Gel Type Resins and Adsorbents", Paper No. 54.

Eryılmaz, Candan, and Ayten Genc.2021.Review of treatment technologies for the removal of phenol from wastewaters. Journal of Water Chemistry and Technology 43.2: 145-154.

Farrier, D.S., Hines, A.L. and Wang, S.W.1979.Adsorption of Phenols and Benzoic Acid from Dilute Aqueous Solution onto a Macroreticular Resin", J. Colloid and Interface Science, 69:233-37.

Jain, S.J. and Snoeyink V.L.1972.Competitive Adsorption from Bisolute Systems on Active Carbon", 45 Annual Conference of the Water Pollution Contr., Fed., Atlandta Geogio, 8 (13).

Kadhum, Shaimaa T., Ghayda Yassen Alkindi, and Talib M. Albayati.2021.Eco friendly adsorbents for removal of phenol from aqueous solution employing nanoparticle zero-valent iron synthesized from modified green tea bio-waste and supported on silty clay. Chinese Journal of Chemical Engineering 36: 19-28.

Kennedy, D.C.1973.Macroreticular Polymeric Adsorbents, Ind. Eng. Chem., Prod. Res. Develop., 12:56-57.

Mattson, J.S.1969.Surface Chemistry of Active Carbon-Specific Adsorption of Phenols", J. Colloid & Interface Sci., 31 (116)

Morris, J.C. & Weber (Jr.) W.J.1963.Kinetics of Adsorption on Carbon from Solution, Proc. Am. Soc. Civ. Engg. J. Sanit. Engg. Div., 89, 31.

Morris, J.C. and Weber (Jr.) W.J.1964.Adsorption of Biochemically Resistant Material from Solution", U.S. Deptt. of Health, Education and Welfare AWCR-9.

Othmani, Amina, Aida Kesraoui, and Mongi Seffen. 2021.Removal of phenol from aqueous solution by coupling alternating current with biosorption. Environmental Science and Pollution Research 28.34: 46488-46503.

Randke, C.J. and Prausnitz J.M.1972.Adsorption of Organic Substances from Dilute Aqueous Solution on Activated Carbon, Ind. Soc. Chem. Fundam, 11 (445),

Randtke, S.J. and Snoeyink, V.L.1983.Evaluating GAC Adsorption Capacity, J. Am. Water Work Assocn., 406-413.

Schaeffer, D.J., Tigwell, D.C., Somani, S.M. and Janardan, K.G.1980.Efficiency and Liquid-liquid Extraction and XAD-4 and XAD-7 Resins in Collecting Organic Compounds from a cake Plant's Effluent, Bull.Environ. Contam. Toxicol., 25:569-78.

Snoeyink, V.L. & Weber, W.J. (Jr). 1967.The Surface Chemistry of Active Carbon", A Discussion of Structure & Surface Functional Groups, Env. Sci. & Technology, 1 (228).

Snoeyink, V.L., Weber, W.J. & Mark, H.B.1969.Env. Sci. Technol., 3:198.

Standard Methods for Examination of Water and Wastewater, APHA, AWWA, WPCF.1987. Published by American Public Health Asso., New York, 16th Ed.

Wang L.K.; Leonard, R.P.; Wang, M.H.; Goupill, D.W.1975.Adsorption of Dissolved organics from Industrial Effluents on to Activated Carbon, J. Applied Chemistry & Biotechnology, 25(7):491-502.

Weber, W. J. and Gould, J.P.Sorption of Organic Pesticides from Aqueous Solution", Advances in Chemistry Series, No. 60, 280, R.F. Gould Ed.

Weber, W.J.1972.Physico-chemical Processes for Water Quality", Wiley-Interscience, New York.

Wojaczynska, M., Kolarz, B.N. and Ruplex, K.1984.Application of the Hannett Equation for Sorption of Aromatic Compounds on Polymeric Sorbents, Reactive Polymers, 3:67-72.

Zogorski J.S. and Hass, J.H.1976. J. Colloid Interface Sci., 55: 329.

23

A Review on Freshwater Cladoceran Diversity of India

V.B. Sakhare and A. I. Shaikh

The Cladocera, commonly known as 'waterfleas' form a primitive group of microcrustaceans. They play an important role in the aquatic food chain and also contribute significantly to the zooplankton dynamics and secondary productivity in freshwater ecosystem. Though most cladocera is filter feeders, they also feed on detritus, algae, small rotifers and copepod nauplii. They keep a check on eutrophication for which this group is also widely used as good indicators of ecosystem health and sentinels of environmental change (Eggermont and Martens, 2011; Padhye and Dahanukar 2015). At the same time, cladocera is the favorite food prey of many invertebrates and many vertebrates. Their characteristic jerky movements make them easily visible to the predators and that is why most of fish larvae were found to prefer cladocera over the prey (Mayer and Wahl,1971).

Cladocerans have a distinct head, a single compound eye, and a large mandible for grinding food particles. Cladocerans feed on algae, small rotifers, and copepod nauplii. Cladocerans reproduce mostly asexually via parthenogenesis. But they can also reproduce sexually based on the environmental conditions.

The distribution of cladocerans is affected by different factors including temperature, rainfall, water quality, nutrients, macrophytes, flood pulse etc. (Ghidini *et al.*,2009; Kiss *et al.,* 2014). Their metabolic rate is variable with water temperature.

The order cladocera belongs to the subclass Branchipoda and includes minute crustaceans generally in the size range of 0.2 to 5.0 mm.

Cladocera have been reported from all over the world. While it is true that several genera and some species are cosmopolitan, certain species are endemic.

In the Indian context the studies on cladocera usually form part of general limnological investigations in lakes, reservoirs and ponds, wherefrom several genera and species are reported from time to time.

Out of nearly 700 species of cladocera occurring globally, only 131 species are reported from India (Sinha,2023). This appears to be under representation of the cladocera diversity of the country given the diverseness of habitats and ecosystems prevailing in the country and also home to four of the global biodiversity hotspots (Sinha,2023).

Of the Indian cladocera, about 75-80 species are known to occur in the country (Sinha,2021), while eighty-seven species were also recognized by Michael and Sharma (1986) and ninety-three species were listed by Sharma and Michael (1987). Diversity of 75 species was observed from state of Assam. Recently the Zoological Survey of India published a comprehensive annotated checklist of 131 cladoceran species, compiled from inland freshwater habitats in India (Sharma and Sharma,2017).

Maharashtra

The cladoceran diversity has been a fascinating subject for a long time. The notable studies on cladoceran diversity of various freshwater bodies of Maharashtra state are those of Rane (2005), Mulani *et al.,* (2009), Jetithor (2010), Ghantaloo *et al.,* (2011), Tidme and Shinde (2012), Kulkarni and Surwase (2013), Patil and Ghorade (2013),Sitre (2013,2014), Padhye and Dumont (2014), Rohankar and Dahegaonkar (2016),Buktar (2017), Patil and More (2018), Dabhade and Chhaba (2019),Tapse (2019), Patil (2020),Yannawar *et al.*, (2022)

Chavan *et al.,* (2006) studied ecology and behaviour of zooplanktons from Manjara project reservoir in Maharashtra and reported *Ceriodaphnia, Moina, Pleuroxus, Chydorus, Bosmina* and *Alona* species. The cladocera contributed to about 15.83% and 15.75% population during first and second year of investigation.

Kadam *et al.,* (2006) reported 5 species of cladocerans from Masoli reservoir of Parbhani district viz., *Daphinia carinata,Ceriodaphnia sp, Moina dubia,Bosmina sp* and *Simocephalus sp.*

Sakhare (2006) studied ecology of Jawalgaon reservoir of Solapur district of Maharashtra and identified 7 species of Cladocera which accounted for about 10.05 %.The maximum cladiran population (210 organisms/Liter) was recorded in April and minimum population (10 organisms/Liter) in November.

Sakhare (2007) documented fisheries and limnology of Yeldari reservoir with 7 species of cladocerans. The maximum cladoceran population (210 organisms/ Liter) was recorded during April and minimum population (10 organisms/ Liter) during November. The cladocerans reported were *Ceriodaphnia cornuta, Moina micrura, Diaphanosoma sarsi, Diaphanosoma excisum, Alona rectangular, Biopertura karna and Indialona ganapati.*

Joshi (2011) made studies on zooplanktons of Rajura lake of Buldhana district of Maharashtra and recorded 6 genera of cladocera i.e., *Macrothrix sp., Bosmania sp.,Cerodaphnia sp.,Daphnia sp.,Dadaya sp.*, and *Chydorous sp.*

Seasonal fluctuations in zooplankton of Powai lake (Mumbai) were studied by Sudhakar *et al.*, (2012). Cladocerans were most abundant during spring with 327 nos/L. This was followed by a second peak of 305 nos/L during winter. In summer, the maximum of cladocera was 282 nos/L. Low abundance of 254 nos/L was observed during monsoon.

Bhadane (2015) identified 32 species of zooplankton from Ramkund of river Godavari in Nashik district. Cladocerans were represented by 9 species with maxima in summer and minima in winter.

Zooplankton diversity of Katepurna reservoir of Akola district was studied by Pawar (2016).Cladocera was represented by 25 species. In first year of investigation, the cladocerans were found maximum during the summer in the month of March,April,May and June and the minimum during the winter season in the month of December,followed by January and February. In second year they were observed maximum in August,September,October,June and July and minimum in March and April.

Sakhare and Vasanthkumar (2016) studied fish fauna, plankton diversity and water quality of few reservoirs from Marathwada region of Maharashtra. In Majalgaon reservoir of Beed district five groups of zooplankton i.e., protozoa, rotifera, cladocera, copepoda and ostracoda were identified. Cladocera was represented by 6 species, with maximum density in summer and minimum in monsoon season. The observed species were *Indialona ganapati, Moina micrura, Diaphanosoma sarsi, Diaphanosoma excisum, Alona rectangular* and *Ceriodaphnia cornuta.* A total of 25 species of zooplankton from Wan reservoir of Beed district which comprised 3 species of protozoa,8 of rotifera,6 of cladocera,5 of copepoda and 3 of ostracoda. Cladoceran members were same as reported in Majalgaon reservoir. Gharni reservoir was constructed in Latur district where cladocera showed summer maxima and monsoon minima. A total of 25 species of zooplankton were recorded from Siddeshwar reservoir of Hingoli district. Out of 25 species, 3 species belonged to protozoa,8 to rotifera,6 to cladocera,5 to copepoda and 3 to ostracoda. Rotifera dominated the zooplankton population (35.49%), followed by cladocera (32.15%), copepoda (15.20%), ostracoda (13.47%) and protozoa (3.69%). In Bhategaon reservoir of Hingoli district rotifera dominated the zooplankton population (31.46%), followed by cladocera (27.65%), copepoda (23.20%), protozoa (11.95%) and ostracoda (5.74%). Cladocerans were reported by 6 species, and all members of this group were present round investigation period. The peak period of cladocera was observed during the month of March and it was minimum during the month of June.

Shrirame *et al.,* (2016) studied seasonal biodiversity and percentage of zooplankton in Ghagardara pond of Kandhar (Maharashtra) and reported 8 species of cladocera. The sequence of order of dominance of the four groups were as Copepods>Cladocerans>Rotifers>Ostracods.

In Girzani lake of Solapur district (Maharashtra) Agarwal and Kumbhar (2017) observed increased percentage of cladocerans in summer followed by monsoon and winter season.

Jetithor and Sakhare (2017) studied plankton diversity in Harni (Katgaon) reservoir of Osmanabad district (Maharashtra). Cladocerans accounted for about 14% of total zooplankton and were represented by 7 species. The maximum population was recorded in March (131 organisms/Liter) and minimum population in the month of November (37 organisms/Liter). Cladocerans were maximum throughout the summer season.

Sakhare (2017) studied zooplankton diversity in Jawalban reservoir of Maharashtra and reported 6 species of cladocera viz., *Diaphanosoma sarsi, Ceriodaphnia cornuta, Moina micrura, Alonella nana, Alona rectangular* and *Indialona ganapati*. Cladocerans accounted for about 19.78 per cent. The peak period of cladocerans was observed in the month of February (158 organisms/ Liter) and it was minimum in the month of August (14 organisms/Liter).

Bhadane (2019) reported *Ceriodaphina sp, Daphina sp, Moina sp* and *Bosmina sp* from Gangapur dam of Nashik district Maharashtra and observed maximum population of cladocera in summer. Cladocera contributed to 4.04% of the total planktonic population of dam.

Sakhare and Jetithor (2019) observed 3 species of cladocera (*Moina micrura, Ceriodaphnia cornuta* and *Indialona ganapati*) from Kundlika reservoir of Beed district.The high and low population densities were recorded in summer and early monsoon season respectively.This higher population density in summer might be due to the temperature acceleration in water body (Manickam *et al.,* 2018).

Amte and Ayare (2020) observed 13 genera of zooplankton which includes rotifers, cladocerans and copepods. Rotifers were observed during all the seasons. Cladocerans were maximum during monsoon season.

A systematic investigation on seasonal variations of cladocera in Udayan sarovar, a perennial pond near Swami Ramanand Teerth Mahavidyalaya at Ambajogai was carried out by Sakhare and Chalak (2020). A total of 9 species of cladocera were identified. The cladoceran diversity fluctuated seasonally and the maximum number of 240 organisms/Liter was recorded during month of April and minimum number of 22 organisms /Liter during the month of September. Chalak and Sakhare (2018) also reported higher cladoceran density in summer and lower in monsoon.

Meshram (2020) studied abundance of cladocera related with the water quality of Chirebandi pond, Fulchur at Gondia (Maharashtra) and recorded only 4 species of cladocerans. The most diversified species was *Moina spp* (354 ind/Lit), *Cereodaphnia spp* (287 ind/Lit), *Diaphanosoma spp* (281 ind/Lit) and *Alona spp* (208 ind/Lit).The total population of cladocerans was recorded as 1130 ind/Lit. Seasonal population density of cladocerans recorded its peak during winter (580 ind/Lit) i.e., 51.33%,during summer (273 ind/Lit) i.e., 24.15%,while least during monsoon (277 ind/Lit) i.e., 24.51%.

Sangve (2020) studied planktonic diversity of three reservoirs of Amravati district (Maharashtra) and reported 5,7 and 3 species of cladocera from Pohra, Wadali and Chhatri reservoirs respectively.

Panpatil (2021) recorded 9 species of zooplankton. Among these,6 species belonged to rotifera,1 species to copepoda and 2 species to cladocera.

Yewale (2021) recorded 40 species of zooplankton from Pentalki lake and reported three groups such as rotifers (22 species), cladocera (14 species) and copepods (4 species).

Studies on zooplankton diversity from Ramadeshwar lake of Nagpur district were conducted by Darvekar and Raut (2022). Cladocerans were represented by 8 species with average composition of 24.15% in summer,25.2% in monsoon and 23.69% in winter.

Madhya Pradesh

The available literature on the zooplankton/cladoceran diversity of freshwater ecosystems of Madhya Pradesh State is of Sharma and Saksena (1981), Rane (1985,1987), Mishra 1989, Bhat *et al.,* (2014), Pandey *et al.,* (2015), Bunkar and Tiwari (2016),Saini and Dube (2018), Khandayat and Singh (2019), and Kushwaha *et al.,* (2022).

Rane (1983) made a comprehensive taxonomic study on freshwater Cladocera of Madhya Pradesh and reported one new species of the genus *Simocephalus* schodler,1858.

Rane and Jafri (1990) recorded a new species of cladocera, Daphnia madhuriae sp. Nov. from Madhya Pradesh.

Mishra and Saksena (1993) studied planktonic fauna in relation to physico-chemical characteristics of Gouri tank at Bhind (Madhya Pradesh) and reported 27 species of zooplankton. Rotifers had 15 species followed by copepoda with 4 species, protozoa and cladocera with 3 species each and ostracoda with 2 species. Cladocerans observed were *Moina branchitata, Ceriodaphnia reticulata* and *Alona intermedia.*

Kadam *et al.,* (2014) studied biodiversity of zooplankton in Pillowa reservoir of Morena district and identified 27 species of zooplankton. Among 27 species,6 species were represented by cladocera viz., *Scaphloleberis kingi, Ceriodaphnia cornuta, Bosmina longirostris, Chydorus sphaericus, Indialona ganapati,* and *Diaphanosoma excisum.* The zooplanktonic forms were exhibited as protozoans were 9.50%, rotifers with 54.84%, cladocerans with 14.43%, copepods with 15.70% and ostracods with 5.54%. The abundance of zooplankton in Pillowa reservoir followed a sequence as Rotifera> Copepoda> Cladocera> Protozoa> Ostracoda.

Shiv *et al.,* (2017) studied zooplankton diversity of River Temar district Jabalpur, Madhya Pradesh and revealed the presence of 34 species of zooplankton belonging to 5 major groups. Among these, 6 species belonged to protozoa,11 species to rotifera,9 to Cladocera,6 to copepoda and 2 species to ostracoda. The group rotifera was the dominant, followed by cladocera and copepoda. The cladoceran population showed minimum in monsoon i.e., in June and maximum in winter.

Wasarat and Chauhan (2018) identified a total of 23 species from Halali reservoir of Madhya Pradesh which included 13 rotifers,7 cladocerans, and 3 copepods.

Rane *et al.,* (2020) reported five zooplankton groups such as rotifera (8 species), cladocera (3 species), copepoda (3 species), ostracoda (1 species) and protozoa (3 species),with a total of 18 different species from Sapna dam of Betul district (Madhya Pradesh).

Singh and Samartha (2021) studied seasonal abundance of certain zooplankton in upper lake (Bhopal). The abundance of zooplankton was studied in pre-monsoon, monsoon and post monsoon season. Zooplankton diversity of upper lake comprised 5 taxonomic groups: protozoa, ostracoda, rotifera, cladocera and copepoda. The greater abundance of zooplankton was recorded during summer season while the lowest abundance was recorded during the monsoon season. Among cladocera, genus Bosmina recorded to be dominant, which is very common in eutrophic lakes having abundant macrophytic vegetation. A total of 6 cladoceran species were recorded viz., *Bosmina sp, Chydorus sp, Daphnia sp, Leydgia sp, Moina sp* and *Moinadaphnia sp.*

Borde and Solanki (2023) studied the zooplankton community of Chikliya pond area of Barwani district and recorded four groups of zooplankton viz., protozoa, rotifera, copepoda and cladocera. They reported 26 species of zooplankton, out of which 10 belonged to protozoa, 7 to rotifera,4 to copepoda, and 5 to cladocera (*Conochilus sp., Cypris sp., Daphnia sp., Macrothirx sp.,* and *Stenocypris sp.*). The density of this group varied from 13 No/Lit to 66 No/ Lit. The minimum was noticed during July and maximum in December.

Karnataka

Studies on various aspects of zooplankton community in freshwater wetlands of Karnataka were carried out by Patil and Gounder (1985), Narayana (1994), Pushpendra and Madhyastha (1994), Kiran *et al.,* (2007), Maganuri *et al.,* (2008), Majagi and Vijaykumar (2009), Rajshekar *et al.,* (2009), Purushothama *et al.*, (2011), Savitha and Yamakanamardi (2012),Veerendra *et al.*, (2012), and Vasanthkumar and Sedamkar 2016).

Limnological studies of two reservoirs of Uttar Kannada district (Karnataka) were undertaken by Kudari *et al.,* (2006). A total of 33 zooplankton species were recorded,out of which 11 species were cladocerans.

Jalilzadeh and Yamakanamardi (2012) studied the seasonal fluctuations of cladocera abundance with special reference to seasonal changes in physico-chemical parameters in lentic ecosystems of Mysore. The physico-chemical variables influencing the abundance of cladocera were dissolved oxygen, NO_3, BOD and conductivity in Bannur lake; COD, turbidity and water temperature in Lingambudi lake. No physico chemical variables revealed correlation with the abundance of cladocera in Hebbal lake.

Kudari and Kanamadizt (2013) studied impact of restoration on zooplankton composition of Kelageri lake, Dharwad. The decrease in the cladoceran species composition immediate after the restoration appears to be due to the removal of macrophytes.

Umadevi (2013) studied the abundance, composition and distribution of zooplankton in relation to water quality parameters of Karanja river of Karnataka.36 species of zooplankton were identified as a total, which included 14 species of rotifera,11 species of cladocera,8 species of copepoda and 3 species of ostracoda.

Harish Kumar and Kiran (2016) prepared a report on diversity of cladocera in sewage fed tank of Bhadravathi taluk, Karnataka and reported 6 species of cladocerans belonging to six families viz., the Daphniidae, Moinidae, Bosminidae, Macrothricidae, Sididae and Chydoridae. It was observed that the rich nutrients coupled with the presence of aquatic weeds favoured abundance of cladocerans.

Joseph (2017) reported zooplankton diversity of Kukkarahalli lake of Mysore. Cladoceran community comprising 15 species belonging to 5 families which were Sididae, Daphniidae, Moinidae, Bosminidae and Chydoridae. The study on the diversity of Cladocera showed that among all the species observed *Diaphanosoma excisum,Moina micrura* and *Biapertura karua* were found to be more in number.

Chhatisgarh

Agarwal *et al.,* (2014) studied diversity and seasonal fluctuation of zooplankton in freshwater reservoir Mongra Bairaj of Rajnandgaon district and reported 18 species of zooplankton belonging to protozoa (9 species), copepoda (5 species) and rotifera (4 species).Among all the zooplankton groups protozoa recorded dominance. Maximum diversity of zooplankton population was recorded in summer season.

Mishra *et al.,* (2014) studied zooplanktons in five ponds of Raipur city (Chhatisgarh) and observed 34 species of zooplanktons belonging to five major groups i.e., protozoa (6), rotifera (22), cladocera (2), copepoda (3) and ostracoda (1).

Shrivastava and Joshi (2014) studied diurnal variation of plankton in Nawagaon wetland of Chhatisgarh. Among zooplankton, crustaceans were found to be more sensitive towards environmental changes as compared to rotifers. Rotifers were observed in the surface water regardless of the illumination while crustaceans were adapted to low light intensity, however their abundance varied accordingly.

Roy *et al.,* (2015) prepared a checklist of commonly available phytoplankton and zooplankton genera of urban and rural ponds of Raipur (Chhatisgarh). Among zooplankton, a total of 24 genera belonging to the classes rotifera (11 genera), copepoda (7 genera), protozoa (3 genera), cladocera (2 genera) and ostracoda (1 genus) have been recorded.

Study of zooplankton in Kello river of Raigarh (Chhatisgarh) was made by Sahu *et al.,* (2016). Zooplankton community of Kello river comprised 17 species belonging to cladocera (2 species), protozoa (1 species), and rotifera (14 species). Rotifera showed dominance in number followed by cladocera and protozoa.

Chandrawanshi (2020) reported ecology of Shivnath river at Madku Deep in Mungali district of Chhatisgarh. The zooplankton abundance was influenced by physical factors of the water body. Diversity of zooplankton tender to be the highest in river during summer. During rainy season copepods and cladocerans were highest and lowest in winter. The zooplankton were recorded qualitatively and quantitatively in the decreasing sequence as protozoa, rotifera, cladocera and copepoda. Total number of 50 species of zooplankton belonging to 32 genera were recorded at Shivnath river.

Verma and Agarwal (2021) studied seasonal variation of zooplankton in Kosrangi dam of Arang block (Raipur district, Chhatisgarh) and reported 19 species of zooplankton, out of which 6 species of protozoa,4 species of rotifera, 5 species of copepoda, 3 species of cladocera and 1 species of ostracoda were recorded. Ostracoda was dominant group throughout the study period. The study of season wise zooplankton analysis showed an average abundance to species in winter season, lower in monsoon and maximum in summer season.

Omkar *et al.,* (2023) studied of variations in physico-chemical parameters and plankton diversity during monsoon and post monsoon season in Taro reservoir of Kawardha, (Chhatisgarh).

Jharkhand

Chandrasekhar and Chatterjee (2002,2004) reported cladoceran diversity from different wetlands of Jharkhand state.

Saba and Sadhu (2015) studied zooplankton diversity of Garga reservoir of Bokaro (Jharkhand) and recorded 11 different species of zooplanktons belonging to 4 different classes namely rotifers (3 species), cladocerans (4 species), copepods (3 species) and ostracod (1 species).Among zooplanktons, cladocera and rotifera showed dominance.

Singh *et al.,* (2021) studied zooplankton diversity and physico-chemical parameters of Pampoo pond of Madhupur (Jharkhand) and recorded 15 genera of zooplankton from the freshwater pond belonging to the four groups namely cladocera, rotifera, copepoda and ostracoda. Among all four groups, maximum abundance of cladocerans (34%) was recorded. Five species of cladocerans, five species of copepods, three species of rotifers and two species of ostracods were observed. The population density order observed as cladocerans>copepods>rotifers> ostracods.

Punjab

Battish (1983) studied ecology and systematics of the cladocera (Daphnidae: Branchidpoda) inhabiting Punjab and reported nine species of cladocera belonging to family Daphnidae.

Thakur and Kocher (2017) recorded nine species of cladocerans from different types of waterbodies of Ludhiana (Punjab). Out of nine species, six were found to belong to family Daphnidae, two to family Moinidae and only one to family chydoridae. Average percent composition of cladocerans in village ponds showed the distribution pattern with predominance of Daphnidae (51.01%) > Moinidae (48.27%) > Chydoridae (0.68%). In fish ponds moinidae family was found to be predominant (62.95%) > Daphnidae (34.61%), > Chydoridae (1.29%). Paddy fields were represented by moinidae family only.

Kaur *et al*., (2018) conducted seasonal study of zooplankton diversity in the polluted water stretch of Buddha Nullah (Ludhiana). Cladocerans were represented by *Simocephalus sp, Moina brachiata, Daphnia sp, Ceriodaphnia sp* and *Alonella sp*. Cladocerans were abundant from March to June and were either absent or present in very negligible numbers during the rest of months. The maximum density of Cladocera during winter season may be due to favorable temperature and availability food, nonoplankton and suspended detritus

Jammu & Kashmir

Several research papers dealing with the freshwater cladoceran diversity of Jammu & Kashmir state are available in literature (Quadri and Yousuf 1980, Battish 1981, Yousuf and Qadri 1981, Khan 1987, Gupta 2002, Sharma *et al.,* 2005, Siraj *et al.,* 2006 & 2007, Sharma and Chandrakiran 2011, Sharma and Kotwal 2011, Ahangar *et al.,* 2012, Shah and Pandit 2013, Pandit *et al*., 2016, Naik *et al.,* 2017, Kotwal 2020 and Kour *et al.,*2021).

Gogoi *et al.,* (2018) studied faunal diversity of cladocera with notes on biogeographically important species in the floodplain wetlands of the Subansiri River basin (India) and reported cladocera assemblage of 55 species belonging to 30 genera and 7 families. The species richness represents 42% and 75% of the total amount of freshwater species reported from India and Assam respectively.

Kotwal (2020) recorded 15 species of cladocera belonging to 8 genera and 4 families from the aquatic habitats of Jammu district. The percental contribution of different families demonstrated a typical trend where Daphniidae> Chydoridae> Moinidae> Macrothricidae. Habitat preference was also recorded in some cladocera.

Sharma and Kour (2021) collected zooplankton samples from a shallow pond located in the Bishnah tehsil of Jammu district and revealed the presence of 13 cladocera species belonging to the families Daphniidae, Chydoridae, Moinidae, Sididae, and Macrothricidae.

Uttar Pradesh

Singh *et al*., (1993) carried out limnological studies for bioenergetic transformation in a Tarai reservoir, Nanak sagar (Uttar Pradesh). The zooplanktonic community of Tarai reservoir comprised rotifera, cladocera and copepoda. In all, it has 20 species belonging to 17 genera, with rotifera, cladocera and copepoda comprising 12,5 and 3 species respectively. On annual basis, cladocera accounted for 38% of the total density.

In freshwater canal of western Uttar Pradesh, the zooplankton community was found to be comprised of 4 species of copepods,5 species of rotifers, and 2 species of cladocera (Arya and Singh,2017). The dominance or peak of zooplankton was recorded during winter months. Numerically, copepoda was the most dominant group throughout the study period, followed by rotifers and cladocerans.

Gangwar and Bhadauriya (2021) studied zooplanktonic community of river Deorania at Bareilly district (Uttar Pradesh) and its relation to seasonal variation. They identified a total of 13 zooplankton species belonging to five classes i.e., Rotifera, Cladocera, Copepods, Ciliates and Mesomycetozoea. The zooplankton diversity was found average abundance of species in rainy season, lower in winter and maximum occurrence in summer season.

Pant *et al.* (1985), Verma (2020) and Tripathi and Shukla (2021) also reported cladoceran diversity from different freshwater wetland of Uttar Pradesh.

Himachal Pradesh

Chauhan (1993 a) studied seasonal fluctuation of claodocerans in Renuka lake.Zooplankton species of Rewalsar lake of Himachal Pradesh have been identified by Chauhan (1993b). In all 29 species were recorded which included 11,4 and 14 species of cladocera, copepoda and rotifera respectively. Maximum (22) species were recorded in August and minimum (11) species in April. The rotatorian were found main contributor to the varation in species composition. The cladoceran species recorded were *Daphnia similis, D.magna, D.longiremis, Diaphanosoma brachyurum, Bosmina longirostris, Sinocephallus expinosus, Moina mucronata, Allonella gibba, Leptodera kindtii, Alonella excisa* and *A.dentistoma.*

Jindal and Thakur (2013) also studied diurnal variations of plankton diversity and physico-chemical characteristics of Rewalsar wetland. Zooplankton were composed of rotifera, cladocera and copepoda. They showed up migration during night hours and down migration during day.

Thakur *et al.,* (2013) studied plankton diversity and water quality assessment of three freshwater lakes of Mandi with special reference to planktonic indicators. They identified 14 species of cladocerans from three freshwater lakes. According to them plankton has the potential to act as bioindicators of pollution status.

Analysis of zooplankton community of Prashar Lake revealed that the maximum contribution was made by rotifera (38%) followed by cladocera (26%), protozoa (25%), copepoda (6%) and ostracoda (5%). During the study period, a total of 27 species of five groups were identified (Sharma and Rama

Kumari,2018). The rotifer was the most dominant group having 10 species representing 38% of the total number of genera. The second dominant group was cladocera having 6 species, contributing 26% of the total genera. The third dominant group was protozoa group, having 6 species contributing 25% of the total number of genera. Copepoda represented with three species. Ostracoda represented with two species. Out of six species of cladocera, *Bosmina sp* and *Daphnia sp* was recorded as most abundant. In many species of cladocera *Daphnia sp* is one of the most common and mostly occurring in all types of environmental conditions. The density of cladocera was reported minimum during monsoon season.

CIFRI (2007) reported cladoceran diversity of Pong Reservoir and later on in 2023 CIFRI also studied fisheries resource of Pong dam. The abundance of zooplankton ranged from 174.97 units/L to 364.98 units/L. The abundance of zooplankton was highest in the lentic zone and lowest in the lotic zone of the reservoir. Six genera from the cladocera were reported. The sector with greatest abundance of cladocera species was the lentic zone (186.65 units/L), followed by the intermediate zone (144.15 units/L) and the lotic zone (96.65 units/L). The study revealed that the highest zooplanktonic abundance was in the lentic sector due to the stable water conditions and lowest in the lotic zone because of the movement of the entry of riverine water into the reservoir ecosystem.

Tamil Nadu

The literature on the cladoceran diversity of few freshwater wetlands of Tamil Nadu state are those of Raghunathan (1983,1985,1986,1990);Venkataraman (1983,1999); Venkataraman and Krishnaswamy (1984a,1984b); Hudec (1987), Muruguan (1989),Balamurugan *et al.,* (1999), (2001);Venkataraman (1999), Venkataraman *et al.*, (2000), Siva Kumar *et al.*,(2001), Suganthi *et al.*,(2014);Manickam *et al.,* (2015) and Vijayapriya *et al.,* 2019.

Murugan (1980) observed 10 different species of cladocera from one small seasonal pond within the Madurai college campus. Two of these, *Kurzia longirostris* and *Ilyocryptus spinifer* were recorded for the first time. At high temperatures cladoerca becomes mature at a smaller size as observed in the field with reference to *M.micrura*.

Sivakumar *et al., (*2001) made qualitative and quantitative analysis of copepods and cladocerans of the freshwater bodies in and around Dharmapuri district. They recorded 4 copepod and 7 cladoceran species and also observed the higher population density of copepods and cladocerans in winter season then in the summer season.

Raghunathan and Suresh Kumar (2002) prepared a comprehensive and consolidated list of 81 species of Cladocera from Tamil Nadu with major representation from family Chydoridae (40) and Daphniidae (17). Compared with other states in India this group has been fairly well studied in Tamil Nadu.

Moorthy and Sultana (2011) identified zooplankton species from different ponds of Arakkonam town of Vellore district and identified eight species of cladocerans.

Sivakami *et al.,* (2015) undertaken study to assess the zooplankton diversity in a lake situated in the suburb of Pudukkottai district.40 species belonging to five different groups were recorded.Out of the 40 species, two each belonged to protozoa and ostracoda,27 to rotifera,5 to Cladocera,3to copepoda and 1 to anostraca.

Krishnamoorthi and Moorthikumar (2020) studied zooplankton diversity in Amravathi dam Tirupur District and reported 25 species of zooplankton. Among them 10 species of rotifers,4 species of copepods,6 species of cladocerans and 5 species of protozoans were recorded.

Bihar

Nasar (1977) recorded eleven species of cladocera from Bhagalpur.All the species recorded were new record for Bihar, except *Moina dubia* and *Ceriodaphnia rigaudi.* Regarding the zoogeographical distribution of the cladocera of Bhagalpur, it appears that it has greater affinity with south-east Asian, Australian and African fauna than the Palaearctic fauna.

Rani *et al.,* (2019) reported zooplankton diversity of various ponds of Patna.A total of 21 species of zooplankton were identified from PWC pond, belonging to 4 species of rotifers,11 species of copepods and 6 species of cladocerans. A total of 16 species were identified in BMP pond,9 species belonging to rotifera, 6 species to copeoda and only one species to Cladocera. A total of 26 species were identified in Patna zoo pond,12 species belonging to rotifera,8 species to copepoda and 6 species to cladocera. The record of various pollution indicator species in all the three ponds infers eutrophic nature of ponds.

Pandit *et al.*, (2020) carried out a study to evaluate the ecology and diversity of the zooplankton of the Ganga river at Arrah. A total of 23 genera of zooplankton belonging to 6 genera of rotifera,5 of protozoa,5 of cladocera,4 of copepoda and 3 of ostracoda were identified with a density from 2 to 213 ind/L. The density of zooplankton declined in post-monsoon and remained maximum in summer. The cladocerans recorded were Bosmina, Chydorus, Daphnia, Daphniosoma and Moina.

Priyama and Thakur (2022) studied zooplankton abundance and species diversity in Shahjangi pond of Bhagalpur district and reported 21 species of zooplankton of four different classes i.e., rotifera, cladocera, copepoda and ostracoda. Rotifera was the most dominant class followed by the cladocera, copepoda and ostracoda. Out of 21 species of zooplankton,10 were rotifers, 6 cladocerans, 3 copepods and 2 ostracods.

On similar lines Chaudhary and Singh (1999), Sharma and Sharma (2001), Malathi (2002), Singh *et al.,* (2012) and Chatterjee *et al.,* (2013) also studied cladoceran diversity from wetlands of Bihar.

Andhra Pradesh and Telangana

The notable contributions on cladoceran diversity of various freshwater bodies in Andhra Pradesh and Telangana are those of Ashan (1982), Jaya (1985), Patil (1986), Siddiqui and Chandrasekhar (1993,1995,1996,1997,1998), Chandrasekhar and Kodarkar (1994); Chandrasekhar (1995,1996, 2014),Dhatrika (2012), Chari *et al*., (2013), Krishna and Kumar (2017),Rao (2017) and Dhatrika *et al*., (2024).

Devi (1997) reported the maximum cladocerans population during post monsoon and premonsoon in Ibrahimbagh reservoir of Hyderabad.

Patil and Panda (2003) studied seasonal population dynamics of freshwater cladocera from a fish tank of Bibinagar (Hyderabad) and recorded twelve species of cladocera. Cladocera was the dominant group forming 38-81% of the average annual population. The reason might be due to the reduced fish population. *Moina micrura* the only cladoceran was found almost throughout the year except May. The peak cladocera were observed in the month of November.

Chandrasekhar (2004) reported 30 species (including two subspecies) belonging to 17 genera spread over five families of cladocera from waterbodies in and around Hyderabad.

Pavan *et al.,* (2016) observed seven species of cladocerans from Bhandam cheryvu of Warangal (T.S.). Cladocera sp. showed high Shannon diversity index value in winter and lowest in monsoon.

Limnological investigations on Hasanparthy Lake (Warangal,Telangana) were undertaken by Balakrishna *et al*., (2017). The observed cladoceran species were *Moina micrua*, *Moina brachiate*, *Ceriodaphnia cornuta*, *Bosmina longirostris* and *Moina microcopa*.

Lingampally *et al.,* (2018) observed two genera (Daphnia and Ceriodaphnia) of cladocera. Cladoceran density was maximum during February and minimum during the month of July. The density of Ceriodaphnia was more compared to

Daphnia throughout the study period. The cladoceran abundance reflects the eutrophic nature of Chakki talab.

Chari *et al.,* (2021) observed 4 cladoceran species from Shanigaran reservoir of Siddipet district (Telangana) and reported highest cladocera population during the pre-monsoon season and lowest during the post-monsoon season.

Rajani *et al.,* (2021) studied zooplankton diversity of Velair freshwater lake, Warangal District (Telangana) and recorded four species of cladocera i.e., *M. branchiata*, *M. macropa*, *Daphnia* sps and *Bosmina longirostris*. Rotifer was dominant (59%), followed by cladocera (21%),copepod (15%) and ostracoda (5%).Cladocera during the study period were observed maximum in monsoon season.

West Bengal

There are many published papers/reports on qualitative and/or qualitative analysis of cladocerans from various freshwater wetlands of West Bengal (Sharma 1978, Venkataraman and Das 1993,1995,1998,2001, Venkataraman *et al.,* 1993, Chakraborty 2004,Biswas 2015, Chakraborty and Mallick 2020 and Sarkar and Pal,2021).

Mukherji and Nandi (2006) documented ecology, biodiversity and management of Rabindar sarovar in Kolkata and identified 8 species of cladocerans.

Chakraborty and Mallick (2021) studied diversity of cladocerans in some selected wetlands of West Midnapore district and revealed the presence of a total of 16 cladoceran species. Chydoridae (56%) was found to be the dominant family followed by moinidae (19%). The study showed that cladoceran species diversity seemed to be positively correlated with the presence of nutrients. The species diversity was associated with the eutrophication status, ecosystem stability and productivity levels of the wetlands. The maximum species diversity was found in the pre monsoon and least in the monsoon period. Thus, increase in temperature indicated a positive correlation with species diversity.

Orissa

Dash *et al.,* (2012) studied diversity of zooplanktons in relation to physico-chemical parameters at hot water spring of Atri. About 11 taxa of zooplanktons were recorded, of which 6 belonged to copepods, 2 to rotifers and 3 to cladocera groups. Zooplankton diversity and abundance were found to exhibit positive relationship with pH, dissolved oxygen, nitrate and phosphate. Hence these variables were the important factors governing the abundance of zooplanktons. The temperature of the hot spring exhibited a negative trend for zooplankton diversity and abundance (Dash *et al.,* 2012)

The percentage composition of various meiofaunal taxa from Kathjudi river were rotifera (30.7%), copepoda (11.2%), nematoda (12.2%), oligochaeta (9.4%),cladocera (17.9%), ostracoda (9.9%) and chironomida (8.6%).The incidence of higher percentage of the rotifers and cladocera is indicative of a healthy environment as these meiofaunules prefer to clean environment (Rath,2015).

Indresha *et al.,* (2018) studied seasonal variations of zooplankton and their correlation with physico-chemical parameters of Kanjia lake of Bhubaneswar. The study revealed the presence of 30 species, out of which 10 species of protozoa (33.33%), 9 species of rotifera (30%),4 species of cladocera (13.33%), 4 species of copepoda (13.33%), 3 species of ostracoda (10%) were recorded. The total densities of zooplankton were found to be maximum during summer and minimum during monsoon season.

Sahu and Rout (2021) recorded 11 species of protozoa,10 species of rotifera,7 species of cladocera and four species of copepoda from river Mahanadi at Cuttack. Peak population of cladocerans was observed during spring or early summer when the temperature was high. There is decline in population during monsoon period.

Rath (1997) and Rath *et al.*, (2004) also documented cladoceran diversity from different rivers of Orissa.

North Eastern States

Northeast states comprise Arunachal Pradesh, Assam, Manipur, Meghalaya, Mizoram, Nagaland and Tripura.

Of the Indian cladocera, about 75-80 species are known to occur in the North Eastern region of the country (Sinha,2021). Maximum diversity of 75 species was observed from the state of Assam (Sinha,2021). Maximum diversity of 75 species accounting for more than 57% of Indian diversity is known from Assam followed by 58 species (44%) and 56 species (~43%) from Meghalaya and Manipur respectively (Sinha,2021).

Saha (1989) conducted an ecological study of two lentic habitats of Agartala and reported 27 species of cladocera, out of which 24 were previously known from India. Two species viz., *Alonella acutirostris* and *Chydorus barroisi* have been recorded for the first time from India and an undesignated species of Bosmina was also recorded which is likely to emerge as a new species.

Sharma and Sharma (2009) reported cladocera as the second most important group of zooplankton of Loktak lake contributing significantly in richness. All the 51 species recorded from Loktak were observed in the samples collected during the first year, while only 47 species were recorded in the following year.

The Chydoridae, the most diverse family of cladocera, contributed dominating (29 species,16 genera) to the faunal diversity in Loktak.

Sharma and Sharma (2010) came across nine interesting species of Cladocera,including four new records from North-Eastern states of India and five new records from Assam i.e., *Streblocerus serricaudatus, Alonella clathratula, Alona guttata tuberculate, A. monacantha tridenta* and *Alona pulchella.*

Sharma and Sharma (2014) studied faunal diversity of Cladocera in wetlands of Majuli (the largest river island) of Assam and recorded 55 cladocera species belonging to 36 genera.

Gogoi (2017) reported 39 species of cladocera belonging to 25 genera in DWR (Hatimora deep water rice field) and 53 species of cladocera belonged to 29 genera were documented in both flood plain lake and oxbow lake.

Daimari (2020) studied ecology of Dhansiri river, Nagaland and reported three groups of zooplankton i.e., protozoa, rotatoria and cladocera.

Sharma and Nogen (2020) conducted limnological survey of Maghuri, Khamti, Gulai and No.11 floodplain lakes (beels) of the Dibru -Saikhowa Biosphere Reserve, Upper Assam and revealed 210 species of zooplankton, belonging to 78 genera and 32 families. Zooplankton richness and abundance followed no definite pattern of monthly or annual variations; Rotifera>Cladocera influenced overall diversity in the three beels.

Salam *et al.*, (2021) studied zooplankton diversity of Lake Kharungpat of Manipur and identified cladocera, copepoda and rotifera as three major zooplankton groups with percentage composition of 47.72,33.87 and 18.41 per cent respectively. The minimum Cladocera density was 188 numbers per Liter in September 2019 and 3350 numbers per Liter in the month of May 2020. Six species were identified under the Cladocera group namely *Daphnia spp, Moina spp, Bosmina spp, Macrothrix spp, Daphanosoma spp*, and *Chydorus spp*.

Pamai and Sharma (2023) studied zooplankton diversity in selected wetlands of Nagaland and revealed biodiverse zooplankton assemblage with 181 species spread over 79 genera and 34 families, belonging to five groups of zooplankton. Rotifera followed by cladocera largely contributed to zooplankton richness while Rhizopoda, copepoda and ostracoda were the other components. Study revealed 32 species new records from Nagaland and 2 species from North-East India, besides 50 species of rotifers has been recently added as new records from the state.

Venkataraman (1995, 1998, 2001), CICFRI(1991), Sharma and Sharma (1999, 2007, 2008, 2010, 2011, 2014), Venkataraman and Das (2000), Sharma (2008,2010), Das and Kar (2016) also reported cladoceran diversity from North-Eastern states of India.

Rajasthan

The notable studies on claodoceran diversity of different freshwater bodies of Rajasthan state are those of Biswas (1964,1971), Nayar (1971),Bhati and Rana (1987),Venkataraman (1990),Gupta (2000),Bagela (2006),Khanna and Yadav (2009), Dube *et al.*, (2010),Sharma *et al.*, (2012),Balai *et al.*, (2014) and Khan and Khan 2017.

Jhingran (1989) mentioned the most abundant cladoceran population in February, followed by July and October in Rangarh reservoir. The main genera recorded were *Ceriodaphnia, Simnocephalus, Chydorus, Daphnia* and *Monia* in order of abundance.

Zooplanktonic communities of two seasonal wetlands of Udaipur were studied by Parihar *et al.,* (2009). Groupwise composition of zooplankton communities indicated following raking: Rotifers> Copepods> Cladocerans> Protozoans>Ostracods.

Sharma *et al.,* (2012) sampled 77 water bodies from South Rajasthan for the cladoceran diversity. In all,54 species cladocerans were reported during the study, belonging to six families viz., the Sididae, Moinidae, Daphnidae, Bosminidae, Macrothricidae and Chydoridae. Rich nutrients, the presence of weeds and shallow waters favoured rich cladoceran diversity.

Bedwal and Mathur (2015) studied physico-chemical status in relation to zooplankton population of Kishangarh lake and recorded 6 species of cladocerans.

Vyas and Chauhan (2017) studied zooplankton diversity and their seasonal variation in Gulabsagar waterbody of Jodhpur. The zooplanktons were represented by 41 species and comprised rotifera (15 species), cladocera (21 species), copepoda (4 species), and ostracoda (1 species). The rotiferans dominated in the population while other were encountered with moderate number.

Meena (2019) reported 30 genera of zooplankton from village pond of Similiya District Kota which were represented by groups of rotifers, cladocerans, copepods, protozoans and ostracods. The annual values of percentage composition among different groups of zooplankton were found to vary considerably.

REFERENCES

Agarwal, P.H. and Kumbhar, A.C.2017. Studies on abundance of zooplankton in Girzani lake of Solapur district, Maharashtra. In: Plankton and Fisheries (Edited by V.B.Sakhare, H.K. Jadhav and S.S. Jadhav),142-146,Discovery Publishing House Pvt. Ltd., New Delhi (India).

Agarwal, R.K., Thiske Sanjay and Mondal, Sunil. 2014. Diversity and seasonal fluctuation of zooplankton in fresh water reservoir Mongra Bairaj Rajnandgaon, District, C.G., India. J. Animal, Veterinary and Fishery Sci., 2(8):1-4.

Ahangar, A., Farooq Mir,M., Saksena, D.N. and Ahangar,M.A.2012.Zooplankton diversity with relation to trophic status in Anchar lake, Kashmir. International Journal of Current Research,4:46-48.

Ahsan, Mohd.1982.Ecology of freshwater zooplankton (Hussainsagar and Saroornagar),Ph.D. thesis, Osmania University, Hyderabad.

Amte,G.K. and Ayare,T.S.2020.Comparative account of planktonic diversity of Gauripada lake, Kalyan, Maharashtra, India. International Journal of Zoology and Applied Biosciences,5(3):163-169.

Arya, M.A. and Singh, Neera. 2017. Zooplankton community and abundance in freshwater canal in Western U.P., Agriways,5(1):10-14.

Bagela, B. 2006. Biodiversity, density and survival of freshwater zooplankton in relation to salinity of water, Ph.D. Thesis, M.L. Sukhadia University, Udaipur (Rajasthan) India.

Balai, V.K., Sharma, L.L. and Ujjania, N.C. 2014.Diversity and seasonal variations of zooplankton in Jaismand Lake, Udiapur, India. International Journal of Animal Research,48:432-437.

Balakrishna, D., Suresh, P., Mahesh, T., Surender Reddy, K. and Ravinder Reddy,T.2017. Limnological studies of Hasanparthy Lake, Warangal District, Telangana, India. International Journal of Zoology and Applied Biosciences,2(6):266-269.

Balamurugan, S., Arivazhagan, R. and Subramanian, P.2001.Seasonal variation of zooplankton population in Ayyampalayam temple tank, Thiruchirapalli, Tamil Nadu, India. J. Adv. Zool., 22(1):52-55.

Balamurugan, S., Gulam, M., Mohuldin, G. and Subramanian, P.1999.Biodiversity of zooplankton in Kavery river stretch, Thiruchirapalli,Tamil Nadu, J. Aqua. Biol., 14 (1&2):21-25.

Battish, S.K.1981.On some chydorid and macrothricid (Crustacea: Cladocera) from Punjab with the description of three new species. Researches on Crustacea,11:17-35.

Battish, S.K.1983.Ecology and systematics of the cladocera (Daphinidae: Branchipoda) inhabiting Punjab, India. Research on Crustacea,12:63-76.

Bedwal,Sujata and Mathur,R.N.2015.Physico-chemical status in relation to zooplankton population of Kishangarh lake (Rajasthan) India. Ecology, Environment & Conservation,21(4):2055-2060.

Bhadane, R.S.2015.Zooplankton diversity of Ramkund of Godavari river, Nashik district, Maharashtra. Bionano Frontier,8(3):128-130.

Bhadane, R.S. 2019.Plankton diversity in the Gangapur dam of Nashik district, Maharashtra. In: Biodiversity and Conservation (Editors V.B.Sakhare, B.Vasanthkumar and S.G. Jetithor),199-203,R.P.Publications,Delhi.

Bhat, M.A., Wanganes, A., and Raina, R.2014.Spatio-temporal variation of the zooplankton community in a tropical wetland (Bhoj wetland),Bhopal, India. J. Ecol. Nat. Environ. 8(8):252-270

Bhati, D.P.S. and Rana, K.S.1987.Zooplankton in relation to abiotic components in the fort moat of Bharatpur. Proc .Nat .Acad.Sci.India,57(13):237-242.

Biswas,M.2015.Seasonal abundance of zooplankton in relation to physico-chemical features in Rabindra Sarobar, Kolkata. International Research Journal of Interdisciplinary & Multidisciplinary Studies,1(5):56-62.

Biswas, S. 1964. A new species of the cladoceran genus Latona Straus 1820 (Crustacea: Cladocera : Sididae) from Rajasthan, India, Proc. Zool. soc. Calcutta, 17 : 149-152.

Biswas, S.1971.Fauna of Rajasthan, India. Part II (Crustacea: Cladocera), Rec. Zool. Surv. India,63:95-141.

Borde, Heeralal and Solanki, Meenakshi. 2023. Study of the zooplankton community of Chikliya pond area, District Barwani, M.P., India.BioGeco,12(2):1098-1105.

Buktar, P.P.2017.Plankton diversity of two reservoirs in Beed district, Maharashtra. In: Plankton and Fisheries (Edited by V.B. Sakhare, H.K. Jadhav and S.S.Jadhav),147-186,Discovery Publishing House Pvt. Ltd., New Delhi (India).

Bunkar, S.K. and Tiwari,A.K.2016.Studies on zooplankton diversity in Barchar dam, Sidhi (M.P.).International Journal of Fauna and Biological Studies,4(1):116-118.

Chakraborty, I.2004.Limnology and zooplankton abundance in selected wetlands of Nadia district of West Bengal, Environ. Ecol., 22:576-578.

Chakraborty, Souradity and Mallick, P.H.2020.Freshwater cladoceran (Cladocera: Branchiopoda) diversity of Lateritic Rarh. Advances in Zoology and Botany,8(3):188-198.

Chakraborty, Souradity and Mallick, P.H.2021.Study on diversity of cladocerans (Cladocera: Branchiopoda) in some selected wetlands of West Midnapore district, West Bengal, India. Environmental and Experimental Biology,19:151-160.

Chalak, A.D. and Sakhare, V.B.2018.Reservoir Ecology, Discovery Publishing House Pvt. Ltd., New Delhi, India.

Chandrasekhar, S.V.A.1994.Biodiversity of zooplankton in Saroornagar lake, Hyderabad. J. Aqua. Biol.,9(1&2):30-33.

Chandrasekhar, S.V.A.1995.Cladocera (Crustacea: Branchiopoda) inhabiting Osmansagar, Hyderebad. Geobios New Reports,14:59-60.

Chandrasekhar, S.V.A.1996.Zooplankton diversity of Medchal Irrigation tank (R.R. District, Andhra Pradesh) with a special reference to rotifera and Cladocera. J. Freshwater Biol.8(4):197-200.

Chandrasekhar, S.V.A.1997.Ecological studies on Saroornagar lake, Hyderabad, Ph.D. thesis, Osmania University, Hyderabad.

Chandrasekhar, S.V.A.1998.Rotatorian and cladoceran fauna of Manjira lake, A.P. with a note on their abundance and indicator value. Proc. Acad. Environ.Biol.,7(1):27-30.

ChandrasekharS.V.A.2004.A study on the cladoceran fauna of Hyderabad and its environs, Andhra Pradesh. Rec.Zool.Surv.India,102 (Part 1 & 2):155-167.

Chandrasekhar,S.V.A.2014.Limnological variations of two Krishna river tributaries in Khammam district, Telangana State .Records od the Zoological Survey of India,114 (3):403-410.

Chandrasekhar, S.V.A. and Chatterjee,T.2002.On a collection of Cladocera from Dimnaand Jublee Park Lakes,Jamshedpur, Jharkhand, Zoos' Print Journal,18(4):1089-1090.

Chandrasekhar,S.V.A. and Chatterjee,T.2004.Freshwater Cladocera of Dhanbad, Jharkhand, Records of the Zoological Survey of India,103(3-4):79-85.

Chandrawanshi,S,2020.Ecology of zooplankton dominated in Shivnath river at Madku Deep in Mungali District C.G.,Ph.D. thesis, Dr. C.V. Raman University, Kota, Bilaspur (C.G.).

Chari, T.J.,Mahender,J.,Kumar, A.S. and Rajashekhar,A.V.2013.Zooplankton diversity abundance and seasonal variation ofNagulakunta water tank Vinijapally Karimnagar Dist. Telangana State, India. International Journal of Science and Research,4(7):1651-1654.

Chari, T.J., Srishylam, B. and Rajashekhar,A.V.2021.Diversity of zooplankton in Shanigaram reservoir, Siddipet District,Telangana,India.J.Aquac.Res.Development,12:637.

Chatterjee, Tapas, Kotov Alexey A., Damme Kay Van, Chandrasekhar, S.V.A. and Padhye Sameer. 2013. An annotated checklist of the Cladocera (Crustacea: Branchiopoda) from India.Zootaxa,3667(1): 89.DOI.10.11646/zootaxa.3667.1.1.

Chaudhary, S. and Singh, D.K.1999. Zooplankton population of Boosra lake at Muzaffarpur, Bihar. Environment and Ecology,17(2):444-448.

Chauhan, R.1993a.Seasonal fluctuation of zooplanktons in Renuka lake, Himachal Pradesh, Uttar Pradesh Journal of Zoology,13(1):17-20.

Chauhan, Ramesh. 1993b. Zooplankton species and their dominance in Rewalsar lake, Himachal Pradesh, Uttar Pradesh J.Zool.13(2):143-147.

Chavan, R.J., Mohekar, A.D. and Hiware, C.J.2006.Ecology and Behaviour of zooplanktons in Manjara Project Reservoir in Maharashtra. In: Ecology of Lakes and Reservoirs (Edited by V.B. Sakhare),151-162, Daya Publishing House, Delhi.

CICFRI.1991. Feasibility studies for fisheries development in Nongmahir reservoir, Meghalaya, Central Inland Fisheries Research Institute, Barrackpore, West Bengal.

CIFRI. 2007. Ecology and Fisheries of Pong Reservoir, Himachal Pradesh, Bulletin No.148 Central Inland Fisheries Research Institute, Barrackpore, West Bengal.

CIFRI.2023. Fisheries resource of Pong dam lake, Himachal Pradesh-An assessment for integrated management. Report prepared for GIZ,pp 145.

Dabhade,D.S. and Chhaba,S.G.2019.Zooplankton diversity around Washim region of Maharashtra. International Journal of Advanced and Innovative Research,6(2):332-336.

Daimari, Puhar.2020. Ecology of Dhansiri river, Nagaland, India. World News of Natural Sciences,30(2):187-193.

Darvekar, Aasidhara and Raut, Vijay. 2022. Studies on diversity, seasonal variation and zooplankton community of freshwater, Ramadeshwar Lake (Hirwa talav) of tehsil Ramtek, District-Nagpur, Maharashtra, India. World Journal of Pharmaceutical and Life Sciences,8(10):91-99.

Das,P. and Kar,D.2016.Biodiversity of zooplankton in a floodplain lake of the Barak Valley, Assam (North-East, India). Int.J.Ad.Res. 4(4):906-910.

Dash, A., Palita, S.K. and Patra, H.K.2012. Diversity of zooplanktons in hot water spring of Atri, Odisha, India. Int. J. Envi. Sci.& Tech., 1(2):111-119.

Devi, Sarla, B.1997.Present status, potentialities, management and economics of fisheries of two minor reservoirs of Hyderabad. Ph.D. thesis, Osmania University, Hyderabad.

Dhatrika, B.2012. Ecological studies and zooplankton diversity in three different lentic water bodies of Warangal District,A.P..Ph.D. Thesis, Kakatiya University, Warangal.

Dhatrika, B., Swaroopa Rani V., and Alaveni, M.2024.Statisticial analysis of zooplankton diversity in Hasanparthy Lake of Hanumkonda District of Telangana State, Uttar Pradesh Journal of Zoology,45(3):51-58. https:// doi. org/10.56557/upjoz/2024/v45i33874.

Dube,P., Garg,V. and Sharma,S.2010.Community structure of zooplankton groups of Kishore Sagar Tank,Kota,Rajasthan.Life.Sci.Bull.,6(3):451-453.

Eggermont, H., and Martens, K. 2011. Preface: Cladocera Crustaceans: Sentinels of environmental change. Hydrobiologia,676,1, https://doi.org/10.1007/510750-011-09089.

Gangwar,S.P. and Bhadauriyua,Gaurav.2021.Study on zooplanktonic community of river Deorania at Barilly District (U.P.) and its relation to seasonal variation. International Journal of Theoretical and Applied Sciences,13(1):12-18.

Ghantaloo, U.S., Sarwade, J.P., Gaikwad, A.F. and Kamble, S.M.2011.Study of zooplankton diversity of Nira left bank canal Sharda Nagar Tal. Baramati Dist. Pune with reference to physico-chemical parameters. Recent Research in Science and Technology,3(3):125-128.

Ghidini, A.R., Serafim-Junior M., Perbiche-Neves G., Brito, L.D.2009.Distribution of planktonic cladocerans (Crustacea: Branchiopoda) of a shallow eutrophic reservoir (Parana state, Brazil).Pan-Am.J.Aquat.Sci.4:29-305.

Gogoi, Budhin. 2017. The cladoceran diversity with special emphasis on feeding ecology of fish in flood plain wetlands of Subansiri river basin, Assam, Ph.D. thesis, Rajiv Gandhi University, Doimukh, Arunachal Pradesh.

Gogoi, Budhin, Francisco Diogo, R. Sousa and Debangshu, N. Das. 2018. Faunal diversity of cladocera (Crustacea: Branchipoda) with notes on biogeographically important species in the floodplain wetlands of the Subansiri River basin, India. Ann. Limnol-Int.J.Lim.,54:36.

Gupta, S.2000.Pollution ecology of two urban water bodies of Jodhpur (India) with special reference to zooplankton. Environmental Protection, (Editor: Vimlesh Chaudhary), Pointer Publisher, Jaipur, India, pp 165-175.

Gupta, S.2002.Studies on the cladoceran diversity of lake Mansar, M.Phil. Dissertation, University of Jammu.

Harishkumar, K. and Kiran, B.R.2016.A report on diversity of Cladocera in sewage fed tank of Bhadravathi taluk, Karnataka. International Journal of Fauna and Biological Studies,3(2):18-20.

Indresha,G.N.,Parida,S.P. and Patra,A.K.2018.Seasonal variations of zooplankton and their correlation with physico-chemical parameters of Kanjia Lake, Nandankanan zoo, Bhubaneswar, Life Sciences Leaflets,99:22-34.

Jalilzadeh, Koorosh and Yamakanamard, S.M.2012.Abundance of Cladocera in three contrasting lentic ecosystems of Mysore, Karnataka. In: Perspectives In Ecology (Eds.V.B.Sakhare and J.S.Mohite),134-143,Oxford Book Company, Jaipur.

Jaya Devi, M.1985. Ecological studies on limnoplankton of three freshwater bodies of Hyderabad. Ph.D. thesis, Osmania University, Hyderabad.

Jetithor,S.G. and 2010.Studies on some aspects of ecology and fisheries of Harni (Katgaon) reservoir in Osmanabad district,Maharashtra,Ph.D. Thesis.Dr.Babasaheb Ambedkar Marathwada University,Aurangabad.

Jetithor,S.G. and Sakhare,V.B.2017.Plankton diversity in Harni (Katgaon) reservoir of Osmanabad district, Maharashtra. In: Plankton and Fisheries (Edited by V.B.Sakhare, H.K.Jadhav, and S.S.Jadhav),108-135,Discovery Publishing House Pvt. Ltd., New Delhi (India).

Jindal, R. and Thakur, R.K. 2013. Diurnal variations of plankton diversity and physico-chemical characteristics of Rewalsar wetland, Himachal Pradesh, India. Recent Research in Science and Technology,5(3):4-9. Retrieved from https:// updatepublishing. com/journal/index. php/rrst/article/view/1034.

Jhingran, A.G.1989.Limnology and production biology of two man-made lakes in Rajasthan (India) with management strategies for their fish yield optimization. Final Report, IDA Fisheries Management in Rajasthan, Central Inland Fisheries Research Institute, Barrackpore, India,1-63.

Joseph, Beenamma.2017. Limnological studies on the species diversity of four groups of zooplankton in Kukkarahalli lake of Mysore, Ph.D. thesis, University of Mysore, Karnataka.

Joshi, P.S.2011.Studies on zooplanktons of Rajura Lake of Buldhana district, Maharashtra, India. Science Research Reoporter,1(3):132-137.

Kadam, C.P., Dandolia, H.S., Kaushik, S., Saksena, D.N. and Shrotriy, V.P.2014.Biodiversity of zooplankton in Pillowa reservoir, District Morena, Madya Pradesh, India. Int.J. of Life Sciences,2(3):263-267.

Kadam, S.U., Gaikwad, J.M. and Babar, Md.2006.Water quality and ecological studies of Masoli reservoir in Parbhani district, Maharashtra. In: Ecology of Lakes and Reservoirs (Edited by V.B.Sakhare),163-175,Daya Publishing House, Delhi.

Kaur, Amritsot, Hundal, S.S. and Aulakh, R.K.2018.Seasonal study of zooplankton diversity uin the polluted water stretch of Buddha Nullah, Ludhiana. Journal of Entomology and Zoology Studies,6(5):2241-2245.

Khan, M.A.1987.Observations on zooplankton composition, abundance and periodicity in two floodplain lakes of Kashmir Himalayan valley. Acta Hydrochim.Hydrobiol.15:174-176.

Khan, S.N. and Khan, S.2017.Variations in the physico-chemical factors and zooplankton diversity during monsoon and post-monsoon period of a Bavori in JDB college campus, Kota, Rajasthan, India. Int. J. Sci. Stud., 5(9):22-28.

Khandayat, Poornima and Singh, Bharat Sharan. 2019. Diversity of zooplankton of Gangulpara dam district Balaghat (M.P.).International Journal of Zoology Studies,4(2):60-62.

Khanna,V. and Yadav,I.2009.Planktonic diversity in the Holy Lake of Pushkar,Ajmer.Nature Environ.Pollut.Technol.8:339-342.

Kiran,B.R., Puttaiah,E.T. and Kainath Devidas.2007.Diversity and seasonal fluctuation of zooplankton in fish pond of Bhadra fish farm, Karnataka. Print Journal,22(12):2915-2516.

Kiss, A., Agoston, A.2014.Microcrustacean (Cladocera, Copepoda, Ostracoda) diversity in three side arms in the Gemenc floodplain (Danube River, Hungary) in different hydrological situations. Acta Zool.Bulg.7:135-141.

Kotwal, Sanjay. 2020. Cladocera diversity in the aquatic habitats of Jammu District (J&K). International Journal of Scientific Research in Biological Sciences,7(3):38-41.

Kour, Sarbjeet, Sharma, Nidhi, Verma Rajan, Kour, Supreet and Kumar, Vipin.2021.Analysis of correlation among zooplankton in two near by ponds of Jammu region, Ecology, Environment & Conservation,27(1):201-208.

Krishna,P.V. and Kumar,H.2017.Seasonal variations of zooplankton community in selected ponds at Lake Kolleru region of Andhra Pradesh, India. International Journal of Current Microbiology and Applied Sciences,6(8):2962-2970.

Krishnamoorthi,A. and Moorthikumar,K.2020.Zooplankton diversity in Amaravathi dam Tirpur District, Tamil Nadu, India. Ecology, Environment & Conservation, 26 (Special Issue):333-336.

Kudari, V.A., Kanamadi, R.D. and Kadadevaru,G.G.2006.Limnological studies of Attiveri and Bachanki reservoirs of Uttar Kannada District, Karnataka, India. Ecology, Environment & Conservation,12(1):87-92.

Kudari, V.A., Kanamadizt Ravishankar,D.2013.Impact of restoration on zooplankton composition of Kelageri lake, Dharwad, (Karnataka State, India).Ecology, Environment & Conservation,19(2):343-350.

Kulkarni, D.A. and Surwase, S.S. 2013.Studies on occurrence, richness and composition of zooplankton in Seena river water at Mohal, Dist. Solapur, M.S., India. International Research Journal of Biological Sciences,2(2):25-28.

Kushwaha, S.K., Mishra, A.K. and Mishra, Shilpi.2022.A study on planktonic status of Bansagar dam, Deolond Shahdol (M.P.) India. International Journal of Advanced Research in Science, Communication and Technology,2(4):416-423.

Lingampally,Vasudha,Solanki,V.R.,Anuradha, D.L. and Sabita Raja.2018.Cladoceran abundance in relation to water quality parameters of Chakki Talab, Bodhan, Telangana. International Journal of Current Engineering and Technology,8(1):1-4.

Maganuri, Neelakanth, P., Ravishankar D.Kanamadi and Vijaykumar A. Kudari. 2008. Limnological study of Ranebennur pond of Haveni district, Karnataka, India. Ecology, Environment & Conservation,14(1):77-84.

Majagi, Shashikanth and Vijaykumar, K.2009.Ecology and abundance of zooplankton in Karanja reservoir. Environ. Monit. Assess.,152(1-4):451-458.

Malathi, D.2002.Ecological studies on lake Hussainsagar Hyderabad with a special reference to zooplankton communities. Ph.D. thesis, Osmania University, Hyderabad.

Manickam,N., Bhavan,P.S.,Santhanam,P.,Muralisankar T., Srinivasan,V., Vijayadevan, K. and Bhuvaneswari,R.2015.Biodiversity of freshwater zooplankton and physico-chemical parameters of Barur Lake, Krishnagiri District, Tamil Nadu, India. Malaya J. Biosci., 2(1):1-12.

Manickam, Narasimman, Periyakali Saravana Bhavan, Perumal Santhanam, Rajagopal Bhuvaneswari, Thirunavukkarasu Muralisankar, Veeran Srinivasan, Annamalai Asaikkuti, Gopalan Rajkumar, Rajendran Udayasuriyan and Madhayan Karthik.2018. Impact of seasonal changes in zooplankton biodiversity in Ukkadam Lake, Coimbatore, Tamil Nadu, India, and potential future implications of climate change. The Journal of Basic and Applied Zoology,79:15.

Mayer, C.M. and Wah, D.H.1997.The relationship between prey selectivity and growth and survival in larval fish.Can.J.Fish.Aquat.Sci.,54:1504-1512.

Meena, Kamlesh. 2019.Ecological studies of a village pond of Similiya, District Kota, Rajasthan, Ph.D. thesis, University of Kota, Kota (Rajasthan).

Meshram, W.J.2020.Abundance of cladocerans related with the water quality of Chirebandi pond, Fulchur at Gondia, Dist. Gondia, Maharashtra, India. Int. Res. Journal of Science & Engineering, Special Issue A9:137-141.

Mishra, Rajesh, Kanungo,V.K., Swati Shrivastava and R.K. Agarwal.2014.Seasonal abundance of zooplanktons in five ponds of Raipur city in Chhatisgarh. International Journal of Fauna and Biological Studies,1(6):78-82.

Mishra, S.R.1989.Zooplanktonic fauna of lentic waters at Bhind (M.P.). Geobios New Reports,8(2):115-118.

Mishra, S.R. and Saksena, D.N.1993. Planktonic fauna in relation to physico-chemical characteristics of Gouri tank at Bhind, Madhya Pradesh. In: Advances in Limnology (Edited by H.R.Singh),57-62,Narendra Publishing House, Delhi.

Moorthy, G.D. and Mazher Sultana. 2011. Identification of zooplankton species collected from different ponds of Arakkonam town, Vellore District, Tamil Nadu. J.Biosci. Res. 2(2):72-75.

Mukherji, Manika and Nandi, N.C.2006. Ecology, biodiversity and management of Rabindar sarovar in Kolkata, West Bengal. In: Ecology of Lakes and Reservoirs (Edited by V.B.Sakhare),36-53,Daya Publishing House, Delhi.

Mulani,S.K.,Mule,M.B. and Patil,S.U.2009.Studies on water quality and zooplankton community of the Panchganga river in Kolhapur city.J.Environ.Biol.,30:455-459.

Murugan, N., 1980.Observations on the natural history of Cladocera of small tropical pond, Ph.D. thesis, Madurai Kamaraj University, Madurai.

Murugan, N.1989. Population dynamics of Moina micrura, Kurz (Cladocera: Moinidae) inhabiting a eutrophic pond of Madurai (South India). Proc. Indian Acad. Sci. (Animal Sci.), 211-222.

Naik, G., Balkhi, M.H., Bhat, F.A. and Rashid, M.2017. Comparative study of present and past cladoceran diversity of Manasbal lake of Kashmir (India). Journal of Fisheries Sciences,11(1):62-68.

Narayana,J.1994.A study on river Cauvery with special reference to zooplankton and macrophytes, Ph.D. thesis, Bangalore University, Karanataka.

Nasar, S.A.K.1977.The zooplankton fauna of Bhagalpur (Bihar).II. Cladocera. Researches on Crustacea,8:32-36.

Nayar, C.K.G.1971.Cladocera of Rajasthan, Hydrobiologia,37, pp 509-518.

Omkar, Sahu, Basant Singh, Vidyabhushan, Uma and Lukesh Kumar Banjare.2023.Study of variations in physico-chemical parameters and plankton diversity during monsoon and post monsoon season in Taro reservoir, Kawardha, Chhatisgarh. The Pharma Innovation Journal,12(121s):1290-1295.

Padhye, S.M. and Dahanukar, N.2015.Determinants of 'water fleas' (Crustacea: Branchiopoda: Cladocera) diversity across seasonal and environmental gradients of a polluted river. Current Science,109(10):1777-1780.

Padhye, S.M. and Dumont, H.J.2014.Species richness of Cladocera (Crustacea: Branchiopoda) in the Western Ghats of Maharashtra and Goa (India), with biogeographical comments. Journal of Limnology, doi:10.4081/jlimnol.2014.1033.

Pamai, Kensiba and Sharma, B.K.2023.Faunal diversity of zooplankton in the selected wetlands of Nagaland, North-East India. Rec .Zool.Surv.India,123(iS2):717-736.

Pandey, Monika, Shukla, Arjun, Mishra, Saket and Rai,Shivani.2015.Distribution and diversity of zooplankton in river Narmada, Madhya Pradesh. International Journal of Current Research,7(12):23471-23474.

Pandit, A.K., Shah, J.A. and Shah, G.M.2016. Research Trends in cladoceran diversity from Kashmir Himalaya. Proceedings of National Academy of Sciences, India,86(2):239-246.

Pandit, D.N., Kumari, P. and Sharma, S.K.2020.Ecology and diversity of zooplankton of the river Ganga at Bihar, India in relation to water quality. Curr. World. Environ., 15(2). DOI: http://dx.doi.org/10.12944/CWE.15.2.18.

Panpatil Pawan and Deshmukh, S.V.2021.Study of zooplankton diversity in Rajura dam at Buldhana district of Maharashtra, Journal of Global Biosciences,10(5):8744-8751.

Pant,M.C., Joshi, A. and Sharma,P.C.1985.Species composition ,temporal abundance and community structure of zooplankton in Lake Sat Tal (U.P.),India.Arch. Hydrobiol.,102(4):519-535.

Parihar, Vijay Laxmi, Sharma, L.L., Sharma, M.S. and Ashwani Kumar.2009. Zooplanktonic assemblages in two seasonal wetlands of Udaipur, Rajasthan. Ecology, Environment & Conservation,15(4):751-755.

Patil, C.S. and Gounder, B.Y.M.1985.Ecological study of freshwater zooplankton of a subtropical pond (Karnataka State) India. Int. Rev. Ges.Hydrobiol.,70:259-267.

Patil, S.G.1986.Records of cladocera (Arthropoda: Crustacea) from Hyderabad. Geobios New Reports,5:78-79.

Patil, S.G. and Panda,P.2003.Population dynamics of planktonic cladocera and copepoda of a freshwater fish tank Bibinagar, Hyderabad, A.P. (Part III).Rec.Zool.Surv.India,101 (Part 3-4):101-107.

Patil, S.S.2020.Study of zooplankton diversity of lower Terna reservoir, District Osmanabad, M.S., India, International Journal of Creative Thoughts,8(4):1899-1903.

Patil,S.S. and Ghorade,I.B.2013.Zooplankton diversity from Godavari river water (Maharashtra). Indian Journal of Applied Research,337-338.

Patil, S.S. and More,V.R.2018.Zooplankton community structure at Kurnur dam in Maharashtra (India) A subtropical freshwater reservoir, International Journal of Life Sciences,6(3):814-820.

Pavan, M., Gowri, P. and Benarjee, G.2016.A study on zooplankton diversity of Bhandan cheruvu, Warangal, T.S., India. International Journal of Innovative Research in Science, Engineering and Technology,5(9):17328-17332.

Pawar,J.R.2016.Zooplankton diversity of Katepurna reservoir Dist.Akola,M.S.,India,Ph.D. thesis,Sant Gadge Baba Amravati University,Amravati.

Priyama Kumari and Ashok Kumar Thakur. 2022. Study of zooplankton abundance and species diversity in Shahjangi pond of Bhagalpur, Bihar (India). International Journal of Trend in Scientific Research and Development,6(3):1895-1899.

Purushothama, R., Sayeswaara, H.A., Gouder, M.A. and Harishkumar, K.2011.Physico-chemical profile and zooplankton community composition in Brahmana Kalasi Tank, Sagar, Karnataka, India.Ecoscan,5(3):99-103.

Pushpendra and Madhyastha, M.N.1994.Seasonal variation and diversity of zooplankton in a small pond near Mangalore. J. Ecobiol., 6(3):197-200.

Quadri, M.Y. and Yousuf, A.R.1980.Influence of physico-chemical factors on the seasonality of Cladocera in lake Manasbal.Geobios,7:273-276.

Raghunathan, M.B.1983.Studies on some planktonic Cladocera of Tamil Nadu. Ph.D. thesis, Madras University, Madras.

Raghunathan, M.B.1985.Studies on freshwater cladocera of Tamil Nadu high altitude waters. Proceedings of National Symposium on Pure and Applied Limnology. Bulletin of Botanical Society, Sagar,60-63.

Raghunathan, M.B.1986.Cladocera (Crustacea) of certain urban water bodies. All India Seminar on water quality around urban ecosystems and management, Chindwara, pp 5-8.

Raghunathan, M.B.1990.Seasonal studies on freshwater cladocera of Chingleput tank, Tamil Nadu. Records of the Zoological Survey of India,86(2):253-259.

Raghunathan, M.B. and Suresh Kumar, R.2002.Cladocera (Crustacea) of Tamil Nadu-Checklist and Bibliography. Zoos' Print Journal,17(12):959-961.

Rajani, V.2021.A study on zooplankton diversity of Velair freshwater Lake, Warangal District, Telangana. International Journal of Creative Research Thoughts,9(2):2079-2082.

Rajshekar, M., Vijaykumar, K. and Parveen, Z.2009.Zooplankton diversity of three freshwater lakes with relation to trophic status, Gulbarga district, North-East Karnataka, South India. Int. J. Sys. Biol., 1&2:32-37.

Rane, Pramod. 1983. A new species of the genus Simocephalus schodler,1858 (Cladocera, Daphnidae) from Madhya Pradesh, India.Crustaceana,45(2):154-156.

Rane, P.D.1985.On a new cladocera Latona narendrai sp.nov. from Madya Pradesh, India. Journal of the Bombay Natural History Society,82(2):387-390.

Rane, P.D.1987.Description of two new species of Cladocera of family Daphniidae from Madya Pradesh, India. Journal of the Bombay Natural History Society,84 (3):638-642.

Rane, P.D.2005. Cladocera: Crustacea. In: Fauna of Melghat Tiger Reserve: Fauna of Conservation area sereies,24:451-500.Zoological Survey of India, Calcutta.

Rane, P.D. and Jafri, S.N.1990.A new species of cladocera, Daphnia madhuriae sp. Nov. from Madhya Pradesh, India, Crustaceana,59(1):62-68.

Rane, Sandeep, Chauhan, Rajendra and Anil Kumar.2020. Biological study of seasonal monitoring and diversity of zooplankton of Sapna dam, Betul district, Madhya Pradesh, India. Journal of Emerging Technologies and Innovative Research,7(10):795-803.

Rani, Sonam, Amrita Kumari, Amita and Sister Stuti, A.C.2019.Fauna diversity of zooplankton of various ponds of Patna, Bihar. Zoology Explore-Journal of Research, XI(1):1-5.

Rao,R.2017.Zooplankton diversity and seasonal variations in Thandava reservoir, Visakhapatnam, India. International Journal of Fisheries and Aquatic Studies,5(1):90-97.

Rath, K.C.1997.Qualitative and quantitative studies on the littoral sedimentary meiofauna of river Kuakhai of Mahanadi river system Orissa, India, Ph.D. Thesis, Utkal University, Bhubenoswar, pp 228.

Rath, K.C.2015.Bioindicators of environmental pollution in Kathjudi river, Odisha, India. J. Exp. Zool.India,18(1):125-132.

Rath,K.C., Rao,D.G. and Sharma,A.L.N.2004.Community structure of benthic meiofauna of river Kuakhai,Orissa,India,Lake-2004.Int.Conf.Conser.Restoration and Management of Lakes and coastal wetlands,Dec.2004,P.8-9.

Rohankar,L.H. and Dahegaonkar,N.R.2016.Studies on diversity of zooplankton in Velgure lake near Aheri, Dist. Gadchiroli, Maharashtra, India. International Journal of Researches in Biosciences, Agriculture and Technology, Special Issue,133-135.

Roy, Koushik, Gupta Sandipan and Nandy Saurav Kumar. 2015. Checklist of commonly available phytoplankton and zooplankton genera of urban and rural ponds of Raipur, Chhatisgarh, India. World Journal of Zoology,10(4):351-357.

Saba, Farhat and Sadhu,D.N.2015.Zooplankton diversity of Garga reservoir of Bokaro Jharkhand, India, International Journal of Bioassays,4(4):3792-3795.

Saha, Ratan Kumar. 1989. An ecological study of two lentic habitats of Agartala India, Ph.D. thesis, University of Calcutta.

Sahu, Seema, Sao Shweta and Singh, R.K. 2016.Study of zooplankton in Kello river Raigarh Chhatisgarh. Journal of Chemistry and Chemical Sciences,6(12):1248-1254.

Sahu,Smrutiranjan and Rout,Prafulla Chandra.2021.Plankton diversity and abundance river Mahanadi at Cuttack.Innovation The Research Concept,6(10):E-13-E-32.

Saini, Deepika and Dube, K.K.2018.Seasonal study of zooplanktonic diversity of River Narmada, Jabalpur Region (M.P.) India. International Journal of Science and Research,7(1):1064-1074.

Sakhare, V.B.2006.Ecology and Jawalgaon reservoir in Solapur district, Maharashtra.In: Ecology of Lakes and Reservoirs (Edited by V.B.Sakhare),16-35,Daya Publishing House, Delhi.

Sakhare, V.B.2007.Reservoir Fisheries and Limnology, Narendra Publishing House, Delhi,pp 187.

Sakhare, V.B.2017.Zooplankton diversity in Jawalban reservoir, Maharashtra. In: Plankton and Fisheries (Edited by V.B.Sakhare, H.K.Jadhav & S.S.Jadhav), 83-107,Discovery Publishing House Pvt. Ltd., New Delhi.

Sakhare, V.B. and Chalak, A.D. 2020.Seasonal variation of cladocera in a perennial pond, Ambajogai (Maharashtra), India. Journal of Fisheries, 8(2):858-860.https://doi.org/10.17017/j.fish.81.

Sakhare, V.B. and Jetithor, S.G.2019. Zooplankton diversity in Kundlika reservoir of Beed district, Maharashtra. In: Environment and Ecology (Editors: C.M. Bharambe, V.B. Sakhare and A.D. Chalak),112-119, R.P.Publications, Delhi.

Sakhare,V .B. and Vasanthkumar,B.2016.Freshwater exotic fishes: Problems and Prospects, Discovery Publishing House Pvt. Ltd., New Delhi, pp 202.

Salam, M.A., Abdul Malik, H.Pokrel, L.P. Mudoi, S.K. Bhagabati and Rajdeep Datta.2021. Status of zooplankton diversity in Lake Kharungpet, Manipur, India, J. Krishi Vigyan,9(2):214-220.

Sangve, K.B.2020.Studies on planktonic diversity in freshwater reservoirs around Amravati district, Maharashtra. Environment Conservation Journal,21(3):137-142.

Sarkar,T. and Pal,J.2021.Zooplankton diversity in the river Jaldhaka, West Bengal, India. Uttar Pradesh Journal of Zoology,42(10):45-53.

Savitha,N., Yamakanamardi,S.M.2012.Studies on abundance of zooplanktons in lakes of Mysore, India. J.Environ. Biol.,33(6):1079-85. PMID:23741805.

Shah, J.A. and Pandit, A.K.2013.Diversity and abundance of cladoceran zooplankton in Wular Lake, Kashmir Himalaya. Res. J. Environ. Earth Sci.5:410-417.

Sharma, B.K. 1978. A note on freshwater cladocera from West Bengal. Bangladesh Journal of Zoology, 6 : 149-151.

Sharma, B.K. and Michael, George. 1987. Review of taxonomic studies on freshwater Cladocera from India with remarks on biogeography. Hydrobiologia,145:29-33.

Sharma, B.K. and Nogen Noroh. 2020. Zooplankton diversity of three floodplain lakes of the Dibru-Saikhowa Biosphere Reserve, Upper Assam, Northeast India.Int.J.Aquat. Biol.,8(1):18-34.

Sharma, B.K. and Sharma, S. 1999. Freshwater cladocerans (Crustacea: Branchiopoda: Cladocera). In: Fauna of Meghalaya: State Fauna Series, 4: 469–550. Zoological Survey of India, Calcutta.

Sharma, B.K. and Sharma, S.2001. Contributions to the cladoceran fauna (Crustacea: Brabchiopoda: Cladocera) of Bihar. Records of the Zoological Survey of India,99(1-4):31-43.

Sharma, B.K. and Sharma, S. 2007. New records of two interesting chydorid cladocerans (Branchiopoda: Cladocera: Chydoridae) from flood plain lakes of Assam, India. Zoos' Print Journal, 22: 2799-2801.

Sharma BK, Sharma S. 2008. Faunal diversity of Cladocera (Crustacea: Branchiopoda) of Deepor Beel, Assam (Northeast India): a Ramsar site. J. Bombay Nat. Hist. Soc. 105: 196–201.

Sharma, B.K. and Sharma, S. 2009. Faunal diversity of Cladocera (Crustacea: Branchiopoda) of Loktak Lake (a Ramsar site), Manipur (N.E. India). Journal of the Bombay Natural History Society, 106: 156-161.

Sharma, B.K. and Sharma, S. 2010. Diversity of Cladocera (Crustacea, Branchiopoda) in Floodplain Lakes of Manipur, Northeastern India. Rec. Zool. Surv. Ind., 110(4): 19-29.

Sharma, B.K. and Sharma,S.2010.Taconomic notes on some interesting cladocerans (Crustacea: Branchipoda: Cladocera) from Assam (N.E. India).Rec.Zool.Surv.India,110(Part 2):39-47.

Sharma, B.K. Sharma S. 2011. Faunal diversity of cladocera (Crustacea: Branchiopoda) of Nokrek Biosphere Reserve, Meghalaya, northeastern India. J. Threat. Taxa. 3: 2120–2127.

Sharma B.K., Sharma S. 2014. Faunal diversity of cladocera (Crustacea: Branchiopoda) in wetlands of Majuli (the largest river island), Assam, Northeast India. Opusc Zool Budapest. 45: 83–94.

Sharma, B.K. and Sharma, S.2017. Crustacea: Branchiopoda (Cladocera). In: Chandra K., Gopi, K.C., Rao, D.V., Valarmathi, K., and Alfred J.R.B.(Eds.). Current status on freshwater faunal diversity of India-An overview, Zoological Survey of India, Kolkata, pp 199-223.

Sharma, K.K. and Chandrakiran, S.2011.Comparative analysis of cladoceran communities from three subtropical freshwater ponds of Jammu. Patterns, Composition and Diversity. The Bioscan,6(2):233-237.

Sharma, K.K. and Kotwal, S.2011.Studies on diversity and dynamics of Cladocera in a subtropical Sungal pond Akhnoor (J&K). The Bioscan,6(4):623-625.,

Sharma, K.K., Kour, S. and Gupta, S.2005.Distributional pattern of cladoceran in subtropical lake, Mansar, J.& K. National Journal of Life Sciences,2(12):23-26.

Sharma, Nidhi and Kour, Sarbjeet. 2021.First Report of three species of the genus Diaphanosoma (Crustacea: Cladocera: Sididae) from Jammu waters (J & K), India. Journal of Threatened Taxa,13(9):19324-19337.

Sharma,R.C. and Rama Kumari.2018.Seasonal variation in zooplankton community and environmental variables of sacred lake Prashar Himachal Pradesh, India. International Journal of Fisheries and Aquatic Studies,6(2):207-213.

Sharma, S. 2008. Notes on some rare and interesting cladocerans (Crustacea: Branchiopoda) from Meghalaya. Rec. Zool. Surv. Ind. 108: 111–122.

Sharma S. 2010. Micro-faunal diversity of cladocerans (Crustacea, Branchiopoda, Cladocera) in rice field ecosystems of Meghalaya. Rec. Zool. Surv. Ind. 110: 35–45.

Sharma, S.P. and Saksena, D.N.1981.Zooplanktonic fauna of some perennial impoundments at Gwalior, India. J. Jiwaji Univer.9:73-92.

Sharma, V., Verma, B.K., Sharma, R., Sharma, M.S. and Gaur, K.S.2012.A report on the freshwater cladocera (Crustacea: Branchiopoda) of South Rajasthan (India). Int.J. Environ.Sci.,3:275-296.

Sharma, Vipul, Sharma, M.S. and Verma, B.K.2012.Freshwater cladocerans of south Rajasthan, India, LAP LAMBERT Academic Publishing,68 pages.

Shiv, Chanchela, Shrivastava, R.K. and Dube, K.K.2017.Studies on zooplankton diversity in River Temar District Jabalpur, Madhya Pradesh, India. International Journal of Interdisciplinary Research and Innovations,5(1):29-33.

Shrirame, N.V., Ugale, B.J., Pawar, U.D. and Gyananath, G.2016.Seasonal biodiversity and percentage of zooplankton in Ghagardara pond of Kandhar, Maharashtra. In: Advanced Studies in Biology (Edited by V.B. Sakhare, C.M. Bharambe and D.B. Sirsath),77-83, Manglam Publishers & Distributors, Delhi.

Shrivastava,K. and Joshi,K.D.2014.Diurnal variation of plankton in Nawagaon wetland of Chhatisgarh,Journal of Inland Fisheries Society,4(6):48-52.

Siddiqui, S.Z. and Chandrasekhar, S.V.A.1993.New distributional records of freshwater Cladocera from Hyderabad . Geobios New Reports,12:105-110.

Singh, C.S., Sharma, A.P. and Deorari, B.P.1993.Limnological studies for bioenergetic transformations in a Tarai reservoir, Nanak sagar (U.P.).In: Advances in Limnology,(Edited by H.R.Singh),29-44,Narendra Publishing House, Delhi.

Singh, Pooja and Smartha, Meenakshi. 2021. Seasonal abundance of certain zooplankton in Upper Lake, Bhopal. Uttar Pradesh Journal of Zoology,42(17):102-108.

Singh, R.K., Pandey, M.K., Kumari, R. and Ranjan, P.2012.Study on the diversity and seasonal variation of zooplankton in Mahendra Nath Pond, Siwan, Bihar. Int. J. Pharm. Biol. Arch.,3:867-870.

Singh, Seema, Erum Usmani, Riya Dutta, Veena Kumari, Salma Praveen, Swati Kumari, Basant Kumar Gupta and Mohommad Arif. 2021. Study on zooplankton diversity and physico-chemical parameter of Pampoo pond of Madhupur, Jharkhand, India. Int.J.Adv.Life.Sci. Res.4(4):34-44.

Sinha, Bikramajit. 2021. Present knowledge of freshwater cladocera diversity in North East India: opportunities and challenges. International Symposium on Aquatic Biodiversity of the Northeast Region, compendium of papers presented on 26-27th October 2021, pp36-38.

Sinha, Bikramjit. 2023. Present knowledge of freshwater Cladocera diversity in North East India: Opportunities and challenges. International Symposium on Aquatic Biodiversity of the Northeast Region, Compendium of papers presented on 26-27th October 2021, pp 1-3.

Siraj, S., Yousuf, A.R. and Parveen, M.2007.Cladoceran community of Sheshbug wetland, Kashmir. Journal of Research and Development,7:67-74.

Siraj, S., Yousuf, A.R., Bhat, F.A. and Parveen, M.2006.Cladoceran community of freshwaters of Kashmir. Journal of Indian Institute of Sciences,76(4):343-350.

Sitre, S.R.2013.Zooplankton biodiversity in Ghotnimbala reservoir of Bhadrawati tahasil of Chandrapur District. International Interdisciplinary Research Journal,3(1):61-67.

Sitre,S.R.2014.Zooplankton fauna assessment of Naik lake of Nagpur city (M.S.) India. International Interdisciplinary Research Journal,4(2):118-123.

Siva Kumar, K., Sujatha, P. and Altaff, K.2001.Studies of freshwater copepods and cladocerans of Dharmapuri Distt Tamil Nadu.J.Aqua.Biol.,16(1&2):5-10.

Sivakami, R., Arumugam, V. and Premkishore, G.2015.An analysis of zooplankton in a lake Pudukkottai District, Tamil Nadu, India. Int. J. Curr. Micrbiol. App. Sci., 4(5):377-389.

Sudhakar, O.,P. Neela Rani, G. Venugopal and George,J.P.2012.Seasonal fluctuations in zooplankton of a freshwater lake. Asian Journal of Microbiology, Biotechnology & Environmental Sciences,14(4):613-616.

Suganthi, A., Ezhumalai, P. and Chezhian,Y.2014.Zooplankton diversity in certain ponds of Cudalore district of Tamil Nadu,India.Biolife,2(4):1316-1319.

Tapse, B.S.2019.Investigation of zooplankton abundance in Sawangi Nullah dam water, Borgaon near Hingna, Maharashtra. International Journal of Current Engineering and Scientific Research,6(1):240-244.

Thakur, Ankita and Kocher, D.K.2017.Diversity and density of cladoceran population in different types of water bodies of Ludhiana, Punjab (India). Journal of Entomology and Zoology Studies,5(3):1568-1572.

Thakur, R.K., Jindal, R., Uday Bhan Singh and Ahluwalia, A.S.2013.Plankton diversity and water quality assessment of three freshwater lakes of Mandi (Himachal Pradesh, India) with special reference to planktonic indicators. Environ. Monit. Assess. DOI 10.1007/ s10661-013-3178-3. The final publication is available at link.springer.com.

Thirumathal, K.2006.Cladocerans of Amaravathy reservoir, Udumalpet, Coimbatore District, Tamil Nadu, India. In: Ecology of Lakes and Reservoirs (Edited by V.B.Sakhare),105-110,Daya Publishing House, Delhi.

Tidme,S.K. and Shinde,S.S.2012.Report on correlation of zooplankton with physico-chemical factors from freshwater temple pond,Journal of Experimental Sciences,3(7):13-16.

Tripathi, R.B. and Shukla, A.2021.Diversity in zooplankton at Sagarrtalab of district Gonda, Uttar Pradesh, India.Int.J.Adv.Res.Biol.Sci.,8(1):7-13.

Umadevi, T.2013.Limnological investigation and zooplankton diversity of Karanja river, Karnataka. International Journal of Science and Research,2(3):134-136.

Vasanthkumar, B. and Sedamkar, E.B.2016.Zooplankton diversity in some lentic water bodies of Karwar (Uttar Kannada). Lake 2016: Conference on Conservation and Sustainable Management of Ecologically Sensitive Regions in Western Ghats. The 10th Biennal Lake Conference.

Veerendra, D.N., Thirumala, S., Manjunatha, H. and Aravinda, H.B.2012.Zooplankton diversity and its relationship with physico-chemical parameters in Mani reservoir of Western Ghats region, Hosanagar taluk, Shivamoga district, Karnataka, India. Journal of Urban and Environmental Engineering,6(2):74-77.

Venkataraman, K.1983.Taxonomy and ecology of cladocera of southern Tamil Nadu.Ph.D. Thesis, Madurai Kamaraj University, Madurai.

Venkataraman, K. 1990.New records of cladoceran of Keoladeo National Park, Bharatpur-III.J. Bombay Nat.Hist.Soc.87:166-168.

Venkataraman, K. 1995. Freshwater Cladocera (Crustacea Branchiopoda) of Tripura, India. Journal of Andaman Science Association, 11, 15–20.

Venkataraman K.1998. Occurrence of Alonella nana (Baird) and Bosmina longirostris (O. F. Muller) (Crustacea: Cladocera) in Sikkim lakes. Journal of the Bombay Natural History Society, 95: 368–369.

Venkataraman, K.1999. Freshwater Cladocera (Crustacea) of Southern Tamil Nadu. Journal of the Bombay Natural History Society, 96, 268–280.

Venkataraman, K.2001. Morphology of Eurycercus lamellatus (Muller), Chydoridae: Cladocera, from the high altitude wetlands of Sikkim, India. Journal of the Bombay Natural History Society, 98:137–142.

Venkataraman, K., Chakraborthy, N. and Sugunan, V. 1993. Indialona ganapati Petkovski (Crustacea: Cladocera) from Hooghly River, West Bengal. Journal of the Inland Fisheries Society of India, 25 (2): 46–47.

Venkataraman, K., Da, S.R. and Nandi, N.C.2000.Zooplankton diversity in freshwater wetlands of Haora district, West Bengal.J.Aqua.Biol.15(1&2):19-25.

Venkataraman, K. and Das, S.R.1993.Freshwater Cladocera (Crustacea: Branchiopoda) of southern West Bengal. Journal of Andaman Science Association, 9 (1/2): 19–24.

Venkataraman, K. and Das, S.R.1995. Occurrence of Palaearctic Cladocera Diaphanosoma brachyurum (Lievin) in West Bengal. Journal of the Bombay Natural History Society, 9:466–467.

Venkataraman, K. and Das, S.R. 1998.On Chydorus faviformis Birge, 1893 from West Bengal (Crustacea: Cladocera: Chydoridae). Journal of the Bombay Natural History Society, 95:364–365.

Venkataraman, K. and Das, S.R. 2000. Cladocera. In: State Fauna Series: Fauna of Tripura,7(4):277-316. Zool. Surv. India, Calcutta.

Venkataraman, K. and Das, S.R. 2001. Freshwater cladocerans (Crustacea: Branchiopoda) of the wetlands of the Indian Botanical garden, Howrah, W. Bengal. Journal of the Bombay Natural History Society, 98(1), 231–236.

Venkataraman, K. and Krishnaswamy, S. 1984a. Occurrence of Moina weismanni Ishikawa, 1896 (Crustacea: Cladocera) from Madurai, South India. Uttar Pradesh Journal of Zoology, 4 (1):115–117.

Venkataraman, K. and Krishnaswamy, S. 1984b. A new record of Diaphanosoma senegalensis Gauthier, 1951 (Cladocera, Sididae) from Madurai, South India. Current Science, 53: 1041–1042.

Verma, A.K.2020.Zooplankton diversity in Guthiataal, wetland of Bahraich (U.P.), India. International Journal of Zoology and Research,2(10):9-18.

Verma, Kumud and Agarwal, R.K.2021.Diversity and seasonal variation in freshwater reservoir Khushrangi (Kosrangi) Raipur District, C.G., India (Part 1).International Journal of Scientific Research in Science and Technology,8(6):206-209.

Vijayapriya, S., Latha, R., Malathi, A., Parameswaran, M., Kiruthika, P. and Banu,G. Sharmila.2019.Assessment of zooplankton diversity in Kosavampatti Lake at Namakkal District, Tamil Nadu. International Journal of Pharmaceutical & Biological Archive. Available at SSRN: https://ssrn.com/abstract=3787256.

Vyas, Naresh and Chauhan, Beena. 2017. Diversity of zoooplanktons and their seasonal variation of density in Gulabsagar water body, Jodhpur (Rajasthan) during 2014-2016. International Journal of Research Studies in Zoology,3(1):34-41.

Wasarat, Mohd. and Chauhan,Rajendra.2018.Freshwater zooplankton diversity in Halali reservoir: A review. International Journal of Scientific Research in Biological Sciences,5(6):196-198.

Yannawar, Sonali, Akkewar, Rani, Yannawar Vyankatesh and Pedewad, Sujit. 2022. Identification of freshwater zooplankton in Godavari river concerning food chain in aquatic ecosystem of Nanded, Maharashtra, India. International Journal of Agricultural and Natural Sciences,15(1):1-14.

Yewale, R.M.2021.Zooplankton diversity in a freshwater lake of Pentalki, Maharashtra. Journal of Emerging Technologies and Innovative Research,8(12):580-583.

Yousuf, A.R. and Qadri, M.Y.1981. Ecology of Diaphanosoma brachyurum Lieven (Cladocera: Crustacea) in lake Manasbal, Kashmir. Journal of Indian Institute of Sciences,63(c):35-42.

24

Growth Performance of Amur Strain of Common Carp in the North Eastern Hills State of Arunachal Pradesh, India

Vivekanand Safi, Tilling Tayo, Nabam Gama, Shivendra Kumar, Mukesh Kumar Singh and Parmanand Prabhakar

ABSTRACT

A multi-location trial was conducted on the growth performance of Amur Common Carp along with Catla and Rohu in the polyculture system in the three locations of the Papumpare district of Arunachal Pradesh. Uniform sizes of fingerlings selected and stocked at the rate of 1no/m^2. Amur Common Carp exhibits a better growth performance and survival rate compared to other species used for the trial. Total mean weight of Amur Common Carp was 683.3g recorded with survival rate of 82%. Catla, Rohu and local Common Carp also showed good growth performance in the demonstration pond with 552.5g, 458.8g and 575.5g mean weights with survival rate of 73%, 65% and 77% respectively in the 10 months duration. The average production of Amur Common Carp with Catla and Rohu was 22.5q/Ha and local Common Carp along with Catla and Rohu was 21.5q/Ha respectively at the end of trial. Among the three species, Amur Common Carp showed a better growth and survival rate than other species in seasonal water bodies, which may be due to the omnivorous feeding habit and resistance to variation to water quality difference.

Keywords: *Amur carp, Common carp, Growth performance, Fish production.*

INTRODUCTION

Aquaculture is the fastest growing food producing sector with annual growth rate of 10.34% and it play a very important role in improving the income generation as well as nutritional security of farmers in many developing countries including India. In the fresh water system, carps are the most widely cultured species throughout the country and carp culture in India is believed to be as old as carp culture in China. Although in early years, carp culture was largely restricted to the North-Eastern States of India, during the last four to five decades, it has spread to almost all parts. The traditional carp culture practice is rapidly transforming into scientific farming in several parts of India. Generally, carp culture in India is a polyculture of 3-6 species of major carps and exotic carps (Vijay Anand, 2019), namely, catla (*Catla catla*), rohu (*Labeo rohita*), mrigal (*Cirrhinus mrigala*), silver carp (*Hypophthalmichthys molitrix*), grass carp (*Ctenopharyngodon idella*) and common carp (*Cyprinus carpio*). Among six species, common carp (*Cyprinus carpio)* cultured in the composite fish farming system is potential species, which plays a vital role in the Inland Fish production especially for the North Eastern Hill states of India as the species has got high adaptability and tolerance towards high range of temperature and dissolved oxygen variation. The species is introduced to India in 1937 as the Prussian strain and in 1957 as the Bangkok strain. Due to their high food value, the farmers of the state have been culturing the fish in rice fields in monoculture or along with other species particularly in Apatani Plateau of Ziro area gaining an average weight 75g to 110g (Hussain *et al.,* 2018).

Locally available strains of common carp; the female gets very early maturity due to which somatic growth is reduced and leads to reduction of pond productivity (Hussain *et al.***,** 2020**)**. The local existing stock of common carp in India has several demerits as the fish attains maturity within six months and breeds naturally in the pond (Jhingran, 1982). Major constraints of existing local strain are early sexual maturation, spawning well before attaining marketable size, recruitment and overpopulation, low growth and yield, in-breeding, ganado somatic index is more in harvested weight, competition for food and space etc (Hussain *et al*., 2020).

Now a days, various strains of common carp (races, landraces, strains, breeds and stocks) have been developed through geographical isolation, adaptation, mutation and selection processes (natural as well as human intervention). Scientist from Karnataka Veterinary, Animal and Fisheries Science University (KVAFSU), Karnataka has developed a superior common carp breed named as Amur which has been reported that percentage increase in weight attained over local was 29.62 in different trial conducted by the University. Amur common carp (*Cyprinus carpio*) is a genetically improved strain of wild common carp of Hungarian origin (Basavaraju *et al.,* 2003). Basavaraju and Reddy (2013) also reported that Amur strain of common carp has greater potential in low-input aquaculture systems due to its better growth than the existing strain. Das (2017) also reported that the improved variety has been found to mature late and grow faster resulting in higher flesh content and fish production in the rural farmers' ponds is expected to increase in the hilly region of North-east India.

To address the existing problem of locally available strain of common carp; Krishi Vigyan Kendra, Papumpare conducted multi-location trials to study the growth performance of Amur Common Carp in Papumpare District of Arunachal Pradesh.

MATERIALS AND METHODS

The trial was conducted in farmers' field at Mani, Midpu and Karsingsa village of Papumpare District of Arunachal Pradesh during May 2022 to February 2023. The fingerlings of Amur common carp brought from NFDB registered hatchery; Pabhai fish farm of Assam and stocked in the ponds after proper acclimatization. The fingerlings were stocked with Rohu (*Labeo rohita*), Catla (*Catla catla*) in a ratio of 35% Catla: 35% Rohu: 30% Amur common carp @ 10,000 fingerlings/ha in one pond and other pond is stocked with Rohu (*Labeo rohita*), Catla (*Catla catla*) in a ratio of 35% Catla: 35% Rohu: 30% local common carp @ 10,000 fingerlings/ha. Water qualities of the ponds were managed by using lime as well as organic manure for a duration of 10 months. Supplementary feeding of rice bran and mustard oil cake mix in the ratio of 1:1 and feed @ 7% of body weight of fishes by broadcasting in the pond water in the initial period of 3 months and later on bag feeding was done @ 3% body weight by using perforated cleaned cement bag filled with feed ingredients rice bran and mustard oil cake with a mixing ratio of 1:1 and fixed in the pond water surface.

RESULTS AND DISCUSSION

Amur common carp showed significant growth performance in the polyculture system of demonstration pond. The average initial weight of fingerling of Amur common carp was 4.2g. The final weight gain by Amur carp were 680.5 (±46) g, 660.8 (±40) g and 710.4 (±30) g at Mani, Midpu and Karsingsa locations respectively (Table 1 & 3). The growth performance of local common carp was 560.6 (±42) g, 575.4 (±46) g and 590.5 (±48) g at Mani, Midpu and Karsingsa locations respectively (Table 2 & 3). In all the locations, intra species growth difference was not observed but the growth performance of Catla has showed comparatively better as of Rohu (Table 1 and 2). The average production of fishes from trial location of polyculture system with existing common carp stock and Amur common carp was recorded as 21.5q and 22.5q respectively (Table 4). Similar observation also recorded by Rajanna *et al.* (2019). Das (2017) also reported that among all the cultured species; Catla (*Catla catla*), silver carp (*Hypophthalmichthys molitrix*), grass carp (*Ctenopharyngodon idella*) and common carp (*Cyprinus carpio*) performed better in composite culture system in the mid altitude conditions.

The overall mean final weight of Amur common carp and local common carp during the harvest was 683.9g and 575.5g respectively in 10 months duration (Table 5). The mean initial weight of local common carp was 5.4g. The survival percentage of Amur Common Carp, local Common Carp, Catla and Rohu were 82%, 77%, 73% and 65% respectively in the demonstration pond (Table 5). It could be inferred from this trial that Amur Common Carp, Catla and Rohu are the most suitable species for polyculture in earthen pond. It was also observed that the growth performance of Amur carp was maximum during the May to November months. The temperature during this period is optimum for growth and development of this fish.

The present trial showed that Amur common carp shows significant growth performance compared to the locally available common carp in Papumpare District of Arunachal Pradesh. The average growth recorded for Amur common carp is 683.9g, which was 18.8 % faster growth than the existing local common carp stock, attaining 575.5g during the culture period (Table 5). The best growth performance of Amur over any other species revealed that the water quality and the primary productivity of the water bodies are suitable to culture this fish. Similar types of observation with a stocking density ranging between 6000 to 8000 number/ha; resulting an average of 26 to 28 percent better growth in Amur common carp than the local common carp was reported by Das (2017), Rajanna *et al.* (2019) and Hussain *et al.* (2020). The reason for better performance of Amur common carp in most of the demonstration ponds could be related to the omnivorous feeding habit and tolerance to wide range of water quality parameters variation in the ecosystem.

Table 1: Growth performance Catla, Rohu and Amur common carp during the trial

Villages	Fish species	Weight (g)		
		Initial wt.	**Final wt.**	**Net gain**
Mani	Catla	3.8±0.6	540.5±36	536.7
	Rohu	4,2±0.4	480.4±30	476.2
	Amur common carp	4.2±0.5	680.5±46	676.3
Midpu	Catla	3.7±0.4	560.7±40	557.0
	Rohu	4.2±0.6	475.4±36	471.2
	Amur common carp	4.2 ±0.5	660.8±40	656.6
Karsingsa	Catla	3.7±0.3	580.2±45	576.5
	Rohu	4.4±0.4	470.6±46	466.2
	Amur common carp	4.2±0.5	710.4±30	706.2

Table 2: Growth performance Catla, Rohu and local Common carp during the trial

Villages	Fish species	Weight (g)		
		Initial wt.	Final wt.	Net gain
Mani	Catla	3.6±0.2	546.6±34	543.0
	Rohu	4.1±0.5	450.6±32	446.5
	Local Common carp	5.6 ±0.6	560.6±42	555.0
Midpu	Catla	3.7±0.4	550.5±42	546.8
	Rohu	4.2±0.6	460.5±37	456.3
	Local Common carp	5.8 ±0.4	575.4±46	569.6
Karsingsa	Catla	3.9±0.3	560.5±48	556.6
	Rohu	4.4±0.4	465.2±42	460.8
	Local Common carp	4.7 ±0.5	590.5±48	585.8

Table 3: Growth performance of Amur & locally available common carp during the trial

Villages	Fish species	Weight (g)		
		Initial wt.	Final wt.	Net gain
Mani	Local Common carp	5.6 ±0.6	560.6±42	555.0
	Amur common carp	4.2±0.5	680.5±46	676.3
Midpu	Local Common carp	5.8 ±0.4	575.4±46	569.6
	Amur common carp	4.2 ±0.5	660.8±40	656.6
Karsingsa	Local Common carp	4.7 ±0.5	590.5±48	585.8
	Amur common carp	4.2±0.5	710.4±30	706.2

Table 4: Average yield (q/ha) of fishes in the study period

System		Mani	Midpu	Karsignsa	Average
Polyculture	With existing common carp	22.4	20.2	21.8	21.5
	With Amur common carp	23.2	21.8	22.6	22.5

Table 5. Growth performance of Amur strain common carp under polyculture in the demonstration pond of Papumpare district of Arunachal Pradesh

Polyculture system				
Parameters	**Amur Common Carp**	**Catla**	**Rohu**	**Local Common Carp**
Initial mean weight (g)	4.2	3.7	4.2	5.4
Final mean weight (g)	683.9	552.5	458.8	575.5
Mean weight gain (g)	679.7	548.8	454.6	570.1
Survival (%)	82	73	65	77

CONCLUSION

Considering the potential character of the breed, the Amur is not yet well-known in Arunachal Pradesh. Therefore, it is the need of hour that multi-location trials and demonstrations are essential to understanding the species' performance in the hill state. If the species performed well in the state, it might benefit the farming community, which struggles with lower fish growth rates and higher production because the state has a long winter season.

REFERENCES

Basavaraju, Y. and Reddy. A. N. 2013. Growth performance of Amur strain of common carp in southern Karnataka. Mysore Journal of Agricultural Science, 47(1): 119-123.

Basavaraju, Y., Penman, D. J. and Mair, G. C. 2003. Stock evaluation and development of a breeding program for common carp (Cyprinus carpio) in Karnataka, India: progress of a research project. NAGA World Fish Center Quarterly, 26(2): 30-32.

Das, S. K. 2017. Performance of an improved breed of common carp - Amur (Hungarian strain) in the North-eastern hill state of Meghalaya, India. Indian Journal of Fisheries, 64 (Special Issue): 33-38.

Hussain, S. M., Debnath, P., Sen, D., Pathak, M., Nabam, J. and Baruah, D. 2018. Integrated Rice Fish Farming System in Arunachal Pradesh: An Over View. Indian Journal of Hill Farming, 31(2):97-101.

Hussain, S. M., Baruah, D. and Pathak, M. 2020. Performance of Amur Common carp Breed in East Siang District of Arunachal Pradesh. Journal of Krishi Vigyan (Special Issue): 244-247.

Jhingran, V. G. 1982. Fish and fisheries of India. Hindustan Publishing Corporation, Delhi.

Rajanna, K. B., Chethan, N., Vijayakumar, S. and Manjappa, N. 2019. Growth performance of amur strain common carp under polyculture in seasonal water bodies of Chamarajanagar district of Karnataka, under Sujala-iii watershed Programme (2015-16). Experimental Zoology India, 22 (1): 67-69.

Vijay Anand, P. E. 2019. The fish farming industry of India. Global Aquaculture Advocate. https://www.aquaculturealliance.org/advocate/the-fish-farming-industry-of India. 2-9.

25

Integrated Fish-cum-Duck Farming System A Profitable Enterprise for Increasing Farmer's Income

Vivekanand Safi, Tilling Tayo, Nabam Gama, Tadang Meena, Budhin Gogoi and Ram Kumar

ABSTRACT

The investigation was carried out on integrated fish-cum-duck farming at farmers' field for three consecutive years 2019, 2020 and 2021 for analysis of income generation. The trials conducted in 3 villages of Papumpare District of Arunachal Pradesh namely Emchi, Jote and Upper Jumi villages. The economics of the integrated farming system as well as farmer's practice has been worked out and it has been found that gross profit to the tune of Rs. 8.15 lakh/ha and Rs. 2.80 lakh/ha were recorded from integrated fish cum duck farming and traditional fish farming practice with a net profit of Rs.5.2 lakh/ha and Rs. 1.77 lakh/ha respectively. This gave an average benefit-cost ratio of 2.81 in integrated fish-cum-duck farming and 1.72 in traditional fish culture practice. Additionally, the consumption of fish, duck meat and duck eggs add to food quality and livelihood security of the resource-poor family. The study has concluded that the integrated fish cum duck farming system could tackle the issues of sustainability, livelihood security and income generation effectively.

***Keywords**: Integrated fish-duck farming, Fish production, Duck production.*

INTRODUCTION

Integrated farming is an innovative approach towards increased fish production. In integrated fish farming two or more commodities are framed together on a common infra-structural base with the objective of optimizing use of available resources. It is based on the concept that "there are no wastes and waste is only a misplaced resource, which can become a valuable material for another product". There are several advantages of integrated farming system viz. increased productivity, greater income, improved cash flow, fuller employment, a better diet for the farmer's family and less biological and economic risk (Pillay, 1990). On the other side, there is the disadvantage that the integrated system is more complex, with a need for more knowledge and better management, since the failure of one system could adversely affect other system (Singh *et al.*, 2014). In integrated farming system, animal manure is shed directly into a fish pond as fertilizer and supports the growth of natural fish food organisms. The fish not only utilize spilled animal feed, but also directly feed on fresh animal excreta which is partially digested and is rich in nutrients. Surplus excreta support the rich growth of planktonic fauna. Fertilizers and supplementary feed are not used, resulting in a drastic cost reduction. These farming systems are relatively confined units with little exchange of water. Manure from livestock which indirectly trigger the growth of mainly photosynthetic organisms resulting in the production high yields fish with low input and limit the dependency on supplementary feed (Saikia *et al.*, 2020 and Sharma *et al.*, 1979; 1998). Ugwumba *et al.* (2010) suggested that the higher the number of viable enterprises integrated into an agricultural system the higher the expected profit. Integrated culture system is quite compatible with the earthen pond culture and relevant to benefit the rural poor (Misra *et al*, 2016, Vohra *et al.*, 2012).

Duck cum Fish farming is very popular in countries like China, Hungary, Germany, Poland and Russia and to a very small extent in India (Ayyappan *et al.,* 1998). The population of India is very high, most of farmers are marginal land holding category. Therefore, integrated fish cum duck farming is only option to raise farm productivity (Edwards *et al.*, 1988). From the viewpoint of input output relationship fish cum duck integration is the best model of integrated fish, livestock and poultry. Ducks are mainly feed on juvenile frogs, tadpoles, dragonfly and insects etc. and keep the pond environment neat and clean for fish. Duck droppings contain all the essential nutrients in the forms of carbon, nitrogen, potassium and phosphorus for the growth of natural fish food organisms. In general, the farmers are using local indigenous variety of ducks for fish-duck integration. On an average, duck dropping contains 0.9 percent nitrogen and 0.4 percent phosphorous. Duck dropping act as good organic manure which helps in production of different variety of phytoplankton & zooplankton in pond. About 300 - 350 ducks are sufficient to fertilize a hectare of water areas. The fish-cum-duck integration system provides meat, eggs in addition to fish. It generates production of additional food and income to the farmer. Approximately 40-50 kg of organic waste is converted into one kg of fish (Saikia *et al.*, 2020). Under this system the nutrients from the duck are recycled in the pond and this allows for escalation of production and income while reducing the affluent along with the dumping of the wastes would have had on the environment (Singh *et al.*, 2014 and Misra *et al.*, 2016). Direct use of livestock wastes is one of the most widespread and conventionally accepted forms of integrated fish farming and the practice increases the efficiency of both duck farming and fish culture through the profitable utilization of animal and feed waste products. The cost of formulated fish feed is usually about 70 per cent of production costs and the use of animal manure considerably reduces operational costs and makes it possible for low-income fish farmers to profitably engage in the enterprise. Banerjee *et al.* (2014) reported that the use of cow dung and duck manure for practicing aquaculture is a viable option for natural biodiversity. Majhi (2018) also explain that raising of ducks over fish ponds fits very well with the fish poly culture as ducks are highly compatible with cultivated fish The ducks feed on such organisms from the pond as larvae of aquatic insects tadpoles, molluscs, aquatic weeds etc. which do not form the food of stocked fish. The duck droppings act as excellent pond fertilizer and the dabbling of ducks at the pond bottom in search of food, releases, and nutrients from the soil which enhances the pond productivity and consequently increases fish production. The ducks in turn get clean and healthy environment to live and quality natural food from the pond for their growth. By this method it has been possible to attain fish production ranging from 3500-4000 kg/ha/yr from the ponds of the farmers in West Bengal without resorting to supplementary feeding. Bhuiyal *et al.* (2014) documented that the integrated farming system improve the efficiency of marginal and small farms that appeared to be the most efficient performers in the integration and arrangement of enterprises. Hence, the present study was aimed to assess the utility of integrated fish and duck farming for self-employment and nutritional security.

Duck integrated farming is becoming more popular as a source of income for people in rural and urban areas and is one option for meeting community demands for nutritious food that is more disease resistant and develops quicker than native chicken become the foundation of life for most individuals who live in agro-ecosystem wetland areas (Sari *et al.*, 2013).

MATERIALS AND METHODS

The experiment was carried out in Emchi, Jote, and Upper Jumi villages of the Papumpare District of Arunachal Pradesh during the years 2019 to 2021. Fingerlings of Catla (*Catla catla*), Rohu (*Labeo rohita*), Mrigala (*Cirrhinus mrigala*), Grass Carp (*Ctenopharyngodon idella*) and Common Carp (*Cyprinus carpio*) were stocked in the ratio of 3:2.5:2.5:1:1 respectively @ 10000 fish fingerling/ha (Mahapatra *et al.*, 2006). Total cost of production includes cost of labour for pond preparation and management, liming, netting etc. and material cost like ducklings, fish fingerling, duck feed, lime, duck house construction materials etc. in the local price. The gross production includes fish production, egg production, and duck meat production. The egg laying capacity of Assam Pati duck calculated as 75-80 nos. per year and duck laying after the age of 8 month.

Preparation of fish pond

Pre-stocking, stocking, and post-stocking management are the three vital categories under which management techniques are used in scientific fish farming. In pre-stocking management removal aquatic weed by hand, eradication of predatory fish, addition of organic manure and liming @ 2000 kg/ha/yr in three instalments to control the pH of pond water. One of the most crucial tasks in stocking management is the transportation of fingerlings and releasing of fingerlings into the pond, stocking of fingerlings done in the morning hours to reduce the mortality of the fishes. Every month samples were taken to assess the growth and health of the fishes as well as duck.

Construction of duck house

Duck house constructed over the pond water with a dimension 8m length, 4m width and 2.5 m height by using locally available materials. The floor of each house made of slated bamboo and roof is made of Toko leaf. The space between slats was just sufficient to make possible the wasted food and duck dropping to fall directly into the pond water. A bridge like structure constructed between the dyke and duck house for easily accessible for the duck and care taker of the farm. Proper disinfection of duck house and equipment was also done before shifting of the duck in the house. 25 days old ducklings of Assam Pati ducks were integrated @ 45 nos./house.

Feeding of duck

Starter feed was provided to the ducklings during the age of 2-20 week @ 45 g/bird/ day in semi closure system whereas a layer mash was provided to the duck above 20 week @ 65 g/bird/day.

RESULTS AND DISCUSSION

The results of the current study showed that integrated fish cum duck farming systems have a number of advantages over conventional fish farming methods. Different fish species, like Catla, Rohu, Mrigala, Grass carp and Common carp harvested from ponds of Emchi, Jote, and Upper Jumi villages of the Papumpare District. Results indicated that Common carp grew more rapidly than other fish species in integrated fish cum duck farming system. For the purpose of results analysis; the average weight of fifteen (15) numbers of fishes was used.

Growth performance

Catla recorded 510±20g weight gain at the time of harvest in integrated fish cum duck demonstration units against 410 ± 30g in the traditional farming practice (Table 1). The initial weight of fingerlings at the time of stocking was 9g. In case of Rohu, the maximum weight was found 430 ± 20 g in integrated fish cum duck demonstration units in comparison to 340±20g in the pond of traditional practicing farmers. The initial weight of fingerlings of Rohu was 9g at the time of stocking. The weight Mrigala recorded 410±15 in integrated fish cum duck demonstration units and 305±30g in the farmers practice ponds. The weight gain by Grass carp was 570 ± 30 in the same demonstration units in comparison to 450 ± 15 recorded in practicing farmer's pond. Common carp recorded highest weight of 790 ± 20g in the integrated pond and 570 ± 20g recorded in traditional practicing farmers. The initial weight of Common Carp was 8g. During the study period, congenial water temperature for fish growth was observed from May to March in all the locations.

Table 1: Growth performance of carried over fish seed after stocking at integrated duck fish farming system

Sl. No.	Species	Initial weight during stocking (g)	Weight of fishes during harvest (g)	
			Integrated duck fish farming	Traditional fish farming
1	Catla	9	510 ± 20	410 ± 30
2	Rohu	9	430 ± 20	340 ± 20
3	Mrigala	8	410 ± 15	305 ± 30
4	Grass Carp	9	570 ± 30	450 ± 15
5	Common Carp	8	790 ± 20	570 ± 20

Average fish yield recorded in integrated fish cum duck farming demonstration units was 20.7q/ha, 21.25 q/ha and 21.8 q/ha during 2019, 2020 and 2021 respectively. This might be attributed to pre-stocking, stocking and post-stocking management practices. Gradual increase in fish productivity in integrated duck fish farming demonstration units over traditional fish farming practice might be due to the residual effect of incorporation of inputs *viz.* lime, manure and feeding materials in the same pond over the years. Similar observations were also made by Manjappa *et al.* (2017) and Kund *et al.* (2010). An increment of fish harvest to the tune of 63.64, 63.46 and 63.33 percent was recorded by adopting integrated fish farming in the year 2019, 2020 and 2021 respectively (Table 2).

Table 2: Average yield (q/ha) of fishes in duck cum fish farming system and traditional fish farming system

Years	Emchi	Jote	Upper Jumi	Balapu	Av. Yield
Average yield (q/ha) of fishes in integrated fish cum duck farming					
2019	21.4	20.6	21.4	19.4	20.7 (63.64%*)
2020	21.8	21.2	21.8	20.2	21.25 (63.46%*)
2021	22.2	21.8	22.4	20.8	21.8 (63.33%*)
Average yield (q/ha) of fishes in traditional fish culture					
2019	13.2	13.2	12.8	11.4	12.65
2020	13.8	12.8	13.2	12.2	13.0
2021	14.2	13.6	13.8	11.8	13.35

[*Fish yield Increase in integrated fish cum duck farming over traditional fish farming practice (%)]

ECONOMIC ANALYSIS

The data on economic component was recorded from all the integrated fish cum duck farming demonstration units is presented in Table 3. The cost of cultivation was observed as Rs. 43200.00 for an average of 1000 m^2 pond including whole expenditure incurred during the trial in compared to the average outcome/return i.e. Rs. 81,120.00 from all the locations after selling of the trial produce like fish, egg and duck which shows about 2.8 fold increment in farms income of the region. The main reason observed for increased income was due to the availability of direct duck dropping along with unused and undigested feed items which was directly used by the fish stock and in other way it acts as a fertilizer to increase the fish pond productivity (Misra *et al.*, 2016). Economic analysis of integrated fish cum duck farming demonstration and traditional farming practice was made to evaluate the sustainability of integrated fish cum duck farming. Average total cost of production over the period of 2019 to 2021 was Rs. 2.95 lakh/ha and Rs. 1.45 lakh/ha in integrated fish cum duck farming and traditional fish farming practice respectively (Table 4). Variation in the cost of production in different years was due to variation in cost of inputs. More cost of production in integrated fish cum duck farming as compared to the traditional fish farming practice is due to duck husbandry, balance feeding, manuring, liming and using chemicals in the integrated system. Mean yield of fishes obtained from these two systems were 21.25 q/ha and 13.0q/ha. Gross profit to the tune of Rs. 8.15 lakh/ha and Rs. 2.80 lakh/ha were recorded from integrated fish cum duck farming and traditional fish farming practice with a net profit of Rs. 5.2 lakh/ha and Rs. 1.77 lakh/ha respectively. This gave an average benefit-cost ratio of 2.81 in integrated fish cum duck farming and 1.72 in traditional fish culture practice. The result reflects that production of fishes and profitability was more in integrated fish cum duck farming over the traditional fish culture practice which is because of adoption of good management practices.

CONCLUSION

From the present study, it can be concluded that with intensification in stocking density of fish and duck resulted in increased fish production up to 20.7 q/ha/year with an annual income of 8.09 lakh/ha/year in 2019, 21.25q/ha/year with an annual income of 8.27 lakh/ha/year in 2020 & 21.8 q/ ha/year with an annual income of 8.09 lakh/ha/ year in 2021, which is 63.64%, 63.46%, 63.33% higher than traditional fish farming practices in three consecutive years respectively. Under reduction of cost of supplementary feed and fertilizers for fish farming strengthening integration of fish cum duck farming which makes this system viable in state environment for employment generation and concerned aspects. As the fish cum duck farms are ageing the farmers become fully experienced and the profit margin would increase. If increased fish production is encouraged, the farmer's income will increase and his poverty level will be reduced.

Table 3: Economics of integrated fish cum duck farming

Expenditure analysis for 0.1 ha		
Sl. No	**Commodity/item**	**Cost involved (Rs.)**
1	Pond preparation	5000
2	Fingerling (n=1000)	1000
3	Duck housing	5000
4	Assam Pati ducklings (n=45)	3200
5	Duck feed	12000
6	Labour cost	12000
7	Miscellaneous	5000
Total cost involved		43200
Outcome statement		
1	Fish (208 Kg) @ Rs 215.00 per kg	44720
2	Eggs (2800 nos) @ 80 eggs/bird for 35 nos female duck@ Rs. 8.00 per egg	22400
3	Duck 40 nos @ Rs. 350.00 per duck	14000
Total outcome per demonstration (0.1 ha)		81120

Table 4: Economics of Integrated fish cum duck farming and traditional fish culture practice

Parameters	Integrated fish cum duck farming				Traditional fish farming practice			
	2019	2020	2021	Avg.	2019	2020	2021	Avg.
Total cost of production (Rs.Lakh/ha)	2.80	2.95	3.10	2.95	1.4	1.45	1.5	1.45
Mean yield of fishes(q/ha)	20.7	21.25	21.8	21.25	12.65	13.0	13.35	13.0
Income from eggs and Duck (Rs. Lakh/ha)	3.64	3.7	3.4	3.58	-	-	-	-
Gross profit (Rs. Lakh/ha).	8.09	8.27	8.09	8.15	2.78	2.80	2.87	2.80
Net returns (Rs. Lakh/ha)	5.29	5.32	4.99	5.2	1.88	1.80	1.67	1.77
Benefit Cost Ratio	2.89	2.80	2.74	2.81	1.98	1.93	1.91	1.72

REFERENCES

Asala, G. 1994. Principles of Integrated Aquaculture. Paper presented at the NFFR/FACU Workshop on Aquaculture Development, Fish Seed Product and Post-harvest Technology.

Ayyappan, S., Kumar, K. and Jena, J. K. 1998. Integrated fish farming practices and potentials. Fishing chimes, 18 (1):15-18.

Bhuiyal, M. S. U., Karim, M. M. and Hossain, S. M. A. 2014. Study on Integrated system model development. Journal of Bangladesh Agriculture University, 12 (2): 325-336.

Banerjee, S., Ruska, N. and Sudip, B. 2014. Study on the zooplankton production in ponds under different fish farming system in West Bengal. Journal of Krishi Vigyan, 3(1):79- 83.

Edwards, P., Pullin, R. S. V. and Gartner, J. A. 1988. Research and education for the development of integrated crop-livestock-fish farming systems in the tropics. ICLARM Studies and Reviews, 16, Manilla, Philippines.

Kund, G. C., Mishra, R. and Sethi, P. K. 2010. An economic analysis of composite fish culture ponds in Sundergarh district, Orissa. Asian Animal Science, 5 (2):139-141.

Majhi, A. 2018. Integrated Duck cum Fish Farming and its Economic Efficiency: A Study in Purulia District, West Bengal. International Journal of Science and Research, 7(3):1390-1395.

Manjappa, N., Patil, R. and Pavadip. 2017. Potential use of village tanks and farm ponds for aquaculture in Karnataka, India - A case study. International Journal of Research on Applied, Natural and Social Science, 5(10): 45-50.

Mahapatra, B. K., Vinod, K., Mandal, B. K. and Bujarbaruah, K. M. 2006. Composite Fish Culture. Technical Bull. No. 20, ICAR-RC NEH, Barapani, Meghalaya, 1-11.

Misra, V. K., Singh, N. D., Mishra, B. P. and Tripathi, A. N. 2016. Study on Integrated Horticulture cum Fish Farming in District West Kameng, Arunachal Pradesh. In: National Conference entilted "Horticulture in North Eastern Region", Organized by College of Horticulture and Forestry, Central Agricultural University, Pasighat, Arunachal Pradesh 16th-18th January 2016, Arunachal Pradesh. 225 pp.

Misra, V. K., Singh, N. D. and Singh, C. P. 2016. Study on integrated fish farming system for livelihood generation and nutritional security in hilly areas of Arunachal Pradesh. In: Proceeding of Lead Papers on Fourth National Symposium on "Transforming Indian Agriculture Towards Food and Nutritional Security" Organized by ICAR Indian Grassland and Fodder Research Institute 20th-21st February 2016, Arunachal Pradesh. 100 pp.

Pillay, T. V. R. 1990. Aquaculture: Principles and Practices. Fishing News Books, A Division of Blackwell Publishing Ltd, Oxford.

Saikia, P., Nath, K. D., Kalita D. N., and Hussain S. M. 2020. Integrated Fish-cum-Duck farming system:A Tool for Increasing Farmer's Income. Journal of Krishi Vigyan, (Special Issue): 162-167.

Sari, F. S., Roesdiyanto and Ismoyowati.2013.Influence of Azolla microphylla and Lemna polyrhiza use, peking duck feed on different protein level on carcass weight and percentage and carcass parts. Scientific Journal Ranchi, 1 (3): 914-923.

Sharma, B. K., Kumar, D., Das, M. K. and Chakraborty, D. P. 1979. Observation on swine dung recycling in composite fish culture Symp. Inland Aquaculture, CIFRI, Barrackpore.

Sharma, A. P., Singh, U.P., Chauhan, R. S. and Singh, V. K. 1998. Duck-cum-fish culture in Tarai region of U.P. In: Ecological Agriculture and sustainable development. Indian Ecological Society, Ludhiana. 1: 258-292.

Singh, U. P., Bisht, H. C. S. and Pandey, N. N. 2014. Poultry cum fish culture as an effective tool for environmentally sustainable and economically viable technology for the mid hills. International Journal of Advance Research, 2(2): 854-860.

Ugwumba, C. O. A., Okoh, R. N., Ike, P. C., Nnabuife, E. L. C. and Orji, E. C. 2010. Integrated farming system and its effect on farm cash income in Awka south agricultural zone of Anambra state, Nigeria. American-Eurasian Journal Agricultural & Environmental Science, 8(1): 1-6.

Vohra, A. R., Narejo, N. T., Naeem, M., Wadhar, G. M., and Dayo, A. 2012. Effect of dry poultry wastes on the physico- chemical and fish growth parameters of exotic carps, Cyprinus carpio at carp hatchery (District Badin), Sindh, Pakistan. Sindh University of Research Journal (Sci. Sec.), 44(2): 239- 244.

26

Vermicomposting of Leaves

V. Kavitha and R. Anandhan

INTRODUCTION

Generation of solid wastes has a universal problem which are facing by most of the countries in present scenario. These wastes are generated due to several human accomplishments as well as natural disaster. It is estimated that in India nearly 700 million tons of organic wastes is generated annually included leaves, husk, sawdust, steam bark, flowers etc. Vermicomposting is a simple biotechnological process of composting, in which certain species of earthworms are used to enhance the process of waste conversion and produce a better end product. Vermicompost is nutritionally rich natural organic fertilizer, which releases nutrients relatively slowly in the soil and improves quality of the plants along with physical and biological properties of soil. It has a more beneficial impact on plants than soil.

During vermicomposting, earthworms eat and grind substrates, convening with anaerobic micro flora to increase the surface area for microbial colonization and enzymatic action. Screening of locally available earthworm species and their suitability for vermicomposting during summer and winter seasons is very much required for effective recycling of available bio-waste for production of quality organic manure – vermicompost. Earthworm's vermicompost is proving to be highly nutritive 'organic fertilizer' and more powerful 'growth promoter' over the conventional composts and a 'protective' farm input (increasing the physical, chemical and biological properties of soil, restoring and improving its natural fertility) against the 'destructive' chemical fertilizers which has destroyed the soil properties and decreased its natural fertility over the years.

Vermicompost is rich in NKP (nitrogen 2-3%, potassium 1.85-2.25% and phosphorus 1.55-2.25%), micronutrients, beneficial soil microbes and also contain 'plant growth hormones and enzymes'. It is scientifically proving as 'miracle growth promoter and also plant protector' from pests and diseases. Vermicompost retains nutrients for long time but the conventional compost fails to deliver the required amount of macro and micronutrients including the vital NKP to plants.

Vermicompost has very 'high porosity', 'aeration', 'drainage' and 'water holding capacity'. They have a vast surface area, providing strong absorbability and retention of nutrients. They appear to retain more nutrients for longer period of time.

The soil treated with vermicompost has significantly more electrical conductivity (EC) and near neutral pH. Animal protein in the feeds of animal and in the food of human being is essential for growth and health respectively. Earthworm meal (vermin meal) is as good as fish meal in terms of quality and has been found to be an efficient substitute for fish meal in domestic animal diets. The potential value of earthworm as a protein source has been established by several authors. Guererro (1981) working on Perionyx excavatus found that the worm is a good source of protein. While Sabine (1983) said that on an average, earthworms have 60-70% crude protein (dry weight). Furthermore, studies had shown that not only earthworms serve as a rich protein source but also form the sources of essential amino acids, especially lysine which exists in many basic foodstuffs.

Bhawalkar (1995) designated earthworm as the "Natural Bioreactor" as they harness the beneficial soil micro flora, destroy soil pathogen and convert organic waste to wealth such as biofertilizers, biopesticides, vitamins, enzymes, antibiotics growth hormones and proteinaceous worm biomass. Beneficial effects of earthworm are mainly due to various activities like burrowing, feeding and excretion that affect physical, chemical and biological properties of soil. Besides improving crop field status, earthworms can be successfully used for composting of weed biomass, crop residue and other bio-waste into organic manure which is popularly termed as vermicompost. Quality of the vermicompost mainly depends upon the type of raw material and type of earthworm species used. The decomposing ability of different species depends highly on agro ecological and environmental conditions of the region.

Considerable work has been carried out on vermicomposting of various organic materials such as animal dung, agricultural waste, forestry wastes, city leaf litter and food wastes. Similarly, industrial wastes such as guar gum industrial waste, paper pulp, and distillery wastes have been vermicomposted and turned into nutrient rich manure. Certain epigeic earthworm species such as *Eisenia foetida, Perionyx excavatus and Eudrilus eugeniae* are voracious feeders of organic wastes.

Vermicompost improves growth and yield of crops, vegetables, flowers and fruits in crops. Earthworms secrete mucus which maintains pH (6.5-7.5) to favour the soil microflora and castings (vermicompost) which are rich in Ca, Mg, K, N, hormones, enzymes and vitamins and certain micronutrients essential for plant growth and promote growth of many important microorganisms like nitrogen fixers and phosphate solublisers. It also has the capacity to control certain diseases in plants.

Extension use of chemical fertilizer and pesticides in Agriculture negatively impact the soil and plants in nutritional products and they accumulate in the food web and environment. Therefore, developing innovative Agricultural practices that employs organic composts and environmentally friendly products will navigate the Agricultural sector toward a greener future and production of wholesome and non-toxic food and products as affordable prices year-round. In addition, innovative Agriculture is required to meet the demands of governments and companies.

Vermicomposting is a green technique that produces vermicompost from different type organic wastes using specific earthworm species. It helps farmers to reduce their use of chemical fertilizer and the overall production costs. Vermicompost is considered an alternative to chemical additives in agricultural crop production that reduces economic cost, while producing healthier organic products for consumer and enriching the environment. Its plays a vital role in non-artificial system such as organic agriculture sustainable Agriculture or environmentally friendly agriculture, owning to its potential to improve the nutritional value of crops and enhance soil fertility.

Improper management of these leaf waste such as burning, dumping in public *et al.* has detoriate the health and environment. Sustainable remediation practices can only way to resolve this problem. Vermicomposting is an important practice of sustainable remediation which has been used in many countries and product that obtained by this process. Compost is rich in plant nutrients and is a micro biologically active organic material formed from the interactions between earthworms and different types of micro-organism. Through the vermicomposting process, environmental risk of leaf waste material reduces transforming into a safer and more stable product suitable for application to soil an also reduces the transporation costs because of the raw organic matter.

During the process of vermicomposting, vermiworms are used to transform organic wastes into a high- quality product from degraded organic matter and the dead bodies of vermiworms. This technique of vermicomposting helps to transform various organic waste (Agricultural waste, Animal manure and domestic wastes) into a nutrient rich compost for the soil and plants. Vermicompost is an essential element in organic Agricultural system, as it contains beneficial and useful properties for plants. It enhances the physical, chemical and biological properties of the soil and increases its organic content.

Earthworms are a major constituent of soil invertebrate fauna and improve soil structure by breaking organic matter and releasing plant nutrient. These add significantly towards nutrient cycling as well as in enhancing biological activity of the soil ecosystem. The activities of the earthworms influence the cycling of the soil nutrients via their influence on biological and physical attributes of the soil. But it must be noted that various collateral factors such as the flora and fauna of the soil, nature of the soil and climate dictates the degree of this influence. Moreover, they maintain soil status through their burrowing and casting activities. Through their burrowing activities, the earthworms construct labyrinth of tunnels which enhances the soil's capacity to take up water and hence, the dearth of earthworm population necessities the application of costly materials such as gypsum or calcium to order to increase soil permeability. Besides, the action of the calciferous gland of the earthworms along with the buffering action of carbonic acid, helps in the maintenance of soil pH. Understood that earthworms have an important impact on soil fertility and soil permeability.

Vermicomposting is a non-thermophilic process which involves a joint action of earthworms and mesophilic microorganisms (Aira *et al.* 2000) for the conversion of organic wastes into a valuable end product called vermicompost. In terms of improving soil health, vermicomposting is better than other ways of managing solid wastes like traditional composting and landfilling. Landfilling is expensive and may lead to leaching of toxic compounds.

As a consequence, the agricultural lands need, continuously, nutritional substance and water to improve and maintain the agronomic fertility. The use of worm casting (vermicompost) with a good fertilizing value, has recently been proposed in experiments for regeneration of degraded land. Earthworm have a great ability to consume all organic waste, reducing their volume by about 50% and expelling the digested materials as castings, which are a useful soil amended and may be easily stored for agricultural use (Tomati *et al.*, 1985). The fertilizing volume of earthworm casting and the beneficial effects on crop have been related to the presence of active mineral nutrients and plant growth regulators with phytohormonal action. Humic substance have been found to possess phytohormonal properties with

Now a days, generation of the substantial quantity of leaf litter produced by trees growing in the forests, sideways the roads, rail lines, in the garden as well as in farm has always been problem. Improper management of these leaf waste such as burning, dumping in public place etc. has detoriate the health and environment. Sustainable remediation practices can only way to resolve this problem. Vermicomposting is an important practice of sustainable remediation, which has been used in many countries and product that obtained by this process called vermicompost. Vermicompost production have been ever reported from leaf of different plant such as Ashoka tree leaf litter (*Polyalthia longifolia*), Teak tree leaves litter (*Tectona grandis*) and Neem tree leaf litter (*Azadirachta indica*) (Jayanthi *et al.*, 2010), Tendu leaf litter (Mushan and Rao, 2012), Mango and Guava leaf (Vasanthi *et al.*, 2013), Rubber leaf litter (Nath and Chaudhuri, 2014;), Teak leaf litter (Nagalakshmi and Prakash 2016) etc.

Action of earthworms on teak leaves composting

Teak trees are widely distributed in India. Hence, their leaves were collected from sundried and chopped into small pieces. For culture of earthworms, various proportions of teak leaves and soil mixtures such as 1:1, 2:1 and 3:1 ratio of experimental media were prepared. Appropriate amount of water was added (60-70% moisture content). L. mauritii worms were introduced into each experiment. Among the proportions, the *L.mauritii* gained maximum biomass and reproduction in 3:1 ratio of leaves and soil mixture. The results were supported by various researchers.

Jesikha and Lekeshmanaswamy (2013) reported that *E.euginea* gained more weight in cow dung and pongamia leaf media on 11th week. Suther (2007) studied the earthworm *E.euginea* growth performance was more in kitchen. Prasad and Saini (2012) reported that high multiplication of *E.foetida* species was observed in 3:1 ratio of Eucalyptus leaves and cow dung mixture. The reason for the multiplication of earthworm during vermicomposting of *Eucalyptus* leaves upon the chemical composition of leaves i,e C:N content of ratio, polyphenol concentrations, terpenoides are enhanced litter palatability of earthworms. It also increased reproductive ability of earthworm (Ketar, C.M., 1993 and Aira *et al.*, 2006).

Jaweria Siddique *et al.* (2005) reported that growth and reproduction of *E.felida* was increased during the vermicomposting of different organic media manures i.e, cow manure sheep / goat manure and dry leaves. Aalok and Tripathi (2010) indicated that *Eucalyptus* leaf litter and municipal solid waste vermicomposts has high soil nutrients than compost. Sabyasachi and Chaudhini, (2017) was reported that soil, cow manure and bamboo leaf litter enhance the growth and number of of hatching production of *Pontoscolex corethrurus* during vercompost. Sonal and More (2017) found that *Eudrilus euginea* can reduced the composting time of banyan leaves, attained maximum biomass as well as produced nutrients rich biofertilizer ie., vermicompost.

Influence of *Adathoda vasica leaves on* biomass and gut microbial population of *L. mauritii*

Adathoda vasica belonging to family *Acanthaceae*, commonly known as Adosa is a small, evergreen shrub found many regions of India and throughout the world, with a multitude of uses in traditional Ayurveda. It is most well-known for its effective in treating respiratory conditions. The leaves and roots contain several alkaloids (chief principle being quinazoline alkaloid, vasicine and vasicinone, vasicinolone and vasicol), which may have a bronchodilator effect of the bronchii. Sundried and powdered cow dung was used as substrate to maintain the adult worms, and cocoons, since cow dung was deemed as a highly suitable natural feed for worms (Lee; 1985). Moisture content about 70-80% was continuously maintained by sprinkling water. This stock cultural was covered with wooden framed iron mesh and maintained at room temperature (27±2°C) inside the laboratory. Every 15 days, the top layer of substrate was removed and added with fresh cattle dung. Non-clitellate worms were used for the present study.

Adathoda leaves was chopped into small pieces and sundried. Cowdung was mixed with Adathoda leaves in three different ratios (E1 - 750g:250g), (E2 - 500g:500g) and (E3 – 250:750g) respectively. After preparation of experimental media, the required quantity of water was added. The media was allowed to pre-decomposition for 15 days. After decomposition, 10 non-ciliated earthworms were introduced into each experimental media. The moisture content was maintained at 60-80% throughout the study period by periodic sprinkling of adequate quantities of water. The containers were kept in the dark under identical ambient conditions (room temperature 25 ± 3°C). Earthworm growth was measured in terms of increase in biomass (weight). *L. mauritii* gained maximum biomass, more cocoon production whereas bacterial and fungal population in gut from the media of 1:3 ratio of cow dung and Adathoda leaf litter. Application of vermicompost in the chilli plant collected from the same media highly enhanced growth and chilli production.

In biology, observation of growth and reproduction are important physiological parameters to observe health of organisms. Several ecological parameters viz. temperature, moisture, adequate dissolved oxygen, pH, available organic matters and also the presence or absence of toxic chemicals governed the distribution, abundance, growth and reproduction of earthworms. Earthworms continue to grow throughout their lives with enlargement of their body segments following emergences from the cocoons (Edwards and Bohlen, 1996).

Godliving (2002) reported that bacteria and fungi degrade wood dust. Dosoretz, Chenn, and Greth (1990) also reported the reduction in carbon content of sawdust when subjected to microbial degradation. It is obvious that all carbon containing compounds undergo essentially oxidation process by the action of microbes, which results in the release of nutrients, CO_2 and humus. Soft plant organic materials are easily decomposed by microbes. However, tougher plant materials do not breakdown readily by soil animals. The breakdown is brought out by microorganisms, and further accelerated when they pass through the guts of earthworms, due to the presence of intestinal microflora and enzymes in the gut (Edwards & Lofty 1975, Lee 1985). Surindrasuthar *et al.* (2014) said that the feasibility of vermitechnology in management of leaf litter originate from urban forestry system. Nagar *et al*, (2017) studied the vermicompost of *Eucalyptus* leaf litter by using *E.fetida* was rich in nutrient and microorganisms which can be used as suitable soil amendment.

Vermicomposting of Banana Leaves

The soil mixed with banana leaves in the ratio of 500g:500g, 400g:600g and 600g:400g. clitellate *L.mauritii* was introduced into each experiment. Maximum biomass and hatchlings production was observed in 400g:600. Whereas high content of macronutrients such as N,P and K and low level of OC,C:N, C:P ratio, pH and EC was observed in the same media. The vermicompost was highly boost up the growth and flower production.

Earthworm can accept organic wastes as feed if it is provided with smaller particle size about 5mm, higher nitrogen content (Edwards and Bohlen,1996), optimal C/N ratio 15:1 to 30:1 (Neuhauser *et al.,* 1980) etc. Quality and availability of food influences the growth and reproductive potential of earthworms. Earthworms consume and require a combination of microbes, nitrogen, cellulose and grit for their maximum growth and reproduction (Flake and Hartenstein, 1984; Edwards and Bohlen, 1998).

Arancon *et al.* (2006) reported that the increase in strawberry plant growth by application of vermicompost was mainly due to the microbial population. Maria Remedios Morales-Corts *et al.* (2018) reported that garden waste compost tea and vermicompost tea was promoted the tomato plant and also increased the shoots, root dry weight, chlorophyll content and stem diameter compared to untreated plants. They were also observed that the compost and vermicompost tea was act as organic fungicides. It had suppressive effect on *R.Solani* and *F. Oxysporum*. Sakthivel Vellaikannu *et al.* (2017) was reported that high content of N, P, K, Ca, Mg, Fe, Zn, Cu and Ec and low level of OC and pH was observed in vermicompost of tea leaves waste mixed with cow dung (30th day) than control. Reduction in pH towards neutrality should be important in retaining nitrogen and seems to promote the nutrients availability to plant (Brady, 1988). Shinde *et al.* (1992) reported that EC values reduced or remain unchanged and ranged between 0.21-0.46 ds/m in the vermicompost.

In recent years, vermicomposting has emerged as a simple, easily adaptable and effective biotechnology for recycling a wide range of organic wastes for agricultural production. The technology is advantageous over thermophilic compost because it contains a considerable amount of organic acids, such as plant growth promoting hormones and humic acids. It also has high water holding capacity, low C:N ratio and low phytotoxicity.

Vermicomposting of cannon ball leaves

Couroupita guianensis, known by a variety of common names including cannon ball tree, is a deciduous tree in the flowering plant family Lecythidaceae. It is native to the tropical forests of Central and South America, and it is cultivated in many other tropical areas throughout the world because of its beautiful, fragrant flowers and large, interesting fruits. Fruits are brownish grey. There are potential medicinal uses for many parts of *Couroupita guianensis*, and the tree has cultural and religious significance in India. Vermibeds were prepared from mixture *of* cowdung and dried leaf litter of *Couroupita guianensis* in the ratio of 3:1, 1:1 and 1:3 ratios. Adult earthworm, *L.mauritii* were added into each media. The *L. mauritii* gained maximum biomass and more hatchling production in 1:3 ratio of cowdung and dried leaves vermibed. Cellulase, amylase, protease, invertase and urease activities in vermicompost were also increased in the same media.

Earthworms voraciously feed on organic wastes while utilizing only a small portion for their body metabolic activities excrete a large part of the consumed materials in a half digested form. The half digested material decompose rapidly and is transformed into vermicompost within a short time since the intestines of earthworms harbour wide range of microorganisms, enzymes, hormones etc. The feeding activity of earthworms and their voiding of microbially and enzymatically enriched casts could have also contributed to higher enzyme activities of vermicompost. Sujatha *et al.* (2003) also reported that the Earthworm castings in the home garden often contain 5 to 11 times more Nitrogen, Phosphorous and Potassium than the surrounding soil. Castings of earthworm also contain abundant sources vitamins, antibiotics and enzymes such as proteases, amylases, lipases, cellulases and chitinases.

Most important feature of vermicompost is that, during processing of the various organic wastes by earthworms, many of the nutrients are changed to available forms that are more readily taken up by plants (Chaudhuri , 2000). Earthworms fragment the substrate in the process of feeding and thereby increase the surface area for further microbial colonization. The enhanced microbial activity accelerated the decomposition process leading to humification thus oxidizing unstable organic matter to stable form. During the passage through the gut of earthworms the surviving microorganisms are voided along with cast. So vermicompost not only provides mineralogical nutrients to soil but also contributes to the biological fertility factor by adding beneficial microbes to soil.

Vermicompost can be considered as the best organic manure than farmyard manure. Thus, vermicomposting can also be considered as a low cost technology system for the processing or treatment of wastes and utilization of earthworms may be an answer as an ecologically sound, economically viable and socially acceptable technology.

Vermicomposting of green Eucalyptus leaf litter by *Eisenia foetida* and *Eudrilus Eugenia*

Eucalyptus belongs to the Myraceae family in the world, including more than 740 species. It is a long tree to a height of 35 to 40 meters, but it grows to a height from 80 to 100 meters in most of the country. Eucalyptus species is remarkable for their rapid growth. Ritu Nagar *et al.* (2017) studied the vermicomposting of Eucalyptus leaf litter with two different species of earthworms i.e. *Eisenia foetida and Eudrilus eugenia.* The initially lower pH was recorded due to production of organic acid and CO_2 by microorganisms due to presence of carbohydrate abundantly. As carbohydrate source depleted, microbial metabolism shift into nitrogenous organic compound which leads to generate ammonia as a result pH turn increase (Ndegwa *et al.* 2000). Initial temperature of 50% leaf waste was higher (36.3°C) than 100 % leaf waste (33.7°C) as compaired to control that was 100 % cattle dung (38.5°C). After completion it was decreased up to around 19°C±1°C. The higher temperature in 50 % leaf was due to presence of more microbial community than 100 % leaf waste because the cattle dung serves the purpose of providing inoculum of microbes which carry out degradation of organic waste. During the process of composting, mesophilic flora predominates with their metabolic activity resulting in the increase in temperature of the organic waste. They are replaced by thermophilic organisms which survive at temperatures greater than 45°C to facilitate composting. When the temperature gets fall, mesophilic microorganism are become again active. According to Mckinley and Vestal (1985), during composting of municipal sewage waste, temperature was raised in between 55-60°C initially then it was down gradually. The temperature of the composting pile was raised due to the biological oxidation of carbon.

The height of biomass showed decrease than initial contents in both 100 % leaf litter and 50 % leaf litter with 50 % cattle dung after decomposing wastes through earthworms. The maximum depletion was 71 % in 100 % leaf litter (70 % in 50 % leaf litter) as compared to 46 % in control (100 % cattle dung).

The greater biomass reduction may be due to microbial richness of substrate materials. In such beddings the microbial respiration may leads to rapid C loss through CO_2 production. The loss of carbon as CO_2 due to microbial respiration and assimilation of simple carbohydrates leads to biomass reductions from waste mixtures. Moreover, carbohydrates and other polysaccharides which are considered major source of carbon are digested rapidly by earthworms and some fraction of digested substances is then assimilated into worm biomass. The greater biomass reduction may be due to microbial richness of substrate materials. In such beddings the microbial respiration may leads to rapid C loss through CO_2 production. Also, earthworm fragments the waste feedstock into fine fractions which results in increased sites for microbial hydrolytic enzyme actions. The biological mutuality between earthworms and associated microbes is primarily responsible for C loss from the organic wastes. Kaviraj and Sharma (2003) reported 20–45% reduction in organic carbon contents of vermicomposted municipal solid waste mixed with CD. Similar observations have been reported by Prakash and Karmegam (2010) during vermicomposting of press-mud from sugar industry.

CONCLUSION

Earthworms are capable of working hard to convert all the organic waste into manure. Earthworms can serve as tool to facilitate these functions. They serve as nature's plowman and form nature's gift to fulfil the nutritional needs of crops. Vermicomposting appears to be the most promising as high value biofertilizer which not only increases the plant growth and productivity by nutrient supply. It reduces the requirement of more land for disposal of wastes. It helps to create better environment. Thus, reduces ecological risk. It results in several benefits to farms, industries, and environment and over all national economy.

REFERENCES

Aalok, A. and Ashutosh Kumar Tripathi, 2010. Composting Vermicomposting of different types of leaves using Earthworm species Eisenia fetida. Dynamic soil, Dynamic plant,4: 139-144.

Aira, M, Monroy, F, Domínguez, J, 2000. Changes in bacterial numbers and microbial activity of pig slurry during gut transit of epigeic and anecic earthworms. J. Hazard. Mater, 162: 1404–1407.

Aira, M., Fernaudo, M. and Doming Z, J. 2006. C to N ratio strongly affects population structure of Eisenia foetida in vermicomposting systems. European Journal of soil Biology, 42:S127-S131.

Arancon, N.Q, Edwards, C.A, and Bierman, P, 2006. Influences of vermicomposts on feld strawberries. Part 2. Effects on soil microbiological and chemical properties. Bioresource Technology, 97: 831-840.

Bhawalkar, M. S., 1995. Sustainable agriculture through vermiculture. Ind. Farmers Times, 13 :13-14.

Brady, N.C.,1988. The nature and properties of soil. Eurasic Publishing House (P) LTD, New Delhi.

Chaudhuri, P.S, Pal, T.K., Bhattacharjee, G, Dey S,K, 2000. Chemical changes during vermicomposting (Perionyx excavatus) of kitchen waste. Tropical Ecol, 41:107–110.

Dosoretz, G. G., Chenn, H. C., and Greth, H. E, 1990. Effect of environmental conditions of extracellular protease activity in lignolytic cultures of phanerochaetae chrusosporium. Applied and Environmental Microbiology, 56: 395-400.

Edwards, C.A. and P.J. Bohlen,1996. Biology and ecology of earthworms. Chapman and Hall, London.

Godliving, M. Y, 2002. Continuous production of lignin degrading enzymes by Bjer kandora adusta immobilized on polyurethane foam. Biotechnology Letters, 24:173-177.

Guerrero R.D, 1981. The culture and use of Perionyx excavatus as a protein resource in the Phillippines. Dorwin Centenary Symp. On Earthworm Ecology (pp 250). Inst. Terrestrial Ecology UK.

Jaweria Siddique, A., Amer Amir Khan, Iftikhar Hussanain and Shamim Akhter, 2005. Growth and reproduction of earthworm (E.fetida) in different organic media. Pakistan J.Zool., 37 (3):211-214.

Jayanthi B, Ambiga G, and Neelanarayanan P,2010. Utilization of mixed leaves litter for converting into vermicompost by using an epigeic earthworm Eudrilus eugeniae. Nature Environment and Pollution Technology, 9(4):763-766.

Jesikha M and M. Lekeshma, 2013. Effect of pongamia leaf medium on growth of earthworm (E.euginea). Inter. Jr. of suen and Research Publications, 3 (1).

Kaviraj and Sharma, S, 2003. Municipal solid waste management through vermicomposting employing exotic and local species of earthworms. Bioresour. Technol, 90:169-173.

Ketkar,C.M, 1993.Use of biogas slurry in Agriculture. In: Biogass slurry utilization. Consortium on rural technology, New Delhi . 76-79.

Lee, K.E., 1985. Earthworms: Their Ecology and Relationship with Soil and Land Use. Academic Press, Sydney. 411.

María Remedios Morales-Corts, Rodrigo Pérez-Sánchez and María Ángeles Gómez-Sánchez,2018. Efficiency of garden waste compost teas on tomato growth and its suppressiveness against soil borne pathogens. Sci. Agric.75(5) :400-409.

Mckinley V.L and Vestal J.R. 1985. Effects of different temperature regimes on microbial activity and biomass in composting municipal sewage sludge. Can. J. Microbiol,31:919-925.

More Sonal and More B.C, 2017. Effects for biomass and Bio fertizer by using earthworm Eudrilus eugeniae ,J.Environ. RES.Denelop. 11(03).

Mushan, L.C. and Rao, K.R, 2012. Physico-chemical analysis of Tendu leaf litter vermicompost processed By Eudrilus eugeniae. DAV International Journal of Science. 1(2):100-102.

Nagalakshmi, P.K. and Prakash, M. 2016. A Microcosm Study of Cast and Gut of an epigeic earthworm Perioynx ceylanensis reared on different substrates. Int. J. Curr. Trend. Pharmacobiol. Med. Sci. 1(2):45-51.

Nagar, AnuragTitor and PraneeshBhati, 2017. Vermicomposting of green Eucalyptus .Inter. Jr. of Ensenia fortida and Eudrilus eugeniae. Inter. Jr. of Environ, Agriculture and Biotechnology, 2(6).

Nath, S. and Chaudhuri P.S., 2014. Growth and reproduction of Pontoscolex corethrurus (Muller) with different experimental diets. Tropical Ecology. 55(3):305- 312.

Neuhauser, E.F., Hartenstein, R. and Kaplan, D.L., 1980. Growth of the earthworm, Eisenia fetida in relation to population density and food rotating. Oikos, 35: 93-98.

Pius Ndegwa and Sidney Thompson K.C. Das ,2000. Effects of stocking density and feeding rate on vermicomposting of biosolids. Bioresource Technology, 71(1):5-12.

Prakash M. and Karmegam N, 2010. Dynamics of nutrients and microflora during vermicomposting of mango leaf litter (Mangifera indica) using Perionyx ceylanensis. International Journal of Global Environmental Issues (IJGENVI), 10: 3-4.

Prasad, A. and Saini, S.K, 2012. Effect of Eucalyptus citridora leaves on leaves on the multiplication of Exotic earthworm species Eisenia foetida, World Journal of Environment Bioscience, 2(1): 36-39.

Sabine, J,1983. Earthworms as a source of food and drugs. In J.E. Satchell (ed) Earthworm ecology: from Darwin to vermiculture. Chapman and Hall, London, UK.

Sabyasachi Nath and Chaudhuri, P.S, 2017. Growth and reproduction of Pontoscolexcorethrus (Muller) with different experimental diets. Tropical Ecology, 55 (3): 305-314.

Sakthivel, Malathi, Jayashri and Saralabai ,2017. Physico-chemical analysis of vermicompost produced from tea leaves waste mixed with cowdung using Eudrilus eugeniae. European Journal of Biotechnology and Bioscience, 5(6).

Shinde, P.H, Naik, R.L, Nazivkar, R.B, Kadam, S.K. and Khaire, V.M, 1992. valuation of vermicompost. Proc. Nat. Sem. Org. Fmg., TNAU, Coimbatore, 54-55.

Sujatha K, Mahalakshmi A and Shenbagarathai R, 2003. Effect of indigenous earthworms on solid waste In: Biotechnology in Agriculture Industry and Environment (Eds. Deshmukh A.M) Microbiology society, Karad, 348-353.

Surindra, Suthar and Shailja, Gairola ,2014. Nutrient recovery from urban forest leaf litter waste solids using Eisenia fetida. Ecological Engineering, 71:660-666.

Suther, S, 2007.Vermicomposting of potential of Perionyx sansibaricus (Perrier) in different waste materials. Biores. Technol., 98:1231- 1237.

Tomati U, Grapelli A and Galli E, 1985. Soil microrganisms and long-term fertility. In: Williams JH, Guidi G, L'Hermite P, editors. Long-term effects of sewage sludge and farm slurries applications. Elsevier Applied Science, London: 14–22.

Vasanthi K , Chairman K and Ranjit Singh AJ A,2013. Vermicomposting of leaf litter ensuing from the trees of Mango (Mangifera indica) and Guava (Psidium guujuvu) Leaves. International Journal of Advanced Research,1(3):33 -38.

Colour Plates

Chapter 3: Biodiversity of The Clupeiform Fishes from Cuddalore, Tamil Nadu

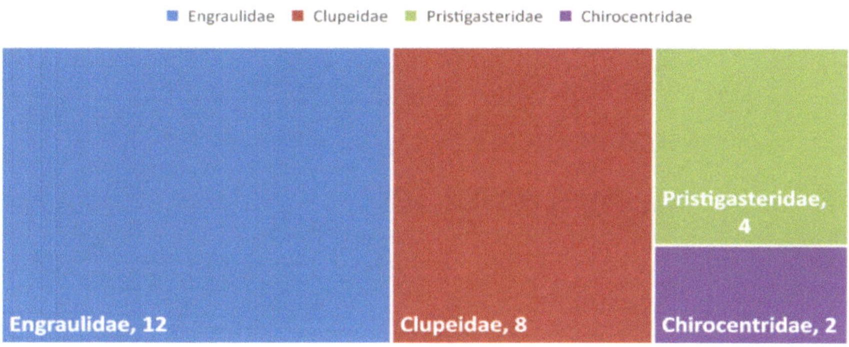

Fig. 3: Key characters of Clupeiform fish Families

Ilisha megaloptera

Ilisha melastoma

Opisthopterus tardoore

Pellona ditchela

Coilia dussumieri

Encrasicholina devesi

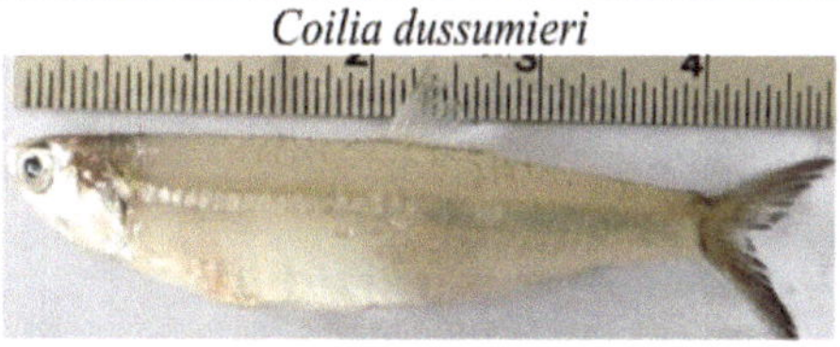

Encrasicholina heteroloba

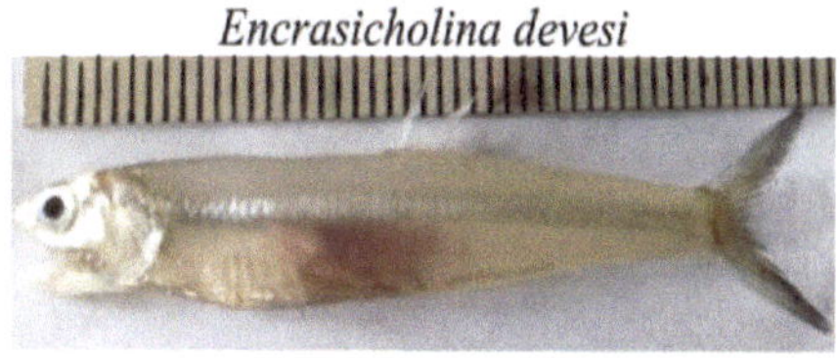

Encrasicholina punctifer

Setipinna phasa

Setipinna taty

Fig. 4: Clupeiform fishes recorded from Cuddalore coast

Setipinna tenuifilis

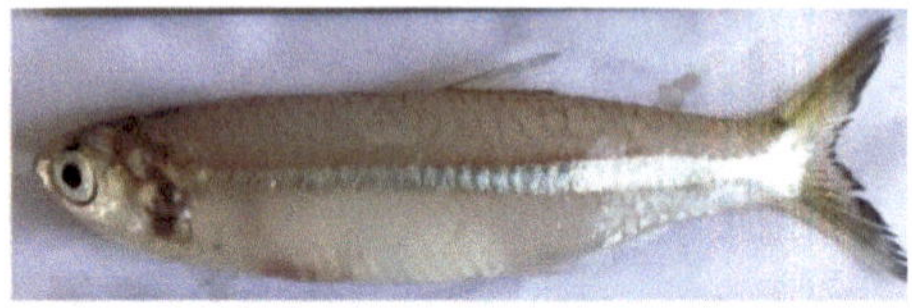

Stolephorus commersonii

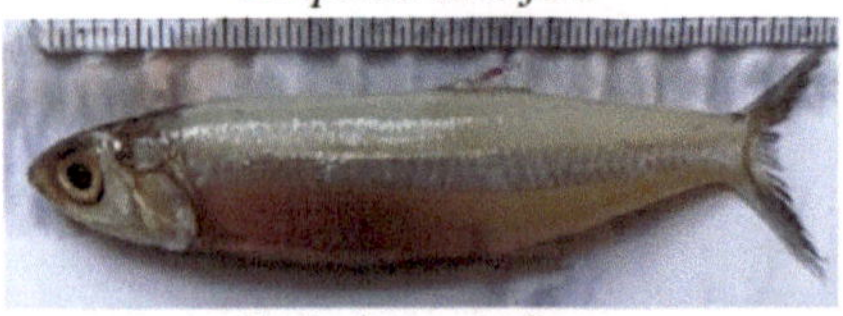

Stolephorus indicus

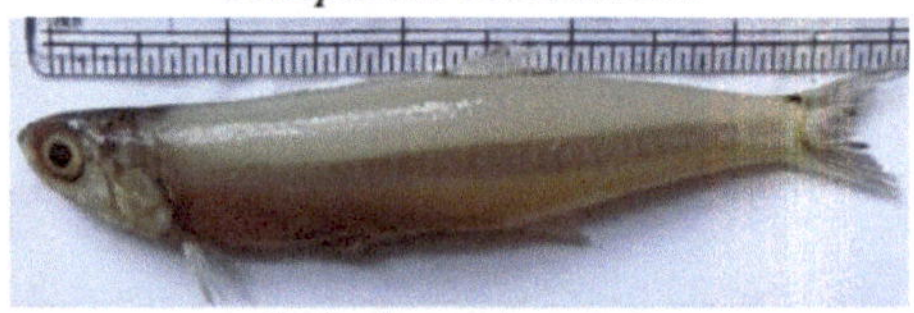

Stolephorus insularis

Thryssa malabarica

Thryssa mystax

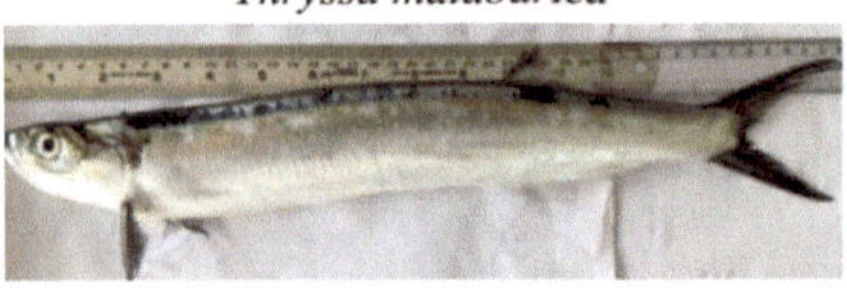

Chirocentrus dorab

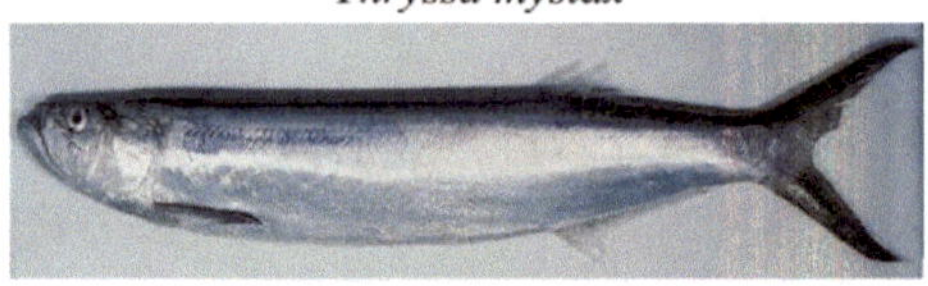

Chirocentrus nudus

Amblygaster sirm

Anodontostoma chacunda

Fig. 5. Clupeiform fishes recorded from Cuddalore coast

Dussumieria acuta

Escualosa thoracata

Sardinella albella

Sardinella fimbriata

Sardinella gibbosa

Sardinella longiceps

Fig. 6. Clupeiform fishes recorded from Cuddalore coast

Chapter 11: Fish Diversity in Maguri Motapung Beel, Tinsukia District of Assam A BriefInvestigation

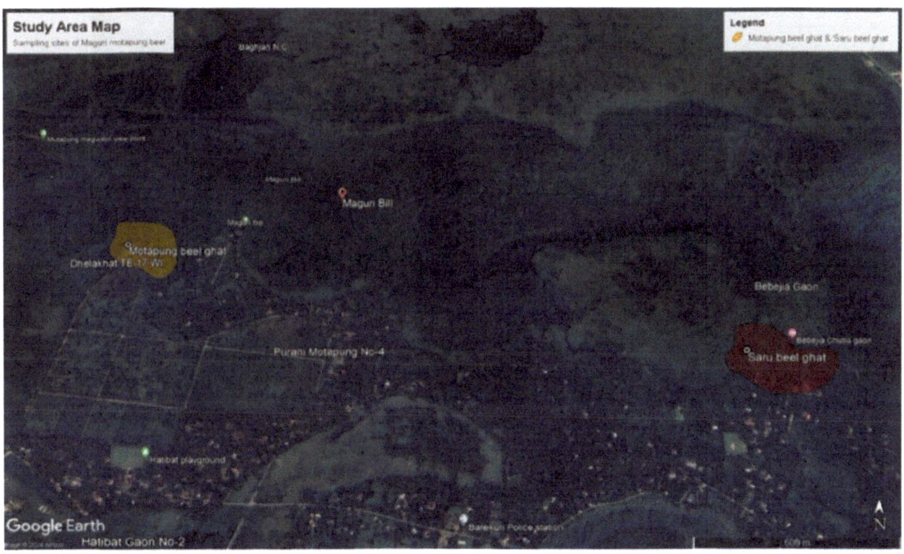

Fig 1a: Study area map (Maguri Motapung Beel)

Fig 1b: Site 1 (Motapung Beel Ghat- Maguri Beel)

Fig 1c: Site 2 (Horu Beel Ghat- Maguri Beel)

Plate-I: Some glimpse of edible & ornamental fish species from the Maguri – Motapung Beel

Macrognathus pancalus

Pseudambassis beculis

Puntius sophore

Wallago attu

Mystus cavasius

Labeo gonius

Mystus nigriceps

Trichoogaster fasciata

Trichogaster lailus

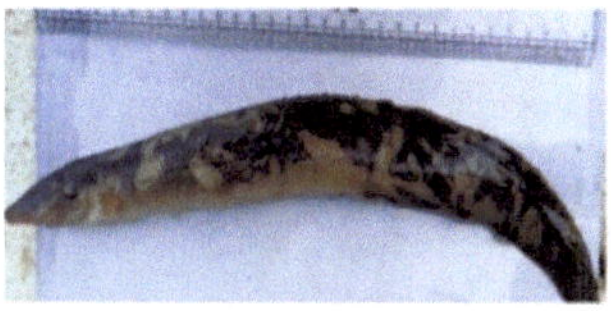
Mastacemcembelus armatus

Cyprinus carpio

Lepidocephalicthys guntea

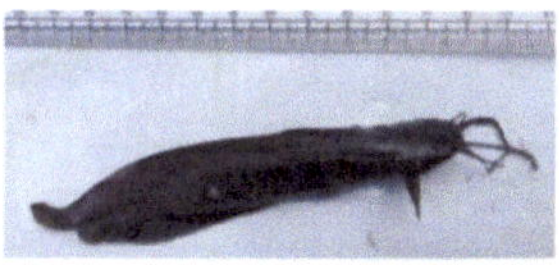
Heteropneustes fossilis

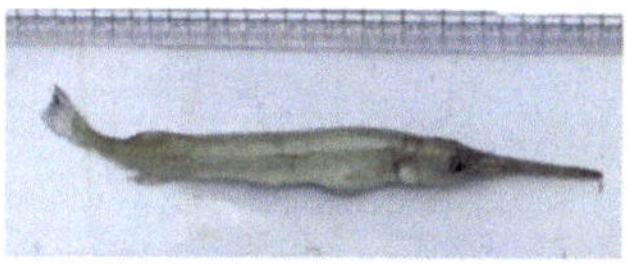
Xenentodon cancila

Rasbora daniconius

Puntius conchoonius

Charax gibbosus

Cirrhinus reba

Notopterus notopterus

Labeo bata

Lepidocephalicthys gongota

Mystus vittatus

Channa punctata

Ctenopharyngodon idella

Chapter 12: A Study on Ichthyofaunal Diversity and Influence of Physico-chemic Parameters on Fish Species of the Maguri-Motapung Beel, Tinsukia District, Assam

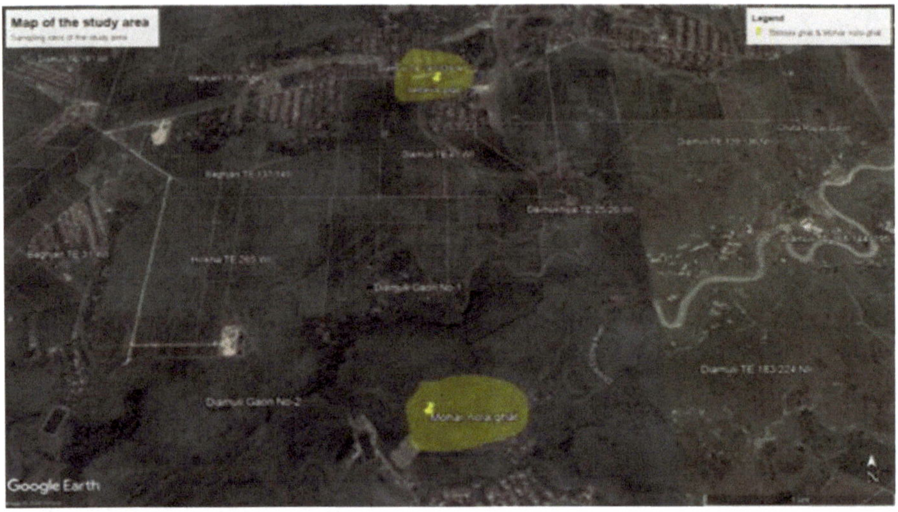

Fig 1: Map of the Study Area

Trichogaster sota

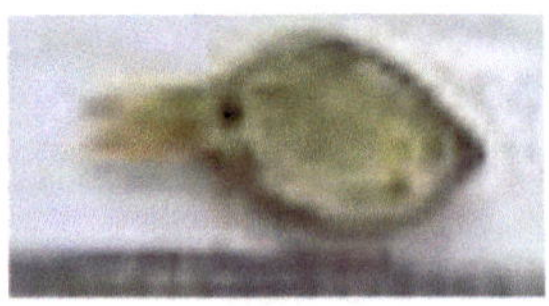

Puntius sophore

Heteropneustes fossils

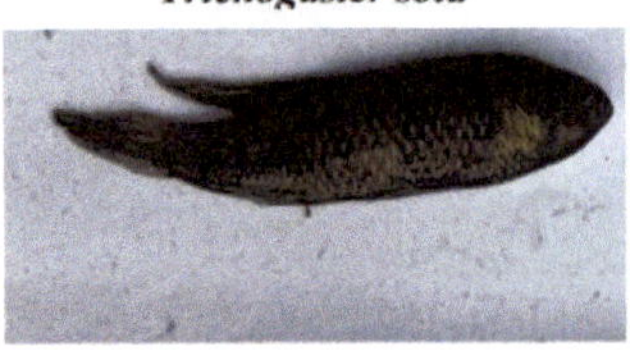

Badis badis

Chanda nama

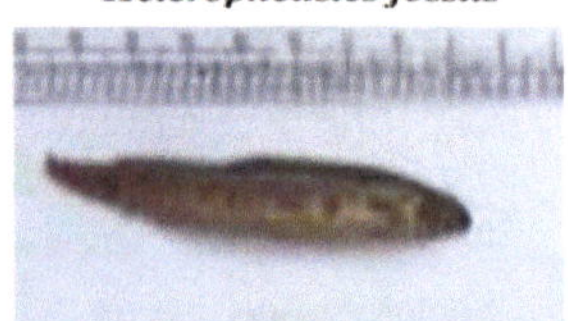

Lepidocephalichthysguntea

Mastacembelus armatus

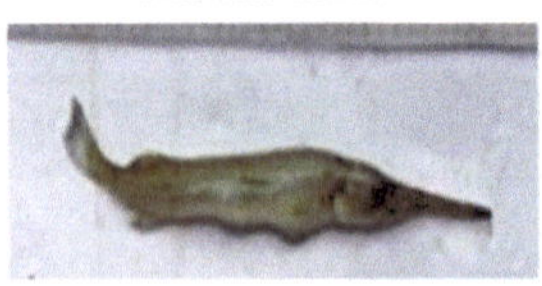

Xenenontodon cancila

Mystus cavasius

Fig. 2: Some glimpse of edible & ornamental fish species from the Maguri – Motapung Beel

Chapter 14: Limnological Studies of Lakes of Nagpur City, Maharashtra, India

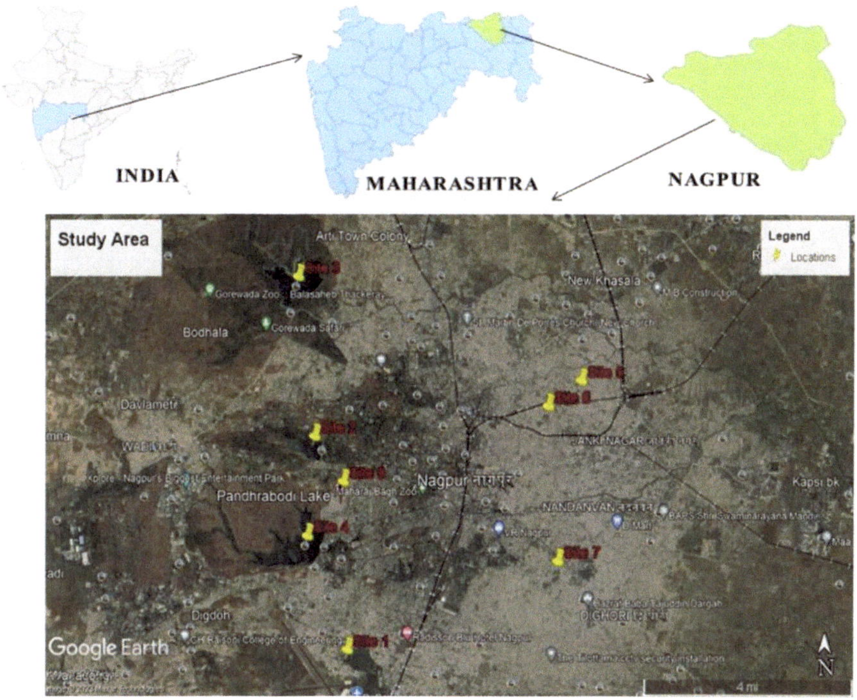

Fig. 1: Study Area Map

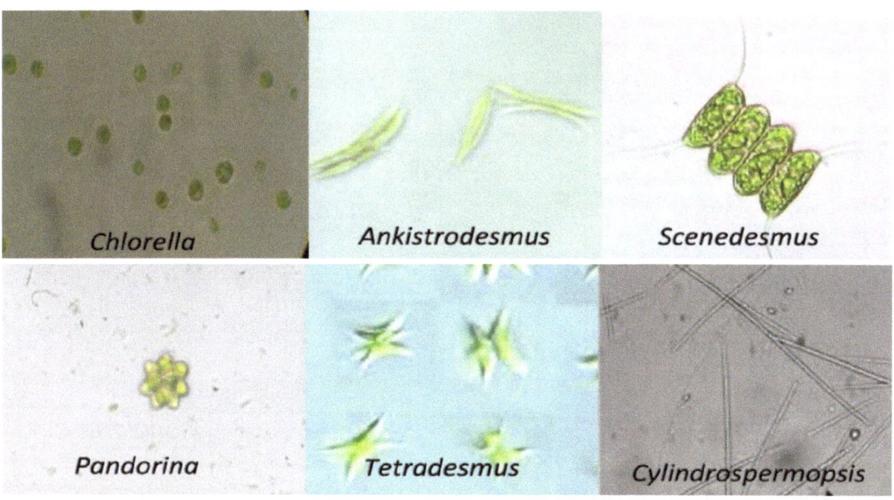

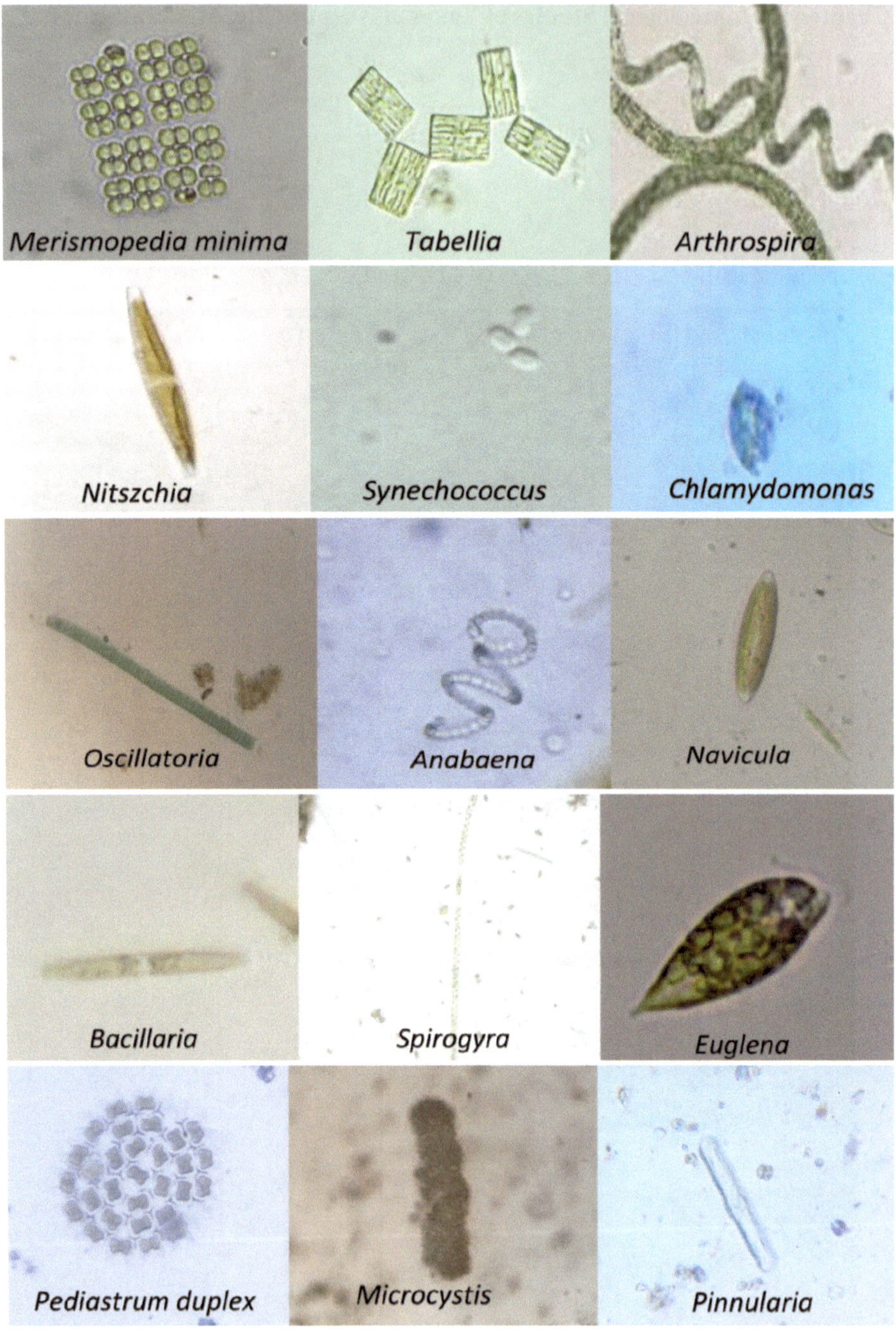
Merismopedia minima
Tabellia
Arthrospira
Nitszchia
Synechococcus
Chlamydomonas
Oscillatoria
Anabaena
Navicula
Bacillaria
Spirogyra
Euglena
Pediastrum duplex
Microcystis
Pinnularia

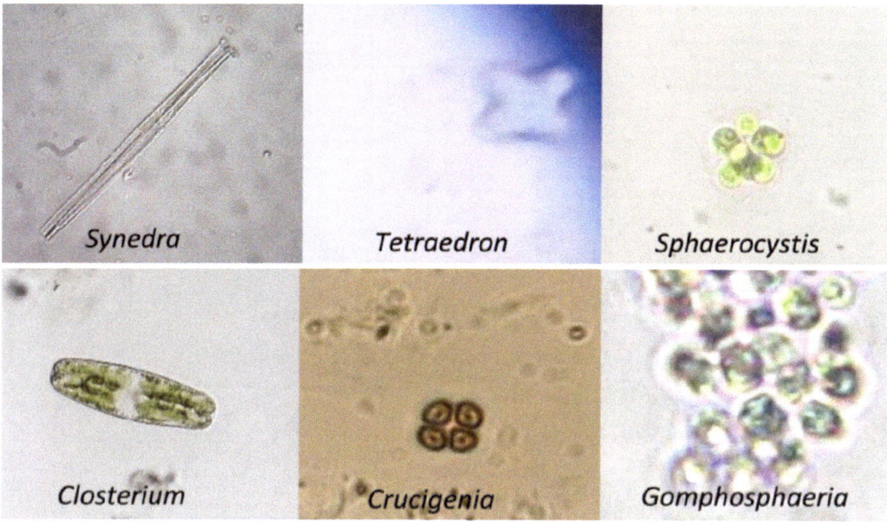

B

Fig. 1: Phytoplankton Observed in Lakes of Nagpur (A and B)

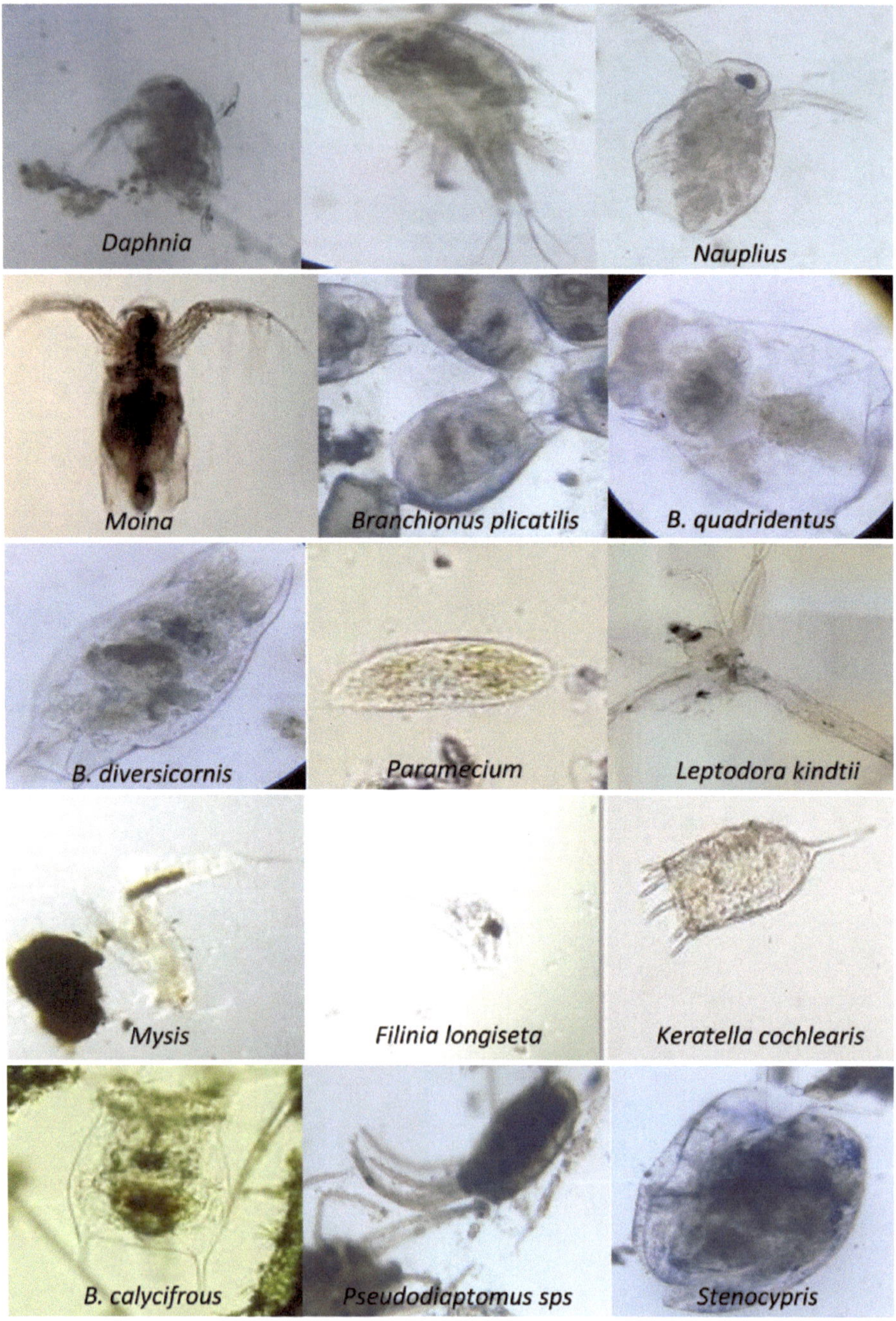
Daphnia
Nauplius
Moina
Branchionus plicatilis
B. quadridentus
B. diversicornis
Paramecium
Leptodora kindtii
Mysis
Filinia longiseta
Keratella cochlearis
B. calycifrous
Pseudodiaptomus sps
Stenocypris

A

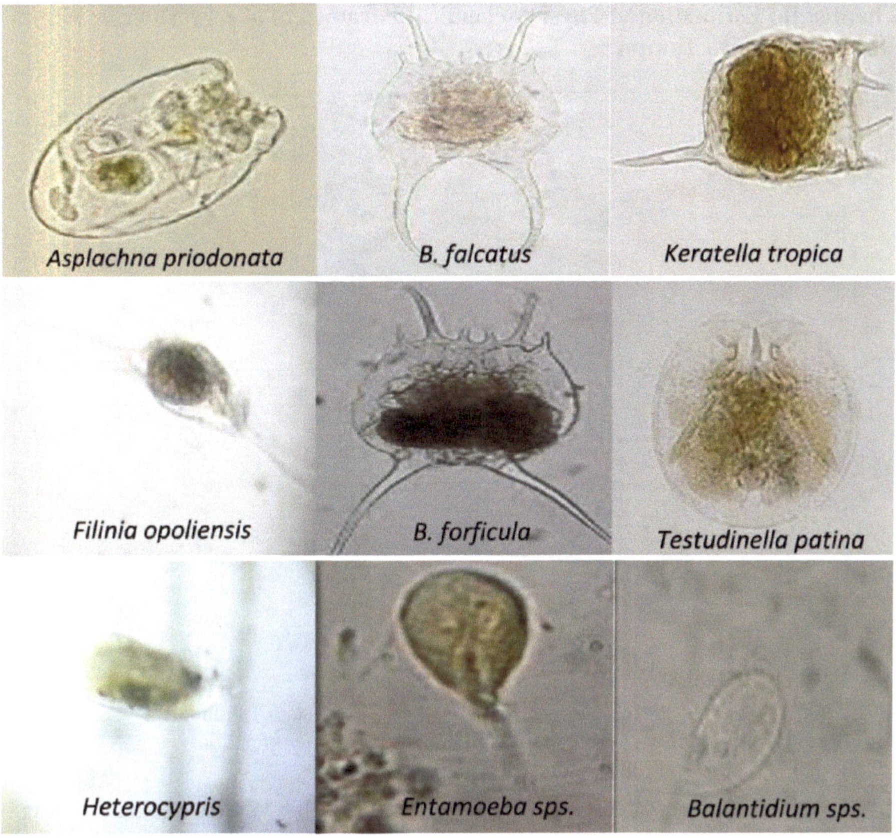

B

Fig. 2: Zooplankton Observed in Lakes of Nagpur (A and B)

Chapter 16: Estimation of Physico-chemical Parameters of a Freshwater Perennial Lake: Satara Tukum

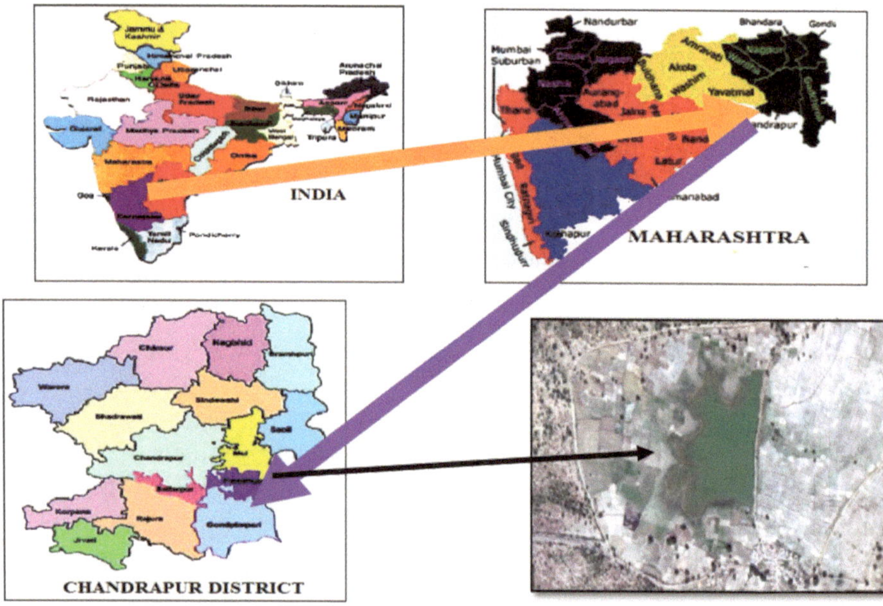

Satara Tukum lake (Aerial View)

Fig. 1: Study Area- Map of Chandrapur district & lake under investigation

Chapter 18: Rotifers Deteriorating Significantly The Water of Madhuganga Lake at Kalmeshwar, District Nagpur

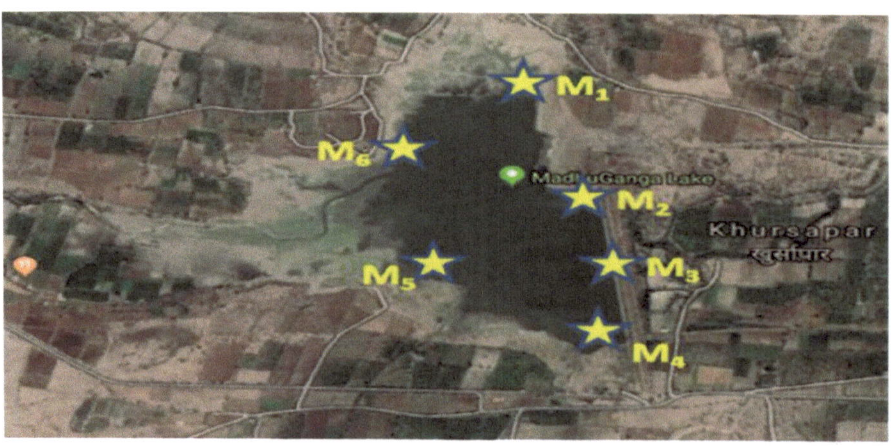

Fig. 1: Location of Madhuganga Lake site , Kalmeshwar, District- Nagpur (Sampling Site)

Index